RF AND MICROWAVE CIRCUIT DESIGN FOR WIRELESS COMMUNICATIONS

For a complete listing of the *Artech House Mobile Communications Library,*
turn to the back of this book.

RF AND MICROWAVE CIRCUIT DESIGN FOR WIRELESS COMMUNICATIONS

Lawrence E. Larson
editor

Artech House
Boston • London

Library of Congress Cataloging-in-Publication Data
RF and microwave circuit design for wireless communications / Lawrence E.
Larson, editor.
     p.   cm.
    Includes bibliographical references and index.
    ISBN 0-89006-818-6 (alk. paper)
     1.Wireless communication systems.   2. Radio circuits.   I. Larson, Lawrence E..
TK5103.2.R4 1996
621.384'12—dc21   97-49987
                                     CIP

British Library Cataloguing in Publication Data
RF and microwave circuit design for wireless communications
    1. Radio circuits—Design   2. Microwave sircuits—Design   3. Microwave
communication systems   4. Wireless communication systems
   I. Title
621.3'8412

ISBN 0-89006-818-6

© 1997 ARTECH HOUSE, INC.
685 Canton Street
Norwood, MA 02062

International Standard Book Number: 0-89006-818-6
Library of Congress Catalog Card Number: 97-49987

10 9 8 7 6 5

▼▼▼

CONTENTS

▼▼▼

PREFACE

The rapid development of commercial markets for analog and digital wireless communications in the last few years has been a blessing to many of the technical professionals hit hard by the defense build-down of the late 1980s and early 1990s. A new market for high-frequency expertise has risen out of the ashes of what was considered a moribund field only a few years ago.

However, the technology required to meet these new markets is very different from that of the older disciplines of DOD RF/microwave electronics—with its emphasis on "performance at any price"—or analog radio engineering—with its emphasis on discrete analog solutions operating at relatively high power levels. This gap between the traditional techniques of RF communication engineering and the more modern disciplines required for wireless communications presents a real difficulty for the practicing engineer. This book was developed with the hope of bridging the gap between the traditional art of radio and RF engineering and the requirements of today's modern wireless communications systems.

Modern wireless communications systems require a thorough understanding of RF and microwave circuit techniques *in addition* to a background in digital communications techniques and familiarity with existing and emerging wireless communications protocol standards. As a result, the first half of this book presents an overview of wireless communications standards, digital communications techniques, and RF system issues related to wireless communications and low-power RF and microwave electronics. The second half of the book covers key circuit areas related to the implementation of wireless communications systems.

This book is organized with the purpose of guiding the reader to the most essential elements of RF and microwave circuit design for wireless communications, and implicitly assumes a knowledge of the basics of active circuit design and communications theory. In any field advancing as rapidly as this one, there is a danger that a narrowly focused book will become rapidly dated. In order to avoid that, each chapter contains a clear exposition of the fundamental aspects of the topic, in addition to more topical material related specifically to wireless communications circuits. It is our hope that the fundamental portions of the book will retain their usefulness well after some of the more topical material has lost its freshness.

In addition, the book emphasizes *monolithic* solutions to microwave/RF circuit design problems. This is because the industry-wide trend is to move to higher and higher levels of integration in order to meet the aggressive cost targets associated with the commercial/consumer market. It is our belief that even if the designer is not actually designing the monolithic circuits, the design of RF systems *using* monolithic circuits will be greatly enhanced by an understanding of monolithic circuit techniques. However, our intent is not to exclude hybrid design techniques, and they are also addressed where appropriate, especially in the chapters on mixer and amplifier design.

Chapter 1 presents an overview of the general area of wireless communications: the market, frequencies, and technologies required to address it. The services addressed by the wireless industry include the ubiquitous cellular telephone and pager, in addition to the emerging markets of PCS, satellite-based telephony and data transmission, low-cost microwave links ,and wireless local area networks. In a field that is changing as rapidly as this, some of the information is evolving even as we finish production of the book, but we have tried to make this material as up to date and thorough as possible.

Chapter 2 of the book presents an overview of wireless communications standards and the RF system design considerations associated with their implementation. These new standards cover voice and data transmission over a variety of frequency bands in the United States as well as Europe and Asia. The chapter makes the point that traditional RF design approaches often fall short of meeting the requirements of modern digital wireless communications products, and presents some illustrative examples where different design methodologies are required to meet these new requirements.

Chapter 3 of this book presents an overview of the general field of wireless RF integrated circuit design, with particular emphasis on the historical development of radio receiver architectures, receiver design, and low-power techniques. This chapter develops the argument that modern communications system designs will rely on direct downconversion and other novel architectural approaches to realize the low-power/low-cost requirements of handheld consumer communications equipment.

Chapter 4 presents a detailed overview of digital RF modulation techniques and their circuit implementations. This material is normally covered in digital communications textbooks. Here, we have developed a one-chapter summary of the key results of that field, with particular emphasis on those modulation schemes that are currently being implemented in modern communications systems. The chapter also adopts the practical approach of outlining the methods of implementing those modulation techniques with a variety of circuit approaches.

Chapter 5 presents a thorough review of the design and fabrication of RF and microwave mixers. Once again, this field is often covered in an entire text. We have chosen here to cover those areas of mixer design particular to the field of wireless communications and have emphasized practical circuit approaches to mixer design, rather than theoretical depth.

Chapter 6 presents a complete overview of frequency synthesizer designs for wireless communications, with particular emphasis on VCO, PLL, and DDS implementations. The circuits often form the "heart" of the communications system in the sense that they provide the modulation and demodulation frequencies for tuning the transmitted and received signals. As a result, their performance and cost are crucial to the realization of acceptable system performance.

Finally, Chapter 7 presents a complete overview of the field of RF and microwave amplifier design. Although the design of amplifiers is a relatively mature field, a variety of new circuit architectures have been developed in the last few years that address the particular requirements presented by the wireless communications field. These requirements include an extremely high dynamic range in conjunction with low-power operation, as well as high efficiency in power output stages in conjunction with low-voltage operation. In addition, monolithic solutions are now being pursued where a hybrid approach would have been standard only a few years earlier.

The development of this book would not have been possible without the efforts of many of my friends and colleagues. In particular, I would like to especially thank all of the authors of the various chapters in the book. They worked tirelessly to produce manuscripts of exquisite quality with a very tight schedule. I would especially like to thank Dr. Paul Greiling of the Hughes Research Labs for his unflagging support of my work over the many years of our association. The long support and friendship of Professor Gabor Temes of Oregon State University is also greatly appreciated. I would also like to thank Drs. Michael Case, David Rensch, Loi Nguyen, Adele Schmitz, Mehran Matloubian, Perry MacDonald, Julia Brown, Ron Lundgren, William Stanchina, Robert Walden, Conilee Kirkpatrick, Cynthia Baringer, William Hooper, and Joseph Jensen of the Hughes Research Laboratories for many enlightening discussions in the area of high-frequency electronics. I would also like to acknowledge the support and encouragement of Mr. Lawrence Blue, Thomas Jackson, Olu George, Steven Rosenbaum, Adam Rachlin, and Ron Coburn from

Hughes Network Systems. I would also like to acknowledge many valuable discussions with Dr. Fred Blum of Sequel. I am indebted to the management of Hughes Electronics—particularly Drs. Arthur Chester, Richard Abrams, and Richard Reynolds of the Hughes Research Laboratories—who have supported my graduate education and career at Hughes.

Finally, I would like to acknowledge my wife Carole, my daughter Hannah, and my son Erik. Their love, support, and patience made the entire project possible.

Lawrence Larson
January 1996

CHAPTER 1
▼▼▼

AN OVERVIEW

Lawrence E. Larson
Head, Telecommunications Technology Department
Hughes Research Laboratories
Malibu, CA

1.1 INTRODUCTION

The rapid development of wireless communications products over the last ten years has led to an explosion in interest in improved circuit design approaches in the radio frequency (RF) and microwave area. However, this field has developed so rapidly that traditional communications and analog circuit design textbooks have struggled to keep up with all of the advances in the field [1, 2]. The purpose of this book is to attempt to bridge the gap between traditional communications/circuit design textbooks, which develop the fundamental principles behind a variety of circuit design approaches, and the most up-to-date advances in the wireless communications field.

In particular, the development of the field is characterized by a shift from purely analog techniques to those employing digital or hybrid digital/analog approaches. The advantage of digital communications techniques are well established both theoretically and practically, and include more efficient use of bandwidth, greater immunity to impairment due to noise, and lower power dissipation [3]. In addition, digital communications techniques can exploit the relentless advance of digital integrated circuit (IC) technology to realize improved functionality at lower power dissipation and lower cost. The migration of RF and microwave technology from analog to digital is briefly summarized in Table 1.1.

Since the widespread introduction of cellular telephone service during the 1980s, there has been rapid growth of those systems as well as mobile paging devices and a variety of wireless "data communications" services, under the broad heading of personal communication services (PCS). In addition to these terrestrial communications systems, the field of satellite-based video, telephone, and data communications services has grown at least as rapidly. The electromagnetic frequencies spanned by all of these systems range from several hundred megahertz to well over 60 GHz. The goal of all of this is to—as George Heilmeier put it in the October 1992 issue of *IEEE Communications Magazine*—"be able to access and communicate...in any medium, or combination of media—voice, data, image, video, or multimedia—anytime, anywhere, in a timely, cost-effective way" [4]. The growth of the market for mobile PCS products is expected to be rapid, as illustrated by the Table 1.2, which estimates market growth for a variety of mobile wireless services [5].

This chapter will briefly summarize the products and services that are anticipated to be developed in the coming years in this diverse field and conclude with some examples of the RF and microwave circuit design approaches that are being planned to meet these challenges.

1.2 MARKETS AND FREQUENCIES FOR WIRELESS COMMUNICATIONS

The majority of interest and activity for wireless communications systems is occurring at frequencies below 2 GHz, although there is increasing interest at frequencies above 2 GHz as well. The reasons that the majority of wireless systems are currently

Table 1.1

Comparison of Traditional Analog Communications RF/Microwave Circuit Approaches to Modern Digital Communications RF/Microwave Approaches

Criterion	Traditional Analog Techniques	Modern Digital Techniques
Level of integration	Hybrid (Low)	Monolithic (high)
Access method	FDMA	TDMA, CDMA
Down/up conversion	Dual	Direct
Power dissipation	High	Low

Table 1.2
Market Projections (in Millions of Users) for PCS Products

Product	1991	1992	1993	1994	1995	1996	1997	1998
Cellular	7.6	10.5	13.6	16.5	20.0	24.0	28.0	32.0
PCS	0.0	0.0	0.1	0.5	1.0	1.5	2.1	2.6
Paging	11.5	13.2	15.2	17.2	19.2	22.0	24.0	26.0
Mobile data	0.25	0.5	1.2	2.1	3.0	4.3	5.7	7.4
Mobile computers	0.0	0.0	0.023	0.075	0.255	0.795	1.515	2.595
PDAs	0.0	0.0	0.002	0.022	0.094	0.289	0.889	1.795

Source: [5].

implemented at frequencies below 2 GHz are both technical and historic. One technical reason for the use of frequencies below 2 GHz is that signal attenuation in the atmosphere and within buildings rises rapidly with frequency. Another reason is that the active circuits that are used to transmit and modulate the signals have historically had little gain at higher frequencies. This latter limitation is changing as advances in IC technology push transistor operating speeds well above 10 GHz. Figure 1.1 is a plot summarizing most of the popular available frequencies available for wireless communications throughout the world [6].

In the United States, the allocation and use of the electromagnetic spectrum is governed by the Federal Communications Commission (FCC). Historically, the spectrum was allocated to users based on a "comparative hearing" process, where the "best qualified" applicant was determined based on a lengthy and expensive

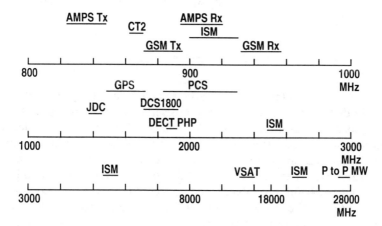

Figure 1.1 Summary of popular wireless transmission frequencies [6].

hearing process [7]. This process had its origins in the 1934 Communications Act, which established the FCC and defined the fundamental law that governed radio and television communications. However, Congress opened up the process of spectrum allocation with the 1993 Omnibus Budget Reconciliation Act, which permitted the FCC to grant spectrum licenses based on a competitive bidding process (an "auction"). The various wireless services encompassed by the FCC spectrum allocation include terrestrial telephony, PCS, satellite communications, and wireless LANs. Each of these areas is briefly discussed below.

Terrestrial Mobile Telephony

Chapter 2 of this book presents an excellent overview of spectrum allocation and data transmission standards for analog cellular, digital cellular, data, and specialized mobile radios (SMRs) (see Tables 2.1 through 2.4). In the United States, the existing advanced mobile phone service (AMPS) analog and IS-54 digital standard covers the frequency range from 824 MHz to 894 MHz. The global system for mobile communication (GSM) is the European digital telephone standard and covers the frequency range from 890 MHz to 960 MHz. GSM is also fast becoming a worldwide standard, with networks in South Africa, Singapore, and Australia to name but a few. Other standards have been proposed as well for future generations of telephones. For example, the Japanese Personal Handy Phone (PHP) is intended to operate at 1.9 GHz.

It is expected that, over the next several years, these telephone services will migrate from analog frequency division multiple access (FDMA) schemes to digital schemes. The resulting increase in cell capacity due to the more efficient bandwidth modulation of digital techniques will provide an added impetus to the implementation of these digital techniques.

PCS

Because of the relative infancy of the markets and services, frequency allocation and access methods for personal communications services (PCS) are highly fragmented. There is also some confusion about the services that will be provided by PCS, although the initial application of PCS in the United States is primarily for mobile telephony. Figure 1.2 shows the frequency allocations for PCS in the United States as of mid-1994, which extend from 900 MHz to well over 2 GHz, and are broken further into narrowband, broadband, and unlicensed services [7]. Narrowband PCS is primarily used for advanced paging services, while broadband PCS is intended for licensed voice, data, or video transmissions. Unlicensed PCS is intended for low-power, limited range devices, primarily for "premises communications."

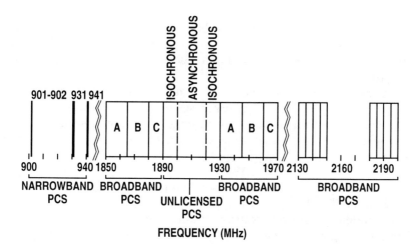

Figure 1.2 United States allocation of PCS frequencies [7].

Satellite Communications

Satellite communications is characterized by traditional voice, video, and data transmission using geosynchronous satellites (mostly at C- and Ku-band frequencies) and fixed earth terminals, and by a variety of proposed *mobile* satellite services (MSS) for handheld voice and data communication. Mobile satellite services are subdivided into geosynchronous (GEO), medium Earth orbit (MEO), and low Earth orbit (LEO), with the latter being subdivided into "little" LEO and "big" LEO. Big LEOs are intended to provide worldwide, or nearly worldwide, voice coverage, as well as limited high-speed data handling capabilities. Little LEOs are intended to provide lower bandwidth, less costly services to local niche markets using inexpensive satellites.

Geosynchronous mobile satellite services include Inmarsat, which is for general voice, data, and video services, and operates at a variety of frequencies from 1 to 6 GHz. OmniTRACs is another GEO MSS, which is used extensively for two-way truck fleet identification and operates in the 12/14-GHz band. In addition, the American Mobile Satellite Corporation has also proposed a GEO MSS system.

Over a dozen LEO systems have been proposed for MSS in the last several years [8]. They include the well-known Motorola Iridium system, as well as the Aries, Ellipso, and Globalstar systems. The Iridium system is a big LEO, and is planned to utilize 66 satellites in LEO, with the capability for global coverage. The Odyssey system, proposed by TRW, is a MEO System of 12 satellites.

Wireless LANs

The impetus for the development of wireless local area networks (LANs) comes from the increasing popularity of laptop computers and personal digital assistants

(PDAs), as well as the high cost of rewiring buildings for the inevitable next generation of wired LANs [9]. Labor and material costs can reach $1,000 per node for installation of copper wire networks, with an additional $200 to $1,000 for each upgrade [10].

So, there are enormous potential advantages if cost-effective, wide-bandwidth LANs can be developed. These networks are generally required to have all of the advantages of traditionally wired networks (high bandwidth, low latency, security, ease of configuration) in addition to several characteristics specific to the wireless environment:

- License-free operation (preferred use of nonlicensed bands);
- Long battery life (low power dissipation);
- Low interference (spread-spectrum techniques);
- Ease of configuration (good propagation characteristics within buildings and between buildings).

The most active frequency bands today for the development of wireless in-building LANs are the so-called industrial, scientific, and medical (ISM) bands at 902 to 928 MHz, 2,400 to 2,483.5 MHz, and 5,725 to 5,850 MHz. These bands were allocated by the FCC in the 1980s for unlicensed transmissions, as long as spread-spectrum transmission is employed and radiated power is less than 1W.

Each band has advantages and disadvantages for wireless LANs. The 2,400-MHz band is available in almost all countries, whereas the 900-MHz and 5,700-MHz bands are strictly limited to U.S. operation at this time. In addition, the 2,400-MHz band is prone to interference from microwave ovens. The 900-MHz band is the most mature technologically, but has a tremendous amount of potential interference from cordless telephones, wireless speakers, and so forth. The 5,700-MHz band has the widest potential bandwidth of any of the bands, but its in-building propagation characteristics are poor, and the analog technologies required to realize the system (GaAs or Si/SiGe) are the least mature. Table 1.3 summarizes the advantages and disadvantages of each of the ISM bands for wireless LAN systems [11].

The 2.4- to 2.5-GHz band has been under particularly active development, with several commercially available chip sets developed for these specific frequencies. The IEEE 802.11 Committee has been meeting for several years with the goal of developing broad hardware and software standards for the use of these frequencies. The overall goal is to support data rates in excess of 1 Mbps and to maintain compatibility with a wide variety of environments, from small rooms to buildings, campuses, and industrial manufacturing facilities. Both direct sequence spread spectrum (DSSS) and frequency hopped spread spectrum (FHSS) are provided for under these standards.

Table 1.3
Wireless LAN Band Alternatives: ISM Bands

Criterion	915 MHz	2.4 GHz	5.8 GHz
Frequencies	902–928 MHz	2.40–2.4835 GHz	5.725–5.825 GHz
Bandwidth	26 MHz	83.5 MHz	125 MHz
FCC licensing	No (Part 15)	No (Part 15)	No (Part 15)
Availability	US/Canada	Worldwide	US/Canada
Development cost	Low (Si)	Low/medium (Si, GaAs)	High (GaAs)
Transmission range	Greatest (power-dependent)	95% of 915 MHz	80% of 915 MHz
Current status	Crowded	Light use	Almost no use
Outlook	High growth	Medium growth	Low growth
Interference sources (U.S.)	Primary users	Primary users	Primary users
	Many LANs	Few LANs	Very few LANs
	Many non-SS	Few non-SS Microwave ovens	Very few non-SS
Additional interference sources (worldwide)	Cellular telephones		Some radar

Motorola has taken a different approach with its Altair System, which operates at an 18-GHz frequency band granted solely to it by the FCC [12]. The advantages of this frequency band include a wide bandwidth as well as lack of potential interference by more common microwave radiators. Recent versions of the system boast a performance of > 10 Mbps, and implement Ethernet and TCP/IP protocols. However, the high-frequency operation of the system also increases its cost.

Higher frequencies than these are being utilized for outdoor "point-to-point" communications systems, with applications for cellular, private, and LAN applications. The most common of these links are from 10 to 38 GHz and have peak transmission rates of 10 to 20 Mbps. An example of a typical 23-GHz product from P-COM is shown in Figure 1.3.

Finally, even higher frequencies are being investigated for applications where wider data bandwidths are required (> 100 Mbps). In particular, the 60-GHz frequency band has been actively pursued for years by a number of groups [13]. The RACE Project in Europe is developing the hardware and software technology for broadband networks at 62 to 63 and 65 to 66 GHz. In addition to the broad-bandwidth capacity at this frequency, the high attenuation of the atmosphere at these frequencies makes frequency reuse more attractive than at lower frequencies [14].

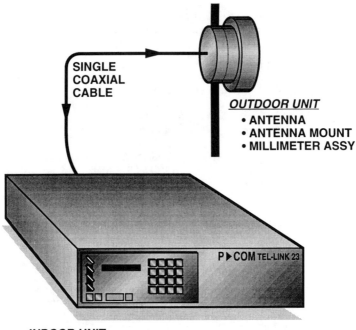

Figure 1.3 A 23-GHz wireless data product from P-COM.

1.3 CIRCUIT TECHNOLOGY OPTIONS FOR WIRELESS RF/MICROWAVE CIRCUIT IMPLEMENTATION

This section will briefly outline the circuit technology options for the implementation of wireless communications circuits.

1.3.1 Semiconductor Devices

The choice of the optimum semiconductor technology for wireless applications is a hotly contested issue, with various proponents arguing passionately for their positions [15]. In the last few years, wireless components have been announced or proposed using silicon bipolar technology, complementary metal-oxide semiconductor (CMOS), Si/SiGe heterojunction bipolar transistors (HBTs), as well as GaAs MESFETs and HBTs [16].

The optimum technology choice is ultimately an economic rather than a technical issue, and whoever can realize components with the lowest cost and quickest time to market is destined to win the argument. However, the economic arguments in favor of one technology or another are even more contentious than the technical arguments, so we will confine ourselves here to the technical merits of the various technologies and let the marketplace sort out the eventual "winners."

The key transistor technology figures of merit for RF and microwave design of wireless communication products are:

1. Short-circuit current gain bandwidth (f_T) (important for both analog and digital circuits);
2. Maximum power gain bandwidth (f_{MAX}) (important for both analog and digital circuits);
3. Minimum noise figure (LNAs, wide dynamic range front ends, mixers);
4. Maximum power-added efficiency (power amplifiers);
5. Linearity (AGC amplifiers, front ends);
6. Yield (relates to price as well as performance).

These various figures of merit can be compared for a number of competing semiconductor technologies and appear in Table 1.4 [17].

This table demonstrates the confusing picture that confronts the designer of wireless communications systems. Unlike digital IC technology, where CMOS is clearly dominant, no one technology dominates (or is likely to dominate) the imple-

Table 1.4
Performance of Various Semiconductor Technologies Addressing Wireless
Communications Applications

Criterion	Si BJT	Si/SiGe HBT	GaAs MESFET	GaAs HBT	BiCMOS BJT
B_{VCEO}/B_{VDS} (V)	4	4	8	15	6
f_T (GHz)	32	55	50	30	13
f_{MAX} (GHz)	35	55	60	70	11
G_{MAX} @ 2 GHz (dB)	24	28	20	19	17
Noise figure @ 2 GHz (dB)	1.0	0.5	0.3	1.5	1.0
IP3/P-1 dB (dB)	9	9	12	16	9
Power-added efficiency @ 3V (%)	–	70	60 @ 5V	70	40

Source: [17].

mentation of wireless communications ICs. In many cases, GaAs technology is used for efficient power amplification, and Si bipolar technology is used for RF signal-processing functions. More recently, Si/SiGe bipolar technology has been introduced as a promising technique for implementing both power amplification and signal processing with the same technology.

1.3.2 Passive Devices

Though often neglected, passive devices often determine the overall size, topology, and performance of a wireless communications transceiver. These elements include inductors, filters, resonators, and capacitors. However, as the size and cost of active devices continue to shrink, the importance of improvements in passive devices performance and size become more urgent. Many of these elements are difficult if not impossible to integrate monolithically.

One of the most important developments in the field of wireless communications was the development of surface acoustic wave (SAW) devices. These devices are typically bandpass filters with very sharp and stable cutoff characteristics. Their operation is based on the interference of mechanical surface waves on a piezoelectric substrate, which have been launched through piezoelectric transducers. The one drawback of SAW devices is their relatively high insertion loss (typically > 20 dB), which creates the need for pre- and postamplification around the circuit. Some low-loss SAW filters have been developed, but they are still in the experimental stage and are not yet commercially available. A drawing of a typical SAW filter appears in Figure 1.4.

In order to eliminate some of the disadvantages of SAW filters (principally the high insertion loss), a number of laboratories have been exploring alternative techniques for the implementation of passive elements. Figure 1.5 shows a dielectric monoblock filter intended for 800-MHz cordless telephone applications [18]. This design consists of a pair of coupled quarter-wave sections, mounted inside a high-dielectric coaxial waveguide. The ceramic material is $BaO-PbO-Nd_2O_3-TiO_2$, with a relative dielectric constant of 90. The filter has a center frequency of 886 MHz, a bandwidth of 2 MHz, and an insertion loss of 2.1 dB. The filter exhibited an out-of-band rejection of 44 dB at 44 MHz from the carrier.

Other important passive elements include dielectric resonators (discussed in Chapter 6), as well as surface-mount transformers and inductors (discussed in Chapter 5), capacitors, and resistors.

1.4 RECENT EXAMPLES OF RF AND MICROWAVE WIRELESS COMMUNICATIONS CIRCUITS

One of the keys to the recent development of wireless communications devices is the increased level of integration of RF and microwave devices. An RF function that

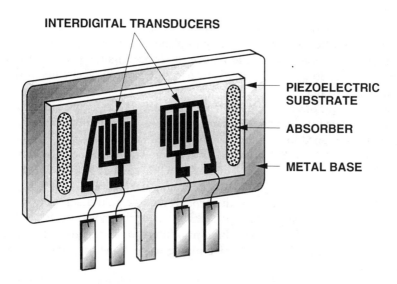

Figure 1.4 SAW filter used for communications transceivers.

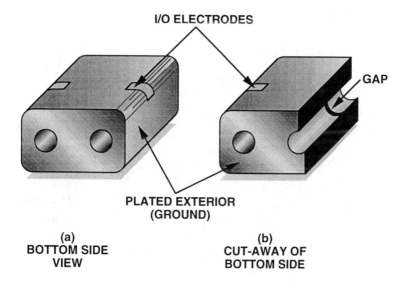

Figure 1.5 Dielectric monoblock filter used for cordless telephone applications [18].

would historically have been implemented with discrete devices and laboriously tuned by hand on the test bench is today implemented in a single monolithic die. Equipment manufacturers are able to reduce the size, weight, cost, and power consumption of the resulting circuits. A wide variety of monolithic circuits of this type have been demonstrated recently.

A good example of a highly integrated transmitter IC for digital cellular applications is demonstrated by Hewlett Packard in the block diagram of Figure 1.6 [19]. This circuit was designed for compatibility with the North American Digital Cellular (NADC) System at 824 to 849 MHz, as well as the GSM European System at 890 to 915 MHz and the Japan Digital Cellular (JDC) System at 940 to 956 MHz.

The circuit was realized in an RF silicon bipolar process, with a 15 GHz f_T and 30 GHz f_{MAX}. Other features of the process include gold metallization, a thick field oxide, and high-density MIS capacitors. Two of the most interesting features of the design are the use of on-chip LC bandpass filters for mixer product suppression and the use of external off-chip inductive tuning to peak the frequency response in the desired band.

ATT recently announced an integrated circuit specifically targeted for the GSM market [20]. A block diagram of the chip set is shown in Figure 1.7. The chip includes two fixed-frequency synthesizers (for transmit and receive IFs), an RF mixer, single IF downconversion with quadrature demodulation and variable gain, as well as a variety of digital control circuitry. The circuit was implemented in a 12-GHz silicon bipolar technology, with 1.5-μm feature size. The addition of a power amplifier and low-noise amplifier are all that are needed to realize the RF portions of a complete GSM transceiver.

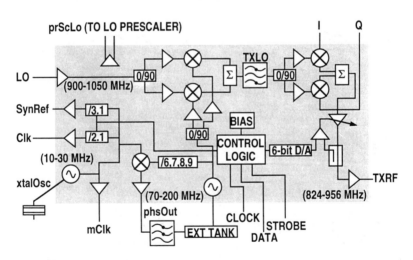

Figure 1.6 Integrated transmitter IC for digital cellular telephone application [19].

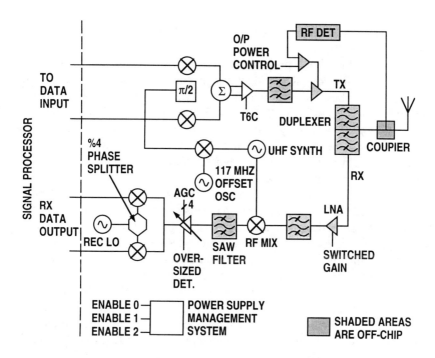

Figure 1.7 GSM chip set recently announced by ATT [20].

1.5 ORGANIZATION OF THE BOOK

This book is organized with the purpose of guiding the reader to the most essential elements of RF and microwave circuit design for wireless communications, and implicitly assumes a knowledge of the basics of active circuit design and communications theory. In any field advancing as rapidly as this one, there is a danger that a narrowly focused book will become rapidly dated. In order to avoid that, each chapter contains a clear exposition of the fundamental aspects of the topic in addition to more topical material related specifically to wireless communications circuits. It is our hope that the fundamental portions of the book will retain their usefulness well after some of the material has lost its freshness.

In addition, the book emphasizes *monolithic* solutions to microwave/RF circuit design problems. This is because the industry-wide trend is to move to higher and higher levels of integration in order to meet the aggressive cost targets associated with the commercial/consumer market. It is our belief that even if the designer is not actually designing the monolithic circuits, the design of RF systems *using* monolithic circuits will be greatly enhanced by an understanding of monolithic circuit techniques. However, our intent is not to exclude hybrid design techniques, and they are also addressed where appropriate, especially in the chapters on mixer and amplifier design.

Chapter 2 of the book presents an overview of wireless communications standards and the associated RF system design considerations associated with their implementation. These new standards cover voice and data transmission over a variety of frequency bands, in the United States as well as Europe and Asia. The chapter makes the point that traditional RF design approaches often fall short of meeting the requirements of modern digital wireless communications products, and presents some illustrative examples where different design methodologies are required to meet these new requirements.

Chapter 3 of this book presents an overview of the general field of wireless RF IC design, with particular emphasis on the historical development of radio receiver architectures, receiver design, and low-power techniques. This chapter develops the argument that modern communications system designs will rely on direct downconversion and other novel architectural approaches to realize the low-power/low-cost requirements of handheld consumer communications equipment.

Chapter 4 presents a detailed overview of digital RF modulation techniques and their circuit implementations. This material is normally covered in digital communications textbooks. Here, we have developed a one-chapter summary of the key results of that field, with particular emphasis on those modulation schemes that are currently being implemented in modern communications systems. The chapter also adopts the practical approach of outlining the methods of implementing those modulation techniques with a variety of circuit approaches.

Chapter 5 presents a thorough review of the design and fabrication of RF and microwave mixers. Once again, this field is often covered in an entire text. We have chosen here to cover those areas of mixer design particular to the field of wireless communications, and have emphasized practical circuit approaches to mixer design rather than theoretical depth.

Chapter 6 presents a complete overview of frequency synthesizer designs for wireless communications, with particular emphasis on voltage controlled oscillators (VCOs), phase lock loops (PLLs), and direct digital synthesis (DDS) implementations. The circuits often form the "heart" of the communications system, in the sense that they provide the modulation and demodulation frequencies for tuning the transmitted and received signals. As a result, their performance and cost are crucial to the realization of acceptable system performance.

Finally, Chapter 7 presents a complete overview of the field of RF and microwave amplifier design. Although the design of amplifiers is a relatively mature field, a variety of new circuit architectures have been developed in the last few years that address the particular requirements presented by the wireless communications field. These requirements include an extremely high dynamic range in conjunction with low-power operation as well as high efficiency in power output stages in conjunction with low-voltage operation. In addition, monolithic solutions are now being pursued where a hybrid approach would have been standard only a few years earlier.

References

[1] H. L. Krauss, C. W. Bostian, and F. H. Raab, *Solid State Radio Engineering*, New York, NY: John Wiley & Sons, 1980.

[2] W. H. Hayward, *Introduction to Radio Frequency Design*, Englewood Cliffs, NJ: Prentice-Hall, 1982.

[3] U. L. Rohde, and T.T.N. Bucher, *Communications Receivers: Principles & Design*, New York, NY: McGraw-Hill, 1988.

[4] G. H. Heilmeier, "Personal Communications: Quo Vadis," in *Int'l Solid State Circuits Conf.*, San Francisco, CA, 1992, pp. 24–26.

[5] Source: The Yankee Group, 1993. From R. Schneiderman, *Wireless Personal Communications—The Future of Talk:* IEEE Press, 1994.

[6] C. Kerramac et al., "New Application Opportunities for SiGe HBTs," *Microwave Journal*, Oct. 1994, pp. 22–37.

[7] N. Colmenares, "The FCC on Personal Wireless," *IEEE Spectrum*, May 1994, pp. 39–46.

[8] J. Lodge, "Mobile Satellite Communications: Toward Global Personal Communications," *IEEE Communications Mag.*, Vol. 29, No. 11, Nov. 1991, pp. 24–30.

[9] T. Freeburg, "Enabling technologies for wireless in-building network communications—four technical challenges, four solutions," *IEEE Communications*, April 1991, pp. 58–64.

[10] J. Lida, "MIS likes unshielded wiring, but worries about support," *MIS Week*, July 1989, p. 31.

[11] D. Bantz, and F. Bauchot, "Wireless LAN Design alternatives," *IEEE Network*, March/April 1994, pp. 43–52.

[12] G. Berline, and E. Perratore, "Portable, affordable, secure: wireless, LANs," *PC Magazine*, Feb. 11, 1992, pp. 291–314.

[13] M. Chelouche, and A. Plattner, "Mobile broadband system (MBS): Trends and impact on 60 GHz band MMIC development," *Electronics and Communications Engineering Journal*, June 1993, pp. 187–197.

[14] R. G. Blake, "The use of millimeterwave for broadband local distribution," *Proc. 17th European Microwave Conf.*, Sept. 1987, pp. 293–298.

[15] L. M. Burns, "Applications for GaAs and Silicon ICs in next generation wireless communications systems," *Proc. of 1994 IEEE GaAs IC Symp.*, pp. 155–158.

[16] J. Neal, "HBT Technology adds power to CDMA chip set," *Microwaves and RF*, Dec. 1994, pp. 205–207.

[17] C. Kermarrec et al., "New application opportunities for SiGe HBTs," *Microwave Journal*, Oct. 1994, pp. 22–36.

[18] M. Matsumoto et al., "A miniaturized dielectric monoblock band-pass filter for 800 MHz band cordless telephone system," *1994 MTT-S Symposium*, San Diego, CA, May 23–27, 1994.

[19] K. Negus et al., "Highly integrated transmitter RFIC with monolithic narrowband tuning for digital cellular handsets," *1994 IEEE Int'l Solid-State Circuits Conf.*, Feb. 1994, pp. 38–39.

[20] T. Stetzler et al., "A 2.7V to 4.5V Single Chip GSM Transceiver RF Integrated Circuit," *1995 IEEE International Solid-State Circuits Conf.*, Feb. 1995, pp. 150–151.

CHAPTER 2

▼▼▼

INTRODUCTION TO WIRELESS COMMUNICATIONS APPLICATIONS AND CIRCUIT DESIGN

Jonathon Y. C. Cheah
Solectek Corporation
San Diego, CA

2.1 INTRODUCTION

Stimulated by the increased consumer and commercial users demand for wireless communications applications, the advancement of sophistication in wireless circuit design has progressed at unprecedented speed in recent years. The last ten years have witnessed a remarkable miniaturization of portable equipment, a similar extension of battery life, and almost an order of magnitude reduction in retail prices. One ubiquitous example of this phenomenon is the North American AMPS handheld cellular telephone. Other examples include handheld portable telephones, paging devices, and wireless local area networks (LANs).

Although not well known among everyday wireless communication consumers, *commercial* wireless communications technology has also experienced a similar impetus in its advancement, both in miniaturization and cost. Commercial very small aperture (satellite ground) terminal (VSAT) is a good example. Before the advent of VSAT technology some ten years ago, the cost of satellite ground terminals for

17

commercial data, voice, and video circuits was at least a few hundred thousand dollars per station, which were customarily housed in specially dedicated buildings. Today, it is not uncommon to see VSATs mounted unobtrusively on top of gas station kiosks and other commercial buildings, at a cost of less than ten thousand dollars per station.

In the portable wireless communication industry, it is well known that the design of the RF transceiver is usually the key element that determines the cost, the size, and the useful battery life of the equipment, as well as how the equipment is used. In *digital* wireless communications, the requirements imposed on the RF transceiver design demand RF engineers to understand and be conversant in digital communications techniques.

For example, in digital communications systems, oscillator phase noise is often specified not only in the traditional dBc/Hz at a fixed-frequency offset profile, but also in noise variance at a determined probability density function. This specification allows the designer to optimize the digital modulation performance while simultaneously optimizing the design in terms of cost and size of the device. Likewise, filters are required to meet phase distortion and group delay requirements, in addition to a prescribed gain versus frequency response.

This chapter will examine some of the tradeoffs involved in the design of the electronic circuits for wireless communications, and how those design parameters can affect ultimate system performance.

2.2 CURRENT AND NASCENT CONSUMER WIRELESS COMMUNICATIONS STANDARDS AND APPLICATIONS

One of the major forces that sustains the continual research and development in the wireless communication arena is the general user acceptance of wireless communications standards. The standards provide a coalescence within the industry to invest in the creation of suitable chip sets for the manufacturing of consumer subscriber units. The resulting high volume drives down the cost of the final product, and the industry is able to take advantage of the same trends that have lowered prices so dramatically in the PC industry. It also encourages investment in the development of communication infrastructure, such as base stations, switch offices and underground copper or optical fiber cables.

It is interesting to note that competing standards create a divisive effect in the market place. Unlike the market for consumer products, where competitive products have had a positive end effect of improved product quality and reduced cost, competing *standards* in the same market confuse the users. The result is a slow market development. In this respect, the rivalry between IS-54 [1] and IS-95 [2] serves as a good example.

The existing major analog cellular telephone standards in the world are listed in Tables 2.1 and 2.2 [3].

Table 2.1
Major Analog Worldwide Cellular Telephone Standards

Mobile Standards	AMPS	TACS	ETACS	NMT 450	NMT 900
Frequency plan (MHz)	824–849	890–915	872–905	453–457.5	890–915
	869–894	935–960	917–950	463–467.5	935–960
Channel spacing (kHz)	30	25	25	25	12.5
Number of channels	832	1,000	1,240	180	1,999
Region	America	Europe	United Kingdom	Europe	Europe
	Australia	Africa		Africa	Africa
	SE Asia	SE Asia		SE Asia	SE Asia
	Africa				

Table 2.2
Additional Analog Worldwide Cellular Telephone Standards

Mobile Standards	C-450	RTMS	Radiocom 2000	NTT	JTACS NTACS
Frequency plan (MHz)	450–455.74	450–455	192.5–199.5	925–940	915–925
	460–465.74	460–465	200.5–207.5	870–885	860–870
			215.5–233.5	915–918.5	898–901
			207.5–215.5	860–863.5	843–846
			165.2–168.4	922–925	918.5–922
			169.8–173	867–870	863.5–867
			414.8–418		
			424.8–428		
Channel spacing (kHz)	10	25	12.5	25/6.25	25/12.5
			12.5	6.25	25/12.5
			12.5	6.25	12.5
Number of channels	573	200	560	600/2,400	400/800
			640	560	120/240
			256	480	280
Region	Germany	Italy	France	Japan	Japan
	Portugal				

However, the current proposed or newly adopted wireless communications standards for personal communications services (PCS) are predominantly based on digital formats. This is because of the proven superiority digital techniques can provide. In contrast, much of the existing wireless RF circuit design techniques were founded upon analog voice systems, such as the AM and FM push-to-talk land or maritime mobile SMR telephones, the omnipresent cordless and cellular telephones, CB radios, and the SSB airline air-to-ground pay phone, to name but a few.

The modern digital standards for handheld and low-power wireless voice communications PCS are shown in Table 2.3 [4].

Table 2.4 shows the equivalent digital voice cellular telephone standards for various existing and proposed systems. These standards also support digital data transmission. Table 2.5 shows a summary of the wireless data communications standards. These standards are used for the transmission of digital data, rather than digitized voice. SMR services currently in operation also provide data communication services.

Apart from the data services indicated in Table 2.5, there are also wireless LAN standards that are being prepared, such as IEEE 802.11 [3–5] in the United States and ETSI/RES10 (otherwise known as HIPERLAN) in Europe. HIPERLAN has ERC allocation of 5.150 to 5.300 GHz and 17.000 to 17.300 GHz for its use. The IEEE 802.11 proposes to support 1 to 2 Mbps channel signaling rate, whereas HIPERLAN targets 20 Mbps.

The IEEE 802.11 is primarily intended to utilize the ISM bands of 902 to 928 MHz, 2.400 to 2.4835 and 5.725 to 5.850 GHz, which are facilitated by the FCC Part 15.247 rule [6] that allows peak transmit radiating power of up to 0 dBw using spread-spectrum techniques. Under this rule, the rigorous regulatory certification process of the intentional radiators is not in force as it is under other FCC regulatory classifications. There are a variety of wireless data products that are already in the U.S. market that take advantage of the FCC Part 15.247 rule. The most common consumer examples are the extended range cordless telephone and cordless stereo speakers using the 902- to 928-MHz band. Unfortunately, there

Table 2.3
Condensed Version of Wireless Communications Technology Standards for Low-Power
Voice Systems

Mobile System	DECT	PHS	CT-2	CT-3	UDPC
Multiple access method	TDMA/ FDMA	TDMA/ FDMA	FDMA	TDMA/ FDMA	TDM/TDMA/ FDMA
Frequency bands (MHz)	1,800–1,900	1,895–1,907	864–868	862–866	Within 1.7–2.3 GHz
Duplex	TDD	TDD	TDD	TDD	FDD
Channel spacing	1,728 kHz	300 kHz	100 kHz	1,000 kHz	350 kHz
Modulation	FSK	π/4 QPSK	FSK	FSK	π/4 QPSK
TX max power	250 mW	80 mW	10 mW	80 mW	200 mW
Speech coding	ADPCM	ADPCM	ADPCM	ADPCM	ADPCM
Bit rate	32 Kbps	32 Kbps	32 Kbps	32 Kbps	32 Kbps
Speech per channel	12	4	1	8	10
Channel bit rate 1152 Kbps	96 Kbps	72 Kbps	640 Kbps	500 Kbps	
Channel coding	CRC	CRC	None	CRC	CRC

Table 2.4
Summary of Digital Voice Cellular Telephone Standards

Mobile System	IS-54	GSM	JDC	IS-95
Multiple access method	TDMA/FDMA	TDMA/FDMA	TDMA/FDMA	FDMA/CDMA
Frequency bands	824–848 869–894	890–915 935–960	810–826 940–956 1,447–1,489 1,429–1,441 1,501–1,513 1,453–1,465	824–848 862–866
Duplex	FDD	FDD	FDD	FDD
Channel spacing	30 kHz	200 kHz	25 kHz	1,250 kHz
Modulation	π/4 QPSK	GMSK BT = 0.3	π/4 QPSK	QPSK
TX max power	600 mW	1W		200 mW
Speech coding	VSELP	RPE-LTP	VSELP	QCELP
Bit Rate	8 Kbps	13 Kbps	8 Kbps	1–8 Kbps
Speech per channel	3	8	3	12
Channel bit rate	48.6 Kbps 270.833 Kbps	42 Kbps	1,228.8 Kbps	
Channel Coding	Rate-1/2 convolution	Rate-1/2 convolution	Rate-1/2,1/3 convolution	CRC

Table 2.5
Summary of Mobile Wide Area Wireless Data Services

Mobile System	ARDIS	MOBITEX	CDPD	IS-95	TETRA
Multiple access method	FDMA/ DSMA	FDMA/SLOTT ED-ALOHA	FDMA/DSMA	FDMA/ CDMA	FDMA/ DSMA
Frequency bands (MHz)	800 band	935–940	824–849 869–894	824–849 869–894	400 band 900 band
Channel spacing	25 kHz	12.5 kHz	30 kHz	1,250 kHz	25 kHz
Modulation	4-FSK	GMSK BT = 0.3	GMSK BT = 0.5	QPSK/SS	π/4 QPSK

is also a plan for the U.S. regulatory agencies to reclaim portions of these frequency bands.

Another wireless communication family is known as specialized mobile radios (SMR). SMRs are used in various applications using often noncontiguous privately licensed frequency bands. As a result, the radio transceiver design and the performance requirements are often more stringent than those of standard consumer products. It is particularly true that the regulatory rules that govern SMR operations are based on traditional AM and FM push-to-talk land mobile radios with respect to the transmit spectrum mask, and thus impose a very tight constraint on the digital communication implementations. The general SMR performance and applications are governed under FCC Part 90 [7] and industrial recommendations can be found in TIA/EIA 603 [8].

There is also an expanded emphasis on wireless telephone usage for replacement of landline telephony local loop infrastructure. This application of wireless voice communications in these applications is commonly known as wireless local loop (WLL). Wireless local loop replaces the underground telephone cable drop to each user and connects the user directly to the public telephone switch network through a cellsite by means of a radio. The advantage of WLL is the deployment of a telephone network virtually overnight in areas where telephone services are not available, such as some cities in third-world countries or rural communities in the western world.

2.3 RADIO TRANSCEIVER TECHNOLOGY REQUIREMENTS

In radio transceiver design, it is often advantageous to have an *a priori* idea of the performance of the state-of-the-art radio transceiver component technology before it is integrated. Mature component data, though not always readily available, allows the systems designers as well as the circuit designers to assess how a required specification can be met in terms of its cost and technical risk. Systems tradeoffs are often dominated by the availability and performance of the concurrent new components in the wireless digital communication industry.

Figure 2.1 shows a typical IS-19 cellular telephone frequency plan and block diagram. Similarly, Figure 2.2 shows that of an IS-55 [9] digital dual-mode cellular telephone. It is interesting to see from Figures 2.1 and 2.2 that, although conceptually both radio transceivers have a similar structure, the actual implementation of these radios is significantly different in the frequency plan and circuit design topology. This is because of various design considerations, such as those related to the number of oscillators needed, the potential spurious frequency mixing products, the availability of mature components, and, most importantly, the unit costs. The resulting design is ultimately driven by the available radio transceiver components technology.

Table 2.6 shows an interesting example of practical product performance limits for an IS-55 dual-mode type cellular telephone using current radio transceiver component technology. The specification items selected are dictated by the limits of state-of-the-art component technology in particular. The limits depicted in Table 2.6 are derived from designs that used the most economical implementation and design techniques viable for commercial production. This table illustrates a baseline for practical achievable performance in radio transceiver design for the consumer market. It can also be used to estimate the performance requirements for other

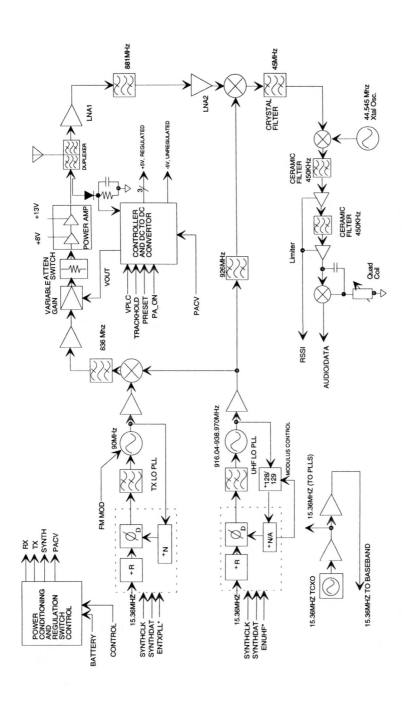

Figure 2.1 IS-19 cellular telephone RF section block diagram. A 45-MHz offset frequency oscillator generates the required receiver and transmitter local oscillator frequencies.

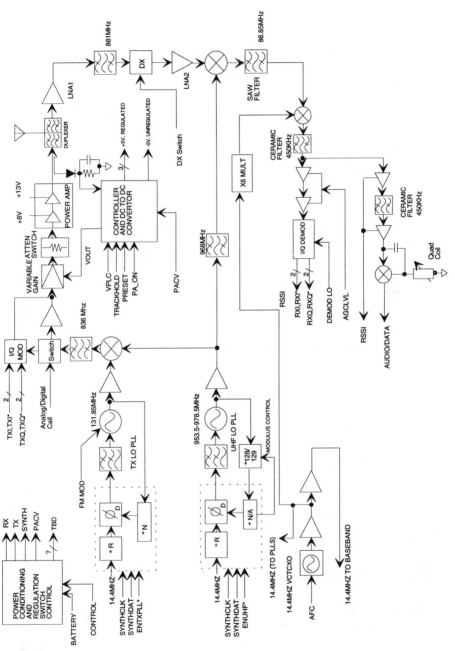

Figure 2.2 IS-55 block diagram. A narrowband IF filter is required for digital operation, as well as an analog-to-digital converter in the baseband.

frequency bands. One can probably assume that if the intended product has the component volume of the cellular market, then the performance and cost of the devices are proportional to the absolute frequency.

Table 2.6
Typical Performance of IS-55 Class of Product Compared with Specifications

IS-55A Specifications	IS-55A Spec	Measured Values
2. Receiver minimum standards:		
2.1.2.1 Sync. acquisition time (digital mode)		
−90 dBm, 8 & 100 km/h	< 130 msec	< 110 msec
−103 dBm, 8 & 100 km/h	< 250 msec	< 130 msec
2.2.2.2 Audio muting		
Audio level	> 40 dB	> 70 dB
Transient	< 25%; < 25 msec	~ 0%; 0 msec
2.2.2.3 Expander		
Above 0 dB	±1 dB	< ±2 dB
Below 0 dB	±2 dB	< ± 1 dB
Attack time	3 ± 0.6 msec	< 3 msec
Recovery time	13.5 ± 2.7 msec	< 4 msec
2.2.2.4 Hum and noise (dB)	> 32 dB	> 44 dB
2.2.2.5 Audio harmonic distortion (%)	5 %	1.30%
2.2.3.2 Supervisory audio tone decoding		
Detection time	< 250 msec	< 200 msec
Removal time	< 259 msec	< 200 msec
2.3 Receiver performance		
2.3.1 Analog		
2.3.1.1 RF Sensitivity (dBm)	−116 dBm	−121 dBm
2.3.1.2 Adjacent channel selectivity	> 16 dB	> 37 dB
Alternate channel selectivity	> 60 dB	> 71 dB
2.3.1.3 Intermod. spurious resp. atten., near	> 65 dB	> 72 dB
intermod, spurious resp atten., Far	> 73 dB	> 76 dB
2.3.1.4 Prot. against spurious resp. interference	> 60 dB	> 82 dB
2.3.2 Digital		
2.3.2.1 RF sensitivity		
Static (3% BER)	−110 dBm	−116 dBm
8 km/H (3% BER)	−103 dBm	< −103 dBm
100 km/H (3% BER)	−103 dBm	< − 103 dBm
2.3.2.2 Adj. and alt. channel desensitization @ −107 dBm		
30 kHz (+) adj. interferer @ −94 dBm	< 3 % BER	< 0.01 %
60 kHz (+) alt. interferer @ −65 dBm	< 3 % BER	< 0.03 %
30 kHz (−) adj. interferer @ −94 dBm	< 3 % BER	< 0.01 %
60 kHz (−) alt. interferer @ −65 dBm	< 3 % BER	< 0.01 %
2.3.2.3 Intermod spurious resp. atten. @ −107 dBm		
120/240 kHz (+) @ −45/−45 dBm	< 3 % BER	
120/240 kHz (−) @ −45/−45 dBm	< 3 % BER	0.3%

Table 2.6 (continued)

IS-55A Specifications	IS-55A Spec	Measured Values
2.3.2.6 Cochannel performance		
FACCH, 8 km/H 14 dB CIR	< 7.3 % WER	0.2 %
FACCH, 100 km/H 12 dB CIR	< 3.3 % WER	0.2 %
SACCH, 8 km/H 14 CIR	< 14 % WER	0.2 %
SACCH, 100 km/H 12 dB CIR	< 9.0 % WER	0.2 %
Class 1, 8 km/H, 17 dB CIR	< 7.1 % WER	0.2 %
Class 1, 100 km/H, 17 dB CIR	< 1.4 % WER	0.2 %
Data Field, 8 km/H, 17 dB CIR	< 3 % BER	0.9 %
Data Field, 50 km/H, 17 dB CIR	< 3 % BER	0.9 %
Data Field, 100 km/H, 17 dB CIR	< 3 % BER	0.9 %
2.3.2.7 Delay interval		
BER Under carrier to noise		
8 km/H, 10.3 usec delay, −103 dBm	< 3 % BER	0.7 %
8 km/H, 20.6 usec delay, −103 dBm	< 3 % BER	0.3 %
8 km/H, 41.2 usec delay, −103 dBm	< 3 % BER	0.07 %
50 km/H, 10.3 usec delay, −103 dBm	< 3 % BER	1 %
50 km/H, 20.6 usec delay, −103 dBm	< 3 % BER	0.5 %
50 km/H, 41.2 usec delay, −103 dBm	< 3 % BER	0.1 %
100 km/H, 10.3 usec delay, −103 dBm	< 3 % BER	1 %
100 km/H, 20.6 usec delay, −103 dBm	< 3 % BER	0.5 %
100 km/H, 41.2 Usec Delay, −103 dBm	< 3 % BER	0.1 %
BER under cochannel interference		
8 km/H, 17 dB Cir, −85 dBm	< 3 % BER	0.4 %
50 km/H, 19 dB Cir, −85 dBm	< 3 % BER	0.5 %
100 km/H, 22 dB Cir, −85 dBm	< 3 % BER	0.5 %
8 km/H, 16 dB Cir, −85 dBm	< 5 % BER	0.5 %
50 km/H, 17 dB Cir, −85 dBm	< 5 % BER	0.3 %
100 km/H, 17 dB Cir, −85 dBm	< 5 % BER	0.8 %
2.4 Conducted spurious emission (dBm)		
No spurious output at antenna terminal out of band	> −47 dBm	> −74 dBm
No spurious output at antenna terminal in RX band	> −80 dBm	> −80 dBm
No spurious output at antenna terminal in Tx band	> −60 dBm	> −60 dBm
2.5 Radiated spurious emission (dBM)		
470 MHz To 1,000 MHz	< −21 dBm	< −69 dBm
3 Transmitter minimum specifications		
3.1 Frequency requirements		
3.1.1 Analog		
3.1.1.2 Frequency stability	± 2.5 ppm	± 1.6 ppm
3.1.1.3 Carrier switching time (usec)		
Attack time	< 2 msec	6 usec
Release time	< 2 msec	4 usec
3.1.1.4 Channel switching time (msec)	< 40 msec	3.5 msec
3.1.2 Digital mode		
3.1.2.1 Reserved		
3.1.2.2 Frequency stability	± 200 Hz	± 10 Hz

Table 2.6 (continued)

IS-55A Specifications		IS-55A Spec	Measured Values
3.2 RF power output requirements			
3.2.1 RF power output			
3.2.1.1 Analog RF power output Pl0-Pl7		+2/–4 dBm	+1/–2 dBm
3.2.1.2 Digital RF power output Pl0-Pl7		+2/–4 dBm	+1/–2 dBm
3.2.2 RF Power transition time			
3.2.2.1 Analog (msec)			
	Pl0-Pl7	< 20 msec	200 usec
	Pl7-Pl0	< 20 msec	800 usec
3.2.2.2 Carrier-on state			
	Inactive	< –60 dBm	< –65 dBm
3.3 Modulation requirements			
3.3.1 Analog			
3.3.1.1 Modulation type And stability (kHz)			
3.3.1.2 Voice modulation			
3.3.1.2.1 Compressor			
	Above 0 dB	± 0.5 dB	± 0.1 dB
	Below 0 dB	± 1 dB	± 0.2 dB
	Attack time	3 ± 0.6 msec	3.58 msec
	Recovery time	13.5 ± 2.7 msec	9.98 msec
3.3.1.2.3 Modulation deviation limiting		< 12 kHz	< 12 kHz
3.3.1.2.4 Audio voice path muting		> 40 dB	> 42 dB
3.3.1.3 Wideband data (kHz)		± 10%	± 1.4 %
3.3.1.4 Supervisory audio tone		± 10 %	± 1 %
3.3.1.5 Signaling tone		± 10 %	± 0.5 %
3.3.1.6 FM hum and noise		< –32 dB	–41 dB
3.3.1.7 Residual amplitude modulation (AM)		5 %	0.5 %
3.3.2 Digital modulation			
3.3.2.1 Modulation type and accuracy			
	Error vector magnitude	12.5 %	8.5 %
	Ten burst error	25 %	9 %
	Origin offset	< –20 dBc	–30 dBc
3.4 Limitations on emissions			
3.4.1 Spectrum noise suppression-broadband			
3.4.1.1 Analog			
Voice plus sat, 20 kHz to 45 kHz from carrier		< 8.7 dBm	–4 dBm
Voice plus sat, > 45 kHz from carrier		< –33 dBm	–45 dBm
Voice plus sat, RX band		< –80 dBm	–90 dBm

2.4 RF COMPONENT REQUIREMENTS FOR TRANSCEIVERS

Radio transceiver circuit designers are often faced with the need to familiarize themselves with the new components that are continually being made available for a particular product family. In particular, the RF functions of a typical transceiver are now being integrated at increasingly high levels onto a single semiconductor die, much the same way that digital IC technology is integrating more and more logic gates on a typical chip. This technological trend is reducing the cost of transceivers even further.

Examination of the performance results shown in Table 2.6 highlights some of the component areas limiting system performance. These include the need for:

1. Duplexers with low insertion loss;
2. Low-noise amplifiers (LNAs) with high intercept point and low current consumption;
3. Mixers with high intercept point and low noise;
4. Filters with low thermal center frequency drift and frequency response tolerance;
5. Low-noise PLLs with fast switching time.

The design considerations of each of these components will be considered more deeply in the following chapters of this book. In particular, the design of mixers is considered in Chapter 5, frequency synthesizers in Chapter 6, and amplifiers in Chapter 7.

2.4.1 Duplexer Requirements

Although the digital mode of the IS-55 specification is indeed half-duplex, and thus the need for a duplexer may be eliminated, the analog mode requirement of IS-55 (and virtually all digital cellular phones) prevents this removal from occurring. Furthermore, extensive GSM experience has indicated that the isolation provided by the duplexer, or alternatively a front-end receive filter, is necessary when a phone is operating in close proximity to one or more other phones. Usually, a duplexer adds an additional 3-dB noise figure in the receiver and 3-dB power loss in the transmit paths at 800 MHz.

In a half-duplex application, the front-end transmit filter requirements are greatly reduced. The > 120-dB transmit-to-receiver noise floor isolation is not required. It follows from this relaxed requirement that critical circuit board layout considerations are correspondingly less stringent. Commercial duplexers available for cellular telephone industry use have the largest physical volume of all other RF components in the transceiver. In handheld portable equipment, the duplexer dimensions and its losses can be a real problem in the enclosure design and the battery life considerations.

2.4.2 Amplifier Requirements

Perhaps one of the hardest specifications a designer of an RF transceiver faces is the balancing act between receiver sensitivity and the intermodulation requirements in conjunction with full transmit power. Current low-noise amplifier components, either in terms of discrete low-noise transistors or low-noise amplifier blocks, have very little margin for tradeoff between LNA gain, receiver sensitivity (or overall noise figure) at 3-dB CNR (carrier-to-noise ratio), and 73-dB intermodulation protection. These limitations are specified by IS-19/TIA-EIA-553 [10, 11] analog cellular telephone specifications and the 3% BER sensitivity at −110-dBm and 52-dB intermodulation limits of the IS-55 dual-mode cellular telephones. At the same time, mismanagement of the transmitter noise floor could certainly doom the overall design.

2.4.3 Mixer Requirements

It is commonly recognized by experienced designers that the selection of the first mixer design topology dominates the receiver chain performance. The mixer performance consideration is accentuated by the need to provide good local oscillator (LO) leakage isolation while maintaining a good balance between the front-end noise figure (NF) and intermodulation isolation. One common design approach replaces the front-end LNA with an active low-noise first mixer. In mobile RF transceiver design, where power drain is not a problem, there are more degrees of freedom in the mixer topology selection, given the fact that the LNA can be biased at higher current and voltage levels to obtain good front-end intercept point performance. In a handheld portable unit, such a luxury is out of the question due to the battery drain budget.

2.4.4 Filter Requirements

Filter designs are the most mature components used among the five items listed above. Filter designs in consumer wireless applications come in a few different forms, dependent on frequency. In the cellular industry, for instance, SAW, dielectric, ceramic, and crystal filters are the most common nonlumped element filters used. The technology for fabricating these elements is quite mature, and they tend to be very inexpensive. However, in modern digital communications, the accuracy of the filter frequency characteristics is frequently a prolific source of product failures due to the more stringent requirements placed on their performance.

The most common cause of severe performance degradation that can be attributable to filter performance in digital communication systems is the passband frequency drift with respect to center frequency of the filter. The passband frequency inaccuracy is usually due to the component production tolerance and thermal characteristics. Most of the digital modulation techniques are extremely sensitive to band-limiting phenomena. In addition, passband drifting also impairs the group delay budget for the modulation. In many cases, active filter equalizations are implemented to counteract this problem. The real tradeoff is in the cost of the design. One can use a broader band filter and satisfy the inadequate adjacent channel rejection requirements by using a tight signal-matched filter with a high signal sample resolution.

In contrast, amplitude distortion seen in the mass-volume-produced commercial filters available today is generally not a serious limiting problem.

2.4.5 Frequency Synthesizer Requirements

The art of PLL design has improved tremendously in the last 10 years. The corresponding chip sets for various PLL structures are now commonplace. Even more esoteric PLL structures such as the fractional-n PLLs, which were not generally in use in the past, can now be seen in some designs. This is because the current challenge in PLL design is to achieve a fast switching time for a small-frequency step size without resorting to multiple PLL designs or using direct digital synthesis (DDS) techniques. Both multiple PLL and DDS options bear the penalty of cost, increased power drain, more printed wiring board (PWB) real estate, and increased EMC problems. Today, existing PLL chip sets are meeting the data communication's demand for switching time only by small margins.

2.5 SYSTEMATIC ANALYSIS OF TRANSCEIVER DESIGN

After mustering all the suitable components for an RF design, the overall transceiver system analysis can be estimated. The cascaded noise figure for an n-stage system is given by [12]

$$NF = NF_1 + \frac{NF_2 - 1}{G_1} + \frac{NF_3 - 1}{G_1 G_2} + \ldots + \frac{NF_n - 1}{G_1 G_2 \ldots G_{n-1}} \qquad (2.1)$$

where NF_1 is the linear noise figure of stage i and G_i is the linear gain of the stage i.

The second-order intercept point (IP2) and the third-order intercept point (IP3) as referenced from the output are given by

$$IP2 = -20\log\left[\sum_i \sqrt{\frac{1}{IP2_i G_{i+1} G_{i+2} \ldots G_n}}\right] \qquad i = 1 \ldots n, \qquad (2.2)$$

$$IP3 = -10\log\left[\sum_i \frac{1}{IP3_i G_{i+1} G_{i+2} \ldots G_n}\right] \qquad i = 1 \ldots n, \qquad (2.3)$$

Using (2.1–2.3), an analysis of the receiver design in Figures 2.2 and 2.3 was carried out, with the results shown in Tables 2.7 through 2.10.

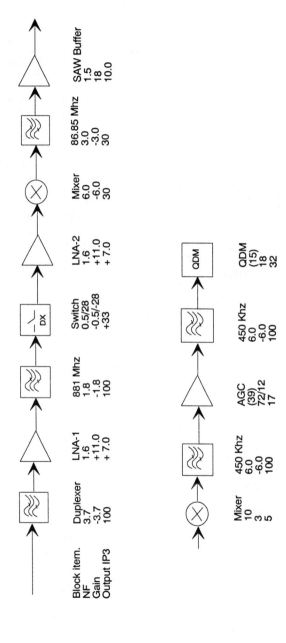

Block item.	Duplexer	LNA-1	881 Mhz	Switch	LNA-2	Mixer	86.85 Mhz	SAW Buffer
NF	3.7	1.6	1.8	0.5/28	1.6	6.0	3.0	1.5
Gain	-3.7	+11.0	-1.8	-0.5/-28	+11.0	-6.0	-3.0	18
Output IP3	100	+ 7.0	100	+33	+ 7.0	30	30	10.0

	Mixer	450 Khz	AGC	450 Khz	QDM
	10	6.0	(39)	6.0	(15)
	3	-6.0	72/12	-6.0	18
	5	100	17	100	32

Figure 2.3 Receiver characteristics used for analysis.

Table 2.7
Small-Signal Analysis Results from Figures 2.2 and 2.3

Receiver Blocks	NF(dB)	Gain (dB)	Output IP3 (dBm)	Cascade NF (dB)	Cascade Gain (dB)	Cascaded IP3 (dBm)	Signal Strength (dBm)
Input							−110.0
Duplexer	3.7	−3.7	100.0	3.7	−3.7	98.46	−113.7
LNA-1	1.6	11.0	7.0	5.3	7.3	7.0	−102.7
881-MHz filter	1.8	−1.8	100.0	5.42	5.5	5.2	−104.5
DX switch	0.5	−0.5	33.0	5.46	5.0	4.69	−105.0
LNA-2	1.6	11.0	7.0	5.63	16.0	6.45	−94.0
UHF mixer	7.0	−6.0	30.0	5.75	10.0	0.44	−100.0
86-MHz filter	3.0	−3.0	30.0	5.87	7.0	−2.56	−103.0
SAW buffer	1.5	15	10.0	5.96	22.0		
VHF mixer	10.0	23.0	5.0	6.02	45.0		
450-kHz filter	6.0	−6.0	100	6.02	39.0	Narrowband Signal	
AGC amp.	(39)	68.0	17.0	6.99	111.0		
450-kHz filter	6.0	−6.0	100.0	6.99	105.0		
QDM	(20)	18.0	32.0	6.99	123.0		

Table 2.8
Summary of Small-Signal Results from Figures 2.2 and 2.3

Performance Parameter	Result
KTB (BW = 24.3 kHz)	−130.14 dBm
NF	6.99 dB
System noise floor	123.15 dBm
Minimum signal E_s/N_0	12 dB
(For 4-phase differential detection @ BER = 10^{-3})	
Implementation loss	1 dB
Required CNR in AWGN	13 dB
Sensitivity @ BER = 1e − 3	−110.15 dBm
Intended minimum signal	−110 dBm
Input IP3	−9.56 dBm
EISA IM method	
IS-55 -2.3.2.3	> 60 dB
IP3 /signal strength	> 20 dB minimum.

Table 2.9
Large-Signal Analysis Results from Figures 2.2 and 2.3

Receiver Blocks	NF(dB)	Gain (dB)	Output IP3 (dBm)	Cascade NF (dB)	Cascade Gain (dB)	Cascaded IP3 (dBm)	Signal Strength (dBm)
Input							−25
Duplexer	3.7	−3.7	100.0	3.7	−3.7	98.46	−28.7
LNA-1	1.6	11.0	7.0	5.3	7.3	7.0	−20.7
860-MHz filter	1.8	−1.8	100.0	5.42	5.5	5.2	−22.5
DX switch	28	−28	33.0	22.58	−22.5	−22.8	−50.5
LNA-2	1.6	11.0	7.0	24.15	−11.5	−1186	−39.5
UHF mixer	7.0	−6.0	30.0	25.01	−17.5	−17.86	−45.5
86-MHz filter	3.0	−3.0	30.0	25.72	−20.5	−20.86	−48.5
SAW buffer	1.5	15	10.0	26.22	−5.5		
VHF Mixer	10.0	3.0	5.0	26.54	−2.5		
450-kHz filter	6.0	−6.0	100	26.59	−8.5	Narrowband Signal	
AGC amp.	(39)	12.0	17.0	47.53	3.5		
450-kHz filter	6.0	−6.0	100.0	47.53	−2.5		
QDM	(20)	18.0	32.0	47.55			

Table 2.10
Summary of Large-Signal Results from Figures 2.2 and 2.3

Performance Parameter	Results
KTB (BW = 24.3 kHz)	−130.14 dBm
NF	47.55 dB
System noise floor	−82.59 dBm
Required signal CNR in AWGN	13 dB
Sensitivity @ BER=$1e-3$	−69.59 dBm
Intended maximum signal	−25 dBm
IP3/signal strength	> 20 dB minimum

2.6 DIGITAL COMMUNICATIONS REQUIREMENTS

The rapid acceptance of digital communication techniques has created an interesting predicament for practicing RF circuit design engineers. The systems parameters suited to digital communication analyses are generally different from conventional RF circuit design parameters or their measurement parameters. Depending on a particular digital communication application, the digital communication system requirements impose different RF transceiver specifications for the same parameter value. Here, two of the fundamental system parameters that are often faced by RF transceivers at the beginning of a design are briefly discussed. They are the bit energy requirement of the system and phase noise specifications. Experience has demonstrated that these two parameters are the most common sources of confusion for RF transceiver designers.

For example, the aggregate receiver phase noise requirement for oscillators is commonly specified as a fixed rms double-side-band noise variance in radians over a signal bandwidth. The signal bandwidth for the phase noise calculation is a function bounded by the product of all the transfer functions of all the filters in the RF transceiver, the signal matched filter, and the highpass transfer function of the automatic frequency control (AFC). This integrated phase noise specification is itself dependent on the required bit error rate (BER) of the system.

The estimation of the required receiver sensitivity is usually based on the minimum E_b/N_0 of the digital communication specification for a set value of BER. Thus,

$$E_b = PT_b \tag{2.4}$$

where, E_b is the energy per bit, P is the power required and T_b is the bit time.

It follows that the theoretical probability of bit error in an additive white Gaussian noise (AWGN) environment with respect to the E_b/N_0 for M'ary PSK can be expressed as

$$\text{Pb} = \frac{1}{2}(1 - \text{erf}(\sqrt{E_b/N_0})) \tag{2.5}$$

In order to estimate the required receiver sensitivity, one needs to know the M'ary nature of the modulation scheme; that is, how many bits per hertz in the signal. For BPSK, the required E_b/N_0 for a 10^{-3} BER in AWGN environment from (2.5) is 6.8 dB. For QPSK, where there are two bits per symbol, the same energy E_b/N_0 value holds for the same BER, but the required symbol power spectral density within the signal bandwidth can then calculated from (2.4) with $T_b = 1/2T$, so the power density required per symbol is in fact twice that of BPSK.

Thus, for example, to achieve BER of 10^{-3} for QPSK symbol rate of 24.3 Ksps, the achievable sensitivity of the receiver using coherent detection can be calculated as

Estimated Receiver Sensitivity = Receiver Noise Floor

$$+ \text{Required SNR} + \text{Implementation Loss} \tag{2.6}$$

where

Receiver Noise Floor = KT + 10log(Bandwidth Hz) + NF(dB) (2.7)

and

Required SNR = E_s/N_0(dB)

$\qquad$ + 10 log (Symbols per second) − 10 log (Bandwidth Hz) (2.8)

$\quad$ On combining these, the receiver bandwidth cancels, leaving

Estimated Receiver Sensitivity = $\ $ kT + NF(dB) + E_s/N_0 (dB)

$\qquad$ + 10 log (Symbols per second)

$\qquad$ + Implementation Loss

$\quad$ = −174 + 7 + E_s/N_0(dB) + 10log(24.3 × 10³) + Implementation Loss

$\quad$ = −123.14 dBm + E_s/N_0(dB) + Implementation Loss (2.9)

assuming KT = −174 dBm/Hz (at room temperature) and system noise figure = 7 dB.

$\quad$ Now, the E_b/N_0 required = 6.8 dB (from (2.5)), and so the QPSK symbol power required is Es/N_0 = 6.8 +3 = 9.8 dB. If the estimated implementation loss = 1 dB, then the actual estimated receiver sensitivity at a BER of 10^{-3} is = −123.14 + 9.8 + 1 = −112.34 dBm.

$\quad$ An interesting comparison with this calculation can be made with an IS-55 digital mode receiver sensitivity requirement of BER = 3%, or $3 \cdot 10^{-2}$, in an AWGN environment, specified at −110 dBm using differential detection. Given that a four-phase differential detection performance curve is typically very close to 2.3 dB degradation equidistant from the AWGN BER versus E_b/N_0 curve for most of the E_b/N_0 range, excluding the nonreducible error floor; this is an interesting exercise for the RF transceiver designer new to digital communications to work out.

$\quad$ The next parameter often encountered by the RF transceiver designer is the phase noise requirement specifications. In a conventional phase noise measurement, as shown in Figure 2.4, phase noise is typically specified as a spectral density in dBc/Hz at a certain frequency offset. A crucial question to answer is: how much noise variance is introduced into the digital system due to this phase noise?

$\quad$ Ignoring the filter skirt residuals, the noise variance (σ^2) is the total phase noise existing within the single sideband of the phase noise profile shown in Figure 2.4 from the highpass 3-dB knee of AFC bandwidth to the lowpass 3-dB knee of the signal matched filter. A simple graphical integration method of Figure 2.4 using a straight-line approximation has been derived below [13,14]:

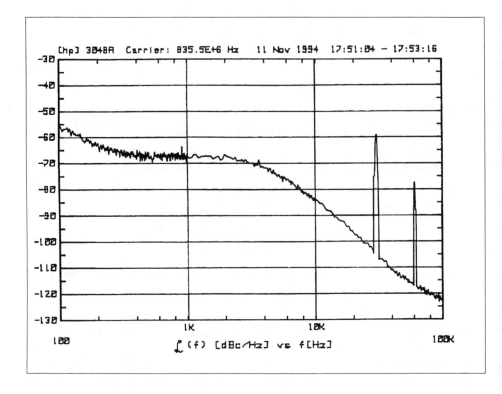

Figure 2.4. Measured phase noise plot of IS-55 transmit carrier. Integration bandwidth is between 200 Hz and 30 kHz.

$$\sigma^2 = \sum_i \left\{ \begin{array}{l} 10^{\frac{\text{dBc}_i}{10}} f_i \log\frac{f_{i+1}}{f_i} \rightarrow x = -1, \\ 10^{\frac{\text{dbC}_i}{10}} \frac{f_i}{x+1} \left[\left[\frac{f_{i+1}}{f_i}\right]^{x+1} - 1 \right] \rightarrow x \neq -1 \end{array} \right\} \tag{2.10}$$

where

$$x = \frac{\text{dBc}_{i+1} - \text{dBc}_i}{10(\log f_{i+1} - \log f_i)}$$

With the value of σ^2 obtained above, one can now estimate the degradation in digital performance by calculating the probability of the bit errors. For BPSK, let $\rho = \sqrt{2}\sigma$. Then, the probability of error can be written as [13–15]

$$P_e = \frac{1}{\sqrt{2\pi}\rho} \int_{-\infty}^{\infty} \exp\left(-\frac{\phi^2}{2\rho^2}\right) \text{erfc}\left(\sqrt{\frac{2E_b}{N_0}}\cos\phi\right) d\phi \qquad (2.11)$$

Other modulation schemes require similar derivations to establish the impact of phase noise on their performance [16].

Using (2.10), the integrated signal sideband phase noise can be directly computed by a graphical integration method. Figure 2.5 shows the accumulation of the sectional integration in terms of σ in degrees rms of the phase noise profile depicted in Figure 2.4.

Using Equation (2.11), it can be shown that the phase noise contribution to the BER performance, as indicated by $\rho = \sqrt{2}\sigma$ for $\pi/4$ QPSK modulation, is not severe even if the phase noise performance is relatively poor. For example, the profile of Figure 2.5 is better than most digital equipment specifications—it shows a double sideband phase noise of 1.98 deg rms. In general, digital communication equipment specifications are overly restrictive in terms of phase noise. The conventional approach has been to use the empirical measurement of a typical system combination of oscillators—in terms of dBc/Hz related to an offset frequency from the carrier—as a profile for the required specifications. Often, the attempt to satisfy the phase

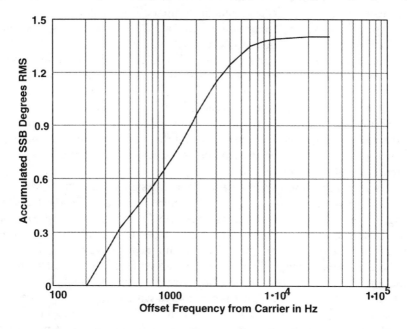

Figure 2.5 Accumulated single sideband phase noise in degrees rms of the phase noise profile shown in Figure 2.4, based on the graphical integration technique described in (2.10).

noise profile resulted in compromises at the expense of channel-switching time, frequency-switching transient response, and, more importantly, per unit cost. These requirements can be relaxed somewhat with an improved understanding of the effect of phase noise on digital system performance.

The introduction of the modulation accuracy [9] concept leads to a more realistic performance specification technique, which encompasses the phase noise contribution as one of many performance elements for the transceiver design. With this technique, the modulated signal is measured against a set percentage of rms signal vector error for its signal constellation. The imperfection in the modulated signal attributable to phase noise, I and Q imbalance, filter imperfections, and circuit noise floor can be determined by a single modulation error vector specification. This approach allows the circuit designer flexibility in circuit design tradeoffs and the communication system designers know explicitly the system performance that the transceiver can deliver.

2.7 ELECTROMAGNETIC COMPATIBILITY IN WIRELESS DEVICES

In RF design processes, one of the most common mistakes committed by RF designers is the omission of up-front electromagnetic compatibility (EMC) considerations. As a result, typical transceiver designs will validate the 20-80 rule; that is, it is common that the last 20% of the RF design takes 80% of the schedule.

Furthermore, consumer and commercial wireless products are not only governed by regulatory rulings, but they are also governed by user expectations. Indeed, the EMC performance requirements in this aspect are generally more stringent than those of a regulatory nature. For example, a consumer would not be satisfied with a mobile cellular telephone installation in his or her car if it means that the FM and AM radio will no longer work properly, even if the mobile cellular telephone passes the regulatory agency requirements. A wireless data modem that causes television interference is equally unacceptable.

To ascertain that the wireless data product design satisfies the EMC requirements in the early phase of the design cycle requires very careful project design and planning. This is because most of the baseband ASICs and noisy digital components may not be available when the tooling commitment for the plastic enclosures, as well as the constraints for other mechanical components, have to be finalized. As a result, the transceiver design is usually debugged without the benefit of the noise contribution information from the baseband circuitry until the final phase of system integration. By then, the RF designer is confronted with a tight project schedule and budget, as well as limited manpower and system workaround flexibility.

One good strategy to mitigate these problems is to use equivalent noisy baseband devices as placeholders for the yet-to-be available baseband ASICs in the initial circuit layouts to present a realistic noisy background for the transceiver circuits.

For example, fast DSPs can be placed to execute endless loop routines, and oscillators with fast-switching interfaces can be placed as dummy placeholders at circuit card locations where the real devices will be eventually go. As a result, EMC problems such as receiver desensitization, broadband frequency emissions, PLL signal contamination as a function of circuit board partition, and EMC shielding requirements can be minimized when options for workaround or design alternatives are easily available.

All wireless communication devices, irrespective of whether they have transceivers or receivers only, are required to pass FCC Part 15.109 Subpart (a) or (b) [17, 18] emission limits dependent on their classifications for use in the United States. This is a requirement for all classifications, and is common to all wireless communication products. It applies to all wireless communications products irrespective of whether a particular product has a transmitter or not, and also irrespective of whether the product can be suitably classified under other relevant emission regulations. FCC Part 15.109 Class A devices are for commercial users and are relaxed by approximately 10 dB compared with FCC Part 15.109 Class B devices, which are for consumer usage. Tables 2.11 and 2.12 show their legal limits.

As can be seen from the FCC Part 15.109 requirements, the emission interference potential to other established domestic appliances, such as broadcast radios

Table 2.11
FCC Part 15.109 Limits (Radiated Emission Limits for
Class B Devices Measured at a Distance of 3 Meters)

Frequency of Emission (MHz)	Field Strength (μV/meter)
30–88	100
88–216	150
216–960	200
Above 960	500

Table 2.12
FCC Part 15.109 Limits (Radiated Emission Limits for
Class A Devices Measured at a Distance of 10 Meters)

Frequency of Emission (MHz)	Field Strength (μV/meter)
30–88	90
88–216	150
216–960	210
Above 960	500

and garage door opener sets, are still considerable, even if the regulatory emission limits are satisfied. It is the duty of the RF transceiver systems designer to impose his or her own specific emission limits as a supplementary requirement to that of the regulatory agency. Outside the United States, one is required to satisfy foreign regulatory limits, and they are generally dominated by VDE and CISPR equivalents of the FCC regulations. These limits are a few decibels more stringent than that of the current equivalent FCC Part 15 rule in most cases. In the European Union countries, the Information Technology Equipment (ITE) family of emission standards are in a state of transition. The real impact of the new EMC requirements can only be assessed after 1996 when they are promulgated.

2.8 CONCLUSIONS

RF design engineers who engage in the design of wireless devices that utilize the maturity of digital communication technology are now facing a new knowledge gap. This gap is a consequence of the disparity of terminology and parlance of the digital communication technology with that of established RF design and measurement techniques. The consequence of this knowledge gap manifests itself in an unusually high design failure rate of the transceivers intended for digital communication use.

However, the explosive demand for wireless applications is just beginning to be felt by designers and the financial investors in these applications. A multitude of traditional analog voice and low-efficiency data communications devices have been updated or replaced by highly efficient digital standards. It is now necessary for practicing RF engineers and technologists to bridge this knowledge and language gap quickly and effectively to be ready for the challenge. In this chapter, some of the introductory facets of this effort have been briefly presented in a narrative manner. Comprehensive tutorials that enable a digital communications engineer to convey his or her requirements fully in the "traditional" RF parameters, or vice versa, are few and are thinly scattered in the open literature. In most instances today, successful designs are only possible when there is a group of close-knit engineers of both disciplines. This should not be necessary when the knowledge horizon of the RF engineers is extended into the digital domain.

References

[1] TIA/EIA Interim Standard, "Cellular System Dual Mode Mobile Station-Base Station Compatibility Standard," IS-54B, April 1992.

[2] TIA/EIA Interim Standard, "Mobile Station-Base Station Compatibility Standard for Dual Mode Wideband Spread Spectrum Cellular System," TIA/EIA/IS-95, July 1993.

[3] J. E. Padgett, C. G. Gunther, and T. Hattori, "Overview of Wireless Personal Communications," *IEEE Communications Mag.*, Vol. 33, No. 1, Jan. 1995, pp. 28–41.

[4] K. Pahlavan, and A. H. Levesque, "Wireless Data Communications," *Proc. of IEEE*, Vol. 82, No. 9, Sept. 1994, pp. 1398–1430.

[5] A. Flatman, "Wireless LANs: Developments in Technology and Standards," *IEEE Computing & Control Engineering Journal*, Vol. 5, No. 5, Oct. 1994, pp. 219–224.

[6] FCC 47 CFR, 10-1-93 Edition, pp. 581–582.

[7] FCC 47 CFR, 10-1-88 Edition, pp. 212–443

[8] TIA/EIA Standard, "Land Mobile FM or PM Communications Equipment Measurement and Performance Standards," TIA/EIA-603, Feb. 1993.

[9] EIA/TIA Interim Standard, "Recommended Minimum Performance Standard for 800 MHz Dual Mode Mobile Stations," EIA/TIA/IS-55, Dec. 1991.

[10] EIA Interim Standard, "Recommended Minimum Standards for 800 MHz Cellular Subscriber Units," EIA/IS-19-B, May 1988.

[11] EIA/TIA Standard, "Mobile Station-Land Station Compatibility Specification," EIA/TIA-553, Sept. 1989.

[12] R. Sagers, "Intercept Point and Undesired Responses," *Trans. IEEE Veh. Technol.*, Vol. VT-32, No. 1, Feb. 1983, pp. 121–133.

[13] J. Y. C. Cheah, K. Karimullah, and M. E. Davis, "Wireless Data Modem , Theory and Implementation," Reading, MA: Addison-Wesley, to be published.

[14] J. Y. C. Cheah, "Analysis of Phase Noise in Oscillators," RF Design, Nov. 1991, pp. 99–104.

[15] J. Y. C. Cheah, "Frequency Reference in VSAT," *IEEE Trans. Communications*, Vol. 42, No. 2/3/4, Feb./Mar/April 1994, pp. 233–236.

[16] R. P. Gilmore, "Specifying Local Oscillator Phase Noise Performance: How good is good enough?" *IEEE International Symp. on Personal, Indoor and Mobile Radio Communications*, Sept. 1991, pp. 166–172.

[17] FCC Gen. Docket No. 89-118, June 29, 1989.

[18] FCC Gen. Docket No. 89-116, June 29, 1989.

CHAPTER 3
▼▼▼

LOW-POWER RADIO-FREQUENCY ICS FOR PORTABLE COMMUNICATIONS

Portions of this chapter originally appeared in the April 1995 *Proceedings of the IEEE Special Issue on Low-Power Electronics* and appear here with permission.

Asad A. Abidi
Electrical Engineering Department
University of California
Los Angeles, CA

3.1 INTRODUCTION

The portable revolution is upon us today. It promises to empower individuals throughout the world by giving them low-cost access to information wherever they may be, thus allowing them to make informed decisions and to be more productive in business and at home without necessarily being tied down to a physical location. It is expected that in the near future, individuals will be equipped with capabilities of local computing and of communications enabling them to perform almost all the tasks that today require the equipment on the office desktop: the telephone, the computer, its connection to a ubiquitous wired network, the fax machine, and so on [1].

The portable revolution has been many years in the making. The personal broadcast radio receiver and cassette player, as pioneered by Sony, was a runaway global success. The user could construct a private audio environment anytime and anywhere, using a device so small and light that its presence was easily forgotten. Portable computers brought about the next wave of change. The personal computer has had such a large impact on the broad working habits of individuals that without it they are lost. The luggable computer has evolved in a matter of one decade into the portable, the notebook, and the subnotebook. With the establishment of a wide area network (WAN) of radio paging transmitters, the personal radio pager has also became very popular since the mid-1980s. Today, the pager network spans the entire continental U.S. and many other parts of the world, enabling the user to receive alphanumeric electronic mail messages. Two-way paging is actively under development. The cellular telephone became widely available shortly thereafter. The user could hook into the international switched-telephone network through the nearest cellular base station with a portable transceiver, which too has scaled down remarkably in size until its weight and volume is the smallest practical [2].

There are many competing visions of how these various portable devices and services will evolve and integrate in the next few years. These are covered extensively in the popular press and in numerous keynote speeches in technical meetings. A common theme is that users will want a multimedia terminal, capable of wireless access to a global network that can transport communications, images, and databases to the user in an on-demand, interactive fashion [3, 4]. Such a terminal will have capabilities of computing, image acquisition and display, and, obviously, communications. It will likely derive as a hybrid of the various portable technologies available today.

What obstacles must be overcome to realize this vision? These include the very highly integrated electronics, effective displays, and a philosophy of design based on low power dissipation to prolong the battery life of the portable device. Sometimes single-battery operation will impose the additional constraint of operation at low voltage, as low as 1V, which will require entirely new ways of doing electronic circuit design.

In the past few years, most designers of mass-market digital integrated circuits (ICs) have been preoccupied with low-power operation [5]. Principles such as operating CMOS logic at the lowest possible supply voltage have become widely known, and power-down modes, gear-shifting of operating clock frequencies, pipelining and parallelism, subthreshold operation, and other such methods that were once the province of specialized areas such as electronic wristwatches and implantable biomedical devices are becoming commonplace [6]. Studies into the fundamental thermodynamic limits to the energy required for computation are being initiated or revived. There is good reason to believe that all this activity will lead to significant improvements in the conventional circuit and system design styles for digital signal processing and computation.

How will this activity affect the communications aspects of the portable device? What similar principles to low-power digital design are there for energy-efficient wireless communications transceivers? These questions do not have simple answers. Low-power communication systems will result from use of the correct architectures, a sensible partition between analog and digital signal processing, low-power circuit techniques everywhere, and a judicious division between active and passive components. There is still not a widely known, integrated vision on this subject. Furthermore, there remains a gap between the IC design used in the portable applications described above and the new designs that will be required over the next few years for advanced portable communicators. Consumers are demanding a great deal more functionality and performance, which is stressing present-day technology to its limits, and wireless communicator design itself is in transition from the classic analog modulation techniques used over the past 50 to 70 years to more sophisticated methods using digital signaling formats and signal-processing methods in transceivers.

This chapter summarizes the key developments in the discipline so far, and from them forecasts wireless IC design trends in the near future.

3.2 KEY SIGNAL-PROCESSING ISSUES IN WIRELESS TRANSCEIVERS

Were it not for the advent of the portable communications revolution, radio technique would almost certainly have become a lost art. The first edition of the last definitive textbook on the subject dates to 1943 [7]. Whereas once radio engineering was synonymous with electronics [8], few university electronics curricula today offer a course on radio communications circuits. Only a few modern textbooks on radio design have been written in the past 25 years [9–13]. Radio communication methods, at least for nonmilitary applications, have remained relatively unchanged since World War II, and the evolutionary improvements in consumer equipment mainly owes to the use of high-frequency discrete transistors [14], smaller passive components, and building-block ICs that improve the long-term reliability and manufacturability of radio and TV receivers. The major impact of IC technology in these consumer items has probably been at baseband, in adding more user features. In contrast, the front-end radio architectures have evolved almost not at all in the past 40 or so years. For instance, ICs have contributed digital volume control, digital frequency tuning, and features to alleviate manual effort on the part of the user, but the RF and intermediate frequency (IF) sections still contain discrete and passive components in rather conventional architectures.

Why is this? It is partly because radio frequencies were too high for the low-cost IC technologies traditionally used in the consumer electronics industry. It is also because advances in component packaging alone have led to rapid downscaling in the size of consumer devices, often obviating the need to rethink the electronics. As a result, only a few individuals in a handful of institutions worldwide have concerned themselves with thinking about these problems. Today, as conventional solutions no longer suffice for the future wireless communications devices, there is a rekindling of interest in this subject, which has led to much rediscovery and some invention. Baseband IC designers are now attempting to apply familiar techniques to wireless, while microwave IC designers are exploring what to them are low-frequency commercial opportunities for their technologies.

To set the stage for further discussion, some of the unique problems of radio receivers and transmitters are first described. Unlike familiar wireline communications, the wireless environment accommodates essentially an unlimited number of users sharing different parts of the spectrum, and very strong signals coexist next to the very weak. The radio receiver must be able to select the signal of interest while rejecting all others. It must do so using less than perfect active and passive components. There are two important problems in the receiver: *image rejection* and *dynamic range*. Image rejection relates to the receiver's ability to select the desired signal from the array of signals occupying the spectrum. Ideally, it might do so with a tunable bandpass filter, whose center frequency could be positioned at will in the RF and whose passband was one channel wide. A filter with this small a fractional passband does not exist. Instead, a practical RF bandpass filer, which may or may not be tunable, will preselect an array of radio channels including the one of interest (Figure 3.1). The other preselected channels are then removed at a lower IF by translating them in frequency with a downconversion mixer and centering the desired channel within a bandpass filter at IF. The other mixer input is a frequency-tunable local oscillator (LO), offset by IF from the desired channel. As the preselected band after downconversion will very likely occupy an interval greater than (0, IF) on the frequency axis, the IF bandpass filter will select both the desired channel and another *image* channel the mixer has translated to −IF. The subsequent detector circuit will be unable to distinguish between the desired and the image channels, and therefore

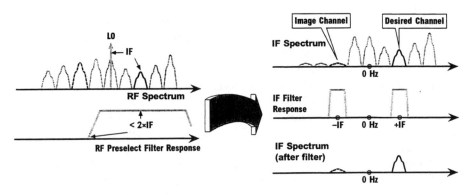

Figure 3.1 The image-rejection problem in radio receivers. The RF preselect filter passband must be determined with prior knowledge of the IF. The preselect filter is responsible for image channel suppression before downconversion.

its output will be the result of the superposition of both. However, if the stopband of the preselect filter lies less than $2 \times$ IF away from the desired channel, it will attenuate the image, so only the desired channel will contribute energy at IF. The receiver designer first studies the available filter technologies and then chooses an appropriate IF, which yields an acceptable image suppression. A high IF relaxes the prefilter passband specification, but it also means that the downconverter signal requires high-frequency amplifiers, which are usually power-inefficient. Further, the IF filter requires a smaller fractional passband. In such cases, following image rejection at this high IF, the channel may be selected after downconversion to a second, lower IF. In such a double superheterodyne, or dual-conversion receiver, the first IF may actually lie at a higher frequency than the incoming RF to make image rejection easier.

The noise level and nonlinearity in the RF amplifier and first mixer usually set the receiver dynamic range. Consider reception of a weak channel surrounded by large undesired channels in the preselection band (Figure 3.2(a)). First, the input-referred noise of the receiver directly adds to the sought signal, corrupting its signal-to-noise ratio (SNR). Second, the large adjacent channels will experience the nonlinearities in the RF amplifier and mixer, and some of the products of the ensuing intermodulation distortion may overlap the desired channel. The IF filter cannot reject these unwanted products, which, like noise, will degrade the received SNR. As the receiver frequency response is normally bandpass, its nonlinearity is measured by applying two tones of equal amplitude closely spaced in frequency (f_1 and f_2) at its input, and measuring the rise in the third-order intermodulation products (at $2f_1 - f_2$ and $2f_2 - f_1$) with the input lever (Figure 3.2(b)). All other intermodulation tones usually lie outside the receiver passband. On a logarithmic plot, the third-order intermodulation level rises at a slope of three relative to the fundamental tone at the output. The two lines intersect at a point called the input-referred *third-order intercept* (IP3). The intercept point is usually extrapolated from measurements at low levels, because the receiver front end will saturate at large inputs. The 1-dB *compression point*, the input level that causes the receiver gain to drop by 1-dB relative to the small-signal gain, specifies the onset of saturation. The input-referred noise level may be included in this plot to define a spurious-free dynamic range (SFDR), although this is rarely used in radio specification. Usually the input noise figure (NF) and the input-referred interception point (IP3) are separately specified.

Similar specifications apply to the transmitter, which operates at much larger signals. Suppose, as is almost always the case, that the transmitter is required to emit a single-sideband suppressed-carrier output; that is, a signal where the LO and the image frequency have been suppressed. However, owing to circuit imperfections, it may also emit small amounts of the carrier and the unwanted sideband, which typically lie in the passband of the subsequent RF filter (Figure 3.3). These unwanted emissions may become interferers for adjacent channels. Nonlinearities in the power amplifier may also produce emissions of intermodulation products at other frequen-

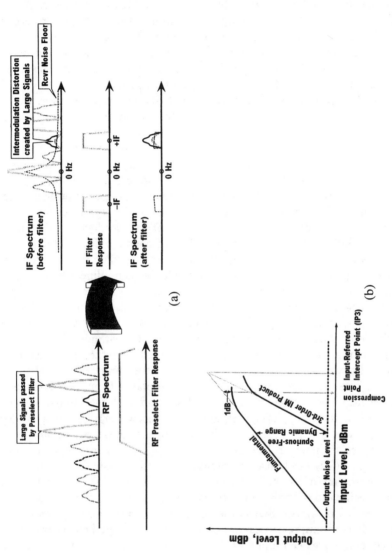

Figure 3.2 The dynamic range problem in the radio receiver. (a) Adjacent large signals may create intermodulation products superimposed on the desired channel. Receiver noise floor is a fundamental limit to sensitivity. (b) Receiver dynamic range is specified (box) in terms of extrapolated third-order intercept point and noise level.

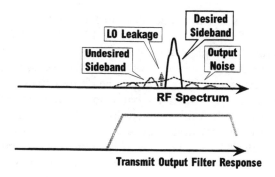

Figure 3.3 Transmitter imperfections (such as mismatches in a quadrature upconverter) result in the appearance of spurii in the emitted spectrum (referred to as "spectral regrowth"). The spurii may not be removed by the output filter, and may superimpose on adjacent channels. Emitted noise might overwhelm weak adjacent channels.

cies. Phase noise in the local oscillator responsible for upconverting to RF will convert to noise added to the signal, and the amplitude of this noise increases with the transmitted signal. This noise could possibly overwhelm nearby weak channels. Transmitter performance is usually specified in terms of the relative levels of unwanted signals to the desired signal and in terms of absolute spectral density of output noise at maximum output power.

To understand the rationale underlying receiver architecture, let us use as an example the familiar broadcast FM receiver. The architecture to be described is the same that Armstrong, the inventor of FM and the superheterodyne, had originally proposed for FM reception. The desired channel consists of a carrier in the 88- to 108-MHz band, modulated by up to ±75 kHz. Neighboring channels are spaced apart by 200 kHz. The receiver must select the desired channel while rejecting nearby channels, and it must be sensitive to a signal of a few tens of microvolts induced on the antenna. A simple FM antenna is wideband, and will pick up signals well outside the broadcast FM band. The low-noise bandpass amplifier in the front end may at best mildly attenuate the out-of-band signals—the actual channel selection must be done elsewhere. The RF amplifier uses an inductive load to resonate with the transistor and tuning capacitances, thereby transforming an inherently lowpass characteristic to bandpass centered on the frequencies of interest. The transistor f_{max} limits the highest frequency at which such a tuned amplifier can still provide a gain greater than unity. This figure of merit is familiar to microwave circuit designers and to device designers, whereas baseband IC designers deal more often with transistor f_T, the capacitance-limited unity current-gain frequency. In bipolar IC processes not optimized for small-signal high-frequency use, f_{max} is comparable to f_T, whereas in the best RF processes it may be twice f_T [15, 16].

It is impossible to select the desired channel at RF, because no tunable filters exist with the required fractional bandwidth of 0.15% (150 kHz/100 MHz). Therefore, following sufficient amplification at RF to overcome the noise level of the following circuits, the signal is mixed down by a variable-frequency local oscillator to a lower IF. Furthermore, a filter to select the desired channel at IF will have a fixed-center frequency and the fractional bandwidth in the passband will be larger. From this perspective, it is desirable to use as low an IF as possible. However, image rejection poses yet another constraint on the choice of IF. Conventional broadcast FM receivers use an IF of 10.7 MHz as a compromise. This IF guarantees that the image always lies outside the FM band (Figure 3.4). It is unlikely, however, that the preselect filter can entirely suppress this image, which will therefore either add noise or AM to the desired signal. However, the subsequent FM detector is inherently insensitive to both these forms of impairment. In this way, a medium-valued IF is made possible by exploiting properties of the detector, and thereby the fractional passband specification for the preselect filter is relaxed. The first mixer downconverts the entire FM band, with the desired channel centered at 10.7 MHz. A varactor-tuned Colpitts oscillator may be used as the first local oscillator. Prior to detection, a cascade of identical fixed-frequency ceramic bandpass filters, each with a 200-kHz passband centered at 10.7 MHz, passes the desired channel while rejecting neighboring channels [17]. In a high-quality receiver, the RF amplifier may be a discrete GaAs MESFET with a tuned load ganged to the LO tuning element, and another ganged tuned circuit may couple the antenna signal into the receiver [17].

To transplant this style of discrete radio circuit design to ICs, one would have to implement LC tuned circuits and filters on silicon. One can indirectly surmise these concerns in the Motorola series of IC design textbooks from the 1960s, which discuss loss in spiral metal inductors fabricated on silicon substrates [18] as well as issues relating to simulated inductors for active filters at IF for radios made with gyrators and capacitors [19]. However, on-chip spiral inductors of useful values

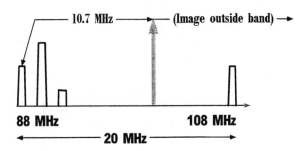

Figure 3.4 The 10.7-MHz intermediate frequency conventionally used in FM receivers is the lowest frequency that will guarantee that the image lies outside the FM broadcast band. The FM demodulator will reject a (non-FM) image channel entering the receiver. The 10.7-MHz passive bandpass filters in the IF strip select the desired channel.

were found to suffer excessive capacitance to the substrate, which lowered their self-resonant frequency to the point that they were not usable beyond the VHF band. It gradually became part of the collective consciousness of IC designers that on-chip tuned circuits are generally impractical.

When useful tuned amplifiers did appear on monolithic integrated circuits, it was not for the VHF-to-UHF range of relevance to consumer applications. Instead, it was military applications at much higher frequencies that drove the development of monolithic microwave integrated circuits (MMICs) during the 1980s. MMICs typically use MESFETs as the active device on semi-insulating GaAs substrates. This technology has enabled miniature radar, remote sensing, and communications at frequencies up to tens of gigahertz. The on-chip wavelengths are so small that monolithic distributed circuits may be built. MMICs take advantage of the semi-insulating substrate in two important ways. Transistors on these substrates have lower parasitic capacitance, which means that they amplify to higher frequencies. It is also possible to build low-capacitance interconnect with air bridge structures, and high-frequency passive components such as spiral inductors required for narrowband tuned circuits. Thus, on MMICs, the board-level design styles used hitherto by radio and microwave engineers could be miniaturized. However, over its many years of existence, GaAs MMIC technology has not had the major impact on consumer electronics that its adherents had hoped for. Makers of consumer electronics favor silicon IC technology wherever feasible because of its low cost, high yields, and the relative ease of mixing analog and digital circuits on a large scale.

It is anticipated that by the year 2000 about 300 million portable consumer wireless devices will be in use [20]. What IC technologies will enable the RF front end of these devices? Do miniaturization and long battery life call for architectural innovations in transceivers? What new circuit design styles will evolve in response? There is much curiosity and speculation on these matters, yet little is generally known about RF-IC design, or on the possible impact of large-scale integration and power-reduction strategies in the front end of wireless transceivers. The next sections present a brief survey of the use of IC technology in wireless receivers and transmitters since the 1970s to date, and from this projects some future trends. RF-ICs are roughly defined as integrated circuits operating in the band of frequencies from 400 MHz to 2500 MHz, which covers most consumer wireless communication devices. As opposed to MMICs, which were almost exclusively fabricated on III-V component semiconductor substrates at small scales of integration, mature silicon technologies will play a large, if not the dominant, role in RF-IC fabrication. It is the author's belief that in response to pressing demands of ubiquitous wireless access, both the underlying semiconductor technology and the design styles will rapidly evolve to realize the single-chip "VLSI radio" in the not too distant future.

3.3 ICs IN BROADCAST RADIO RECEIVERS

The two-way wrist radio has fascinated the popular imagination since its introduction in the popular American cartoon strip *Dick Tracy* in 1946 [21]. This is the ultimately unobtrusive piece of consumer electronics. Let us now see how feasible it is to build an FM receiver of this size with microelectronics technology. For the average user to accept such a radio, its selectivity and sensitivity must be comparable to that of tabletop models. If, in the FM receiver described in the previous section, all the transistors in the receiver electronics were to be integrated onto one silicon chip, the radio would still need a considerable number of off-chip tunable inductors and ceramic filters, and in spite of state-of-the-art miniature packaging, the components could not plausibly all fit into a wristwatch. Neither would the power dissipation be commensurate with the life of a wristwatch battery.

The first generation of silicon bipolar ICs developed in the late 1970s for the IF and baseband portions of broadcast receivers more or less contained the transistors of conventional receivers assembled onto one or more ICs [22–24]. However, integration did afford freedom to use transistor-rich circuits for higher performance. Circuit techniques such as double-balanced mixers using Gilbert analog multiplier, PLLs as FM demodulators, and use of balanced on-chip signal paths to attain greater immunity to pickup and common-mode noise became widely used as a result. By eliminating many of the coupling coils and other noncritical discrete components found in older radio circuits, ICs contributed to lowering the cost of assembling and aligning the final product. In the RF section, though, the receivers still used the conventional 10.7-MHz IF superheterodyne architecture implemented with shielded discrete-component circuits.

In the early 1980s, Kasperkovitz (at Philips) [25] made the first significant explorations into alternative architectures for highly integrated radio receivers. He realized that to reduce receiver size and power dissipation, it was very important to eliminate the many off-chip passive components. If certain passive inductors, capacitors, and resistors could not be eliminated, they could at least be packaged in surface-mount outlines for very small size. However, neither the volume of the IF ceramic filters could be readily scaled down, nor could their characteristic impedance be scaled much above 50Ω. Each filter requires an on-chip analog driver of comparable impedance. Alternatively, the filter may be realized on-chip as an active bandpass circuit of sufficient selectivity and dynamic range. Although a gyrator-capacitor-based active filter is possible in principle, small phase shifts in the gyrator transistors at the 10.7-MHz IF can seriously upset the filter passband shape. Active resonators are also known to suffer from a larger internal noise level than their passive counterparts, and this discrepancy worsens with increasing pole-Q and pole frequency [26]. Kasperkovitz solved the problem with an *architectural innovation*, by dramatically lowering the IF from 10.7 MHz to 70 kHz. This makes it a great deal easier to implement an IF active channel-select filter, which now need only be lowpass. Further, at a given dynamic range, the power dissipation in an active filter also scales down with the IF [26]. The low IF eliminates the off-chip channel-select filter and reduces power dissipation, both very desirable properties. But what of the image frequency, the principal reason for the choice of 10.7 MHz?

At a 70 kHz IF, the image frequency lies halfway to the adjacent FM channel (Figure 3.5(a)). The image, therefore, is the interchannel noise in the FM band. As an RF preselect filter cannot possibly reject an image this close to the desired signal, it will pass unattenuated to worsen the received signal-to-noise ratio (SNR) by 3 dB. Another consequence of this choice of IF is that after the first downconversion, an instantaneous frequency deviation in the received FM signal of more than 70 kHz will alias around dc to produce distortion. This is avoided by compressing the frequency deviation to ±15 kHz with a negative feedback frequency-locked loop prior to downconversion (Figure 3.5(b)). The FM monostereo radio [27–29] requires, in addition to the single-chip receiver, only 15 small capacitors and 2 inductors, and this collection of parts readily fits inside a wristwatch. A miniature earphone is plugged into the watch and the earphone lead serves as the antenna. The radio when active drains 8 mA from a 4.5V supply.

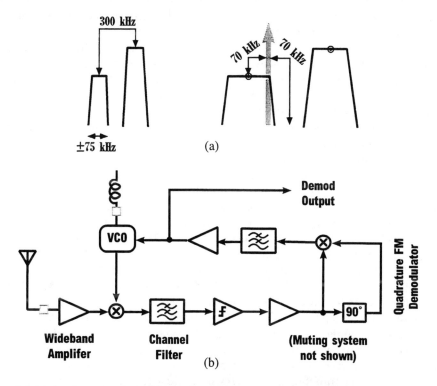

Figure 3.5 (a) An alternative choice of IF in the FM band, which places the image in the gap between adjacent channels. The IF strip, including the channel filters, now operates at a 70-kHz frequency. (b) An FM receiver using a 75-kHz IF. The channel filter is an active-RC implementation on-chip. A frequency feedback loop compresses the incoming frequency swing. Except for a tuning inductor, no high-frequency off-chip components are required.

Sony, one of the world's leading makers of miniature radios, has also recently modified its integrated FM radio-receiver architecture from the conventional 10.7-MHz IF [30] to low IF [31]. In the new architecture, a high first IF of 30 MHz is used, so any out-of-FM band image falls in the stopband of a fixed 80 to 110 bandpass preselect SAW filter after the antenna. A wideband IF amplifier boosts the received signal level of *all* FM channels *without* any filtering—amplification at 30 MHz is not a problem on this modern silicon bipolar IC process. Further, at a 30-MHz IF, the entire broadcast FM band falls to one side of the LO frequency, which means that any channel in the FM band may be selected after the first downconversion. Following this, another mixer converts to a low second IF of 150 kHz, and thereafter an on-chip ninth-order active-*RC* lowpass filter rejects adjacent channels. The low second IF, however, will pass an image FM channel as well as the desired channel, and as there is no filtering at all at the first IF, an *image-rejection mixer* is used for the second downconversion (Figure 3.6). The variable-frequency first LO tunes the desired channel, while the second LO, which must produce quadrature outputs for the image-rejection mixer, is at a fixed frequency.

The image-rejection mixer provides a trigonometric solution to a difficult filtering problem [32]. The desired channel and its image are frequency-converted into two paths by mixers driven by quadrature phases of an LO. The mixer outputs are then phase-shifted 90 degrees with respect to one another. The sum of these two signals will select the desired channel and suppress the image while, vice versa, the difference will select the image. The extent of image suppression depends on the gain matching of the two paths, and on the phase accuracy of the LO quadrature outputs. For these reasons, this concept has only become practical with IC technology, where the two paths are well-matched on-chip and track each other over temperature. Image suppression on the Sony chip is limited to about 40 to 45 dB by residual gain mismatch in the two paths. An allpass active RC-CR filter produces 90-deg phase-shifted versions of the downconverted input. This receiver drains about 15 mA from voltages as low as 0.9V in either FM or AM mode. All the necessary transistors are integrated on-chip—the RF amplifier portion, however, uses an off-chip load inductor, and the local oscillator needs an off-chip LC tuned circuit.

The circuit techniques that enable sub-1V operation are also interesting (Figure 3.7). The antenna signal drives the emitter of a common-base NPN, which forms the tail of a differential pair. The input resistance of the common-base stage matches the antenna impedance. The signal develops at one inductively loaded collector of the differential pair, while the other collector dumps a fraction of the signal current into the supply in response to an automatic gain control (AGC) differential control voltage. Following amplification by a resistively loaded differential pair in cascade, the balanced RF signal is level-shifted into the first mixer, a simplified double-balanced Gilbert-cell with resistors instead of current sources in the tails.

As inductors do not drop a dc voltage, they make convenient loads for stacked transistor circuits operating at a low supply voltage. The signal on the transistor

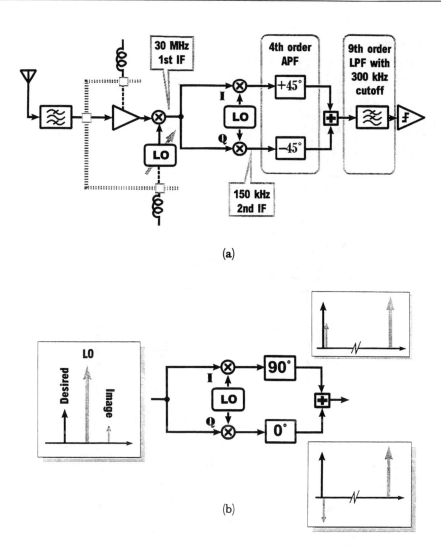

Figure 3.6 (a) An alternative architecture for a single-chip FM receiver, with a 30-MHz IF chosen for strong image rejection and to translate all the channels in the FM band to IF on to one side of dc on the frequency axis. The image-rejection mixer at 150 kHz selects the desired channel while rejecting the undesired one 300 kHz away. Aside from a noncritical RF preselect filter, the remaining filters are active on-chip, including the phase shifts in the two arms of the mixer. (b) The image-rejection downconversion mixer. The image and desired tones are at positive and negative frequency offsets from the local oscillator and are discriminated in the mixer by a relative inversion of polarities in the two arms.

RF Amplifier

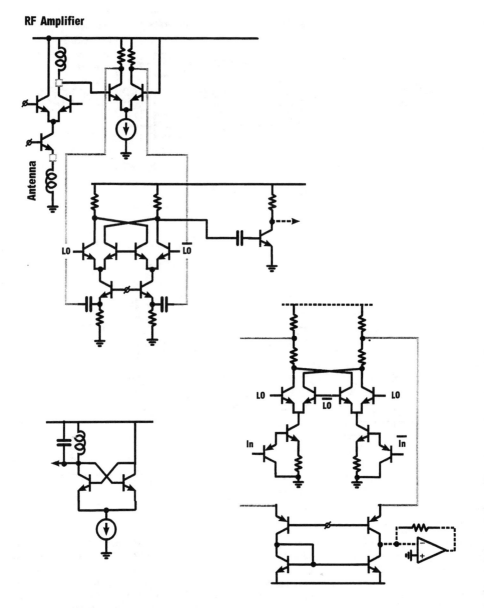

Figure 3.7 Low-voltage circuits capable of operation with a 0.9V supply. Note how the antenna carries the RF amplifier bias current, and the use of on-chip capacitors for level-shift in the RF signal path. As the signal is downconverted in frequency, active level-shift and op amp circuits appear. The cross-coupled differential pair negative resistance is popular for LC oscillators.

end of the inductor will swing above the supply. This is the case in the local oscillator, which implements a negative resistance of $-1/g_m$ with a cross-coupled differential pair, causing an LC tuned circuit across it to oscillate. The oscillation amplitude is limited by the differential pair nonlinearity to a small multiple of kT/q.

As the signal propagating through the receiver downconverts in frequency, low-frequency circuit techniques and devices with a lower f_T are used. For instance, the 30-MHz IF signal couples into the second mixer through PNP emitter followers. At the 150-kHz second IF, common-base PNPs and current mirrors level-shift down the signal, which is further amplified in an op amp.

This receiver is a significant example of how architectural rethinking combined with appropriate IC design styles has resulted in a very different solution to the well-established broadcast FM receiver.

3.4 ICs IN WIRELESS PAGING RECEIVERS

Miniature wireless communicators to page people on the move were first developed in the late 1950s. The Bell System's Bellboy™ paging receiver [33] anticipated many of the concepts underlying today's pagers. The system operated at a 150-MHz RF, addressing a receiver by frequency modulating the carrier with a unique set of three tones, which the intended receiver recognized at baseband by the simultaneous response of three passive reed resonators. The superheterodyne receiver operated at 4V using a total of only ten transistors [34], a notable early example of low-power and low-voltage circuit design. The low IF of 6 kHz meant that simple, capacitively coupled 10-kHz lowpass (rather than bandpass) filters could select the desired channel. Also, all low-frequency amplifiers after the RF section used transistors biased at small currents. The two stages in the cascode RF amplifier were inductively coupled to share, or reuse, the same bias current, a power-saving method found even in today's MMICs.

Paging receivers have been in continuous evolution since then. They are supported by a sophisticated nationwide wireless infrastructure. The modern pager uses digital signaling at rates anywhere from 500 to 1,200 bps, encoding binary data with a simple positive or negative offset of the carrier frequency—the binary frequency-shift keyed (FSK) modulation. In spite of market pressure to reduce the battery drain and miniaturize the unit, paging receivers until the early 1980s used conventional radio architectures without exploiting the powerful simplifications implied by this signaling scheme. Such a double-superheterodyne receiver [35] might consist of a *first upconversion* of the received signal to suppress the image channel with an RF crystal bandpass filter, and then a *downconversion*, followed by channel selection with a ceramic bandpass filter. An analog frequency discriminator demodulated the FSK. Much as in broadcast FM receivers, this architecture required tuned amplifiers and passive filters, which constrained further miniaturization of the paging receiver.

Vance, at ITT Standard Telecommunication Laboratories, first realized that by taking the idea of low IF to its limit with *zero IF*, an FM receiver could be scaled down to one chip with only one or two passive RF components. A quadrature downconversion mixer could discriminate positive and negative frequency modulation centered around dc [36]. There is now *no* image to be rejected, and a lowpass filter suppresses adjacent channels. If applied to analog FM as in broadcast signals, however, this scheme suffers from the dc offsets in the amplifiers and their flicker noise, which will seriously corrupt the SNR at midchannel. Furthermore, the limiting amplifier that analog FM receivers use in place of AGC does not respond to dc inputs, because there are no zero crossings. In paging receivers, however, the spectral energy clusters in two lobes on either side of dc owing to the relatively large modulation index, and zero IF (or direct conversion) is exactly the right solution [37, 38]. This remarkably simple receiver (Figure 3.8) consists only of a quadrature demodulator, lowpass filters in each arm, limiters, and a D-type flip-flop detector. When integrated on an early bipolar chip, it drains 2.5 mA from 1.8V when active, although in standby the current drain falls to a mere 50 μA. Large-value off-chip capacitors are used for ac coupling and for the lowpass filters. A second low-frequency digital CMOS IC performs all the user interface functions. Data is encoded by offsetting the carrier frequency by ±4.5 kHz, so capacitive coupling with a corner frequency below 4 kHz blocks out receiver dc offsets and lower frequency flicker noise from corrupting the SNR. A 10-kHz lowpass filter suppresses high-frequency out-of-band noise and adjacent channels. Vance also pointed out that data in a binary FSK signal downconverted in quadrature to dc may be recovered by a simple flip-flop when the limited output from one arm of the downconversion mixer is applied to the D-input, and from the other arm to the clock input. Although exceptionally simple, this flip-flop detector makes instantaneous decisions and therefore has poor immunity to a single noise spike. More sophisticated detectors must be used to get acceptably low bit error rates (BER) at typical received SNRs. For instance, the inputs of two flip-flops may be cross-coupled to the quadrature channel outputs, and the decision may be derived from the analog average of the two outputs [37]. The optimum binary FSK detector in the presence of Gaussian noise correlates the downconverted signal with the two possible offset frequencies, integrates the output, and declares a valid bit when one of the integrator outputs crosses a threshold [39]. Even the most complex detector will dissipate a small power, because it operates at the low baseband data rate.

Pager ICs from Philips originally used a frequency-offset receiver principle [40, 41] whereby the local oscillator frequency is adaptively offset from the received carrier by 2 kHz, thus converting the FSK tones to 2.5 kHz and (aliased to) 6.5 kHz. This avoids a quadrature downconversion to differentiate between positive and negative frequency, but it requires a fairly sophisticated automatic frequency control. A frequency discriminator detects data. In addition to the local oscillator crystal (operated here at its fifth overtone), this chip requires three off-chip tuned

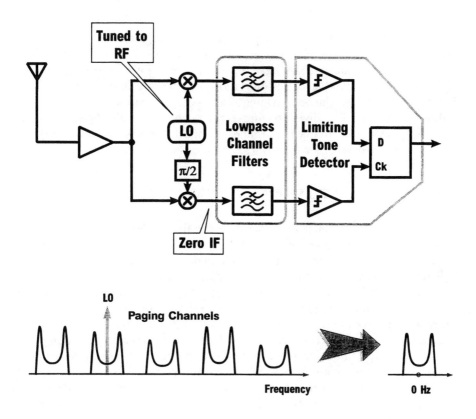

Figure 3.8 A direct-conversion single-chip receiver for FSK demodulation. There is no image channel here, and all channel selection filtering is on-chip at baseband. Note also that no AGC is required, only limiters, because all information is contained in the zero crossings. The signal path must be capacitively coupled to suppress the undesirable dc offsets in the receiver electronics.

circuits. Philips later recognized the simplicity of a zero-IF receiver for this application [42]. In this single-chip receiver, the RF amplifier requires one off-chip inductor, the quadrature phase-shift circuits in the mixer another; both inductors are combined into one off-chip signal path. The channel-select filters are entirely on-chip. They consist of a third-order active RC lowpass filter, followed by a seventh-order gyrator-based lowpass filter with a 15-kHz cutoff. Most of the signal amplification occurs at these low frequencies.

NEC's paging receivers evolved from the supeheterodyne [35, 43] to direct conversion [44] in an effort to reduce receiver volume and parts count. Others have developed similar zero-IF bipolar integrated front-ends [45, 46]. A notable feature of the chips is that much of the die area is taken up by the capacitors for ac coupling the signal path and for the on-chip lowpass filters. Most pager ICs operate at supplies of 2V to as low as 1V and are implemented in silicon bipolar technology. In addition, full-featured pagers require 20,000 gate-equivalent digital ICs for the user interface [47], low-voltage EEPROMs for customization and software, and capability to drive a liquid crystal display. The basic paging receivers can fit within a wristwatch [48]. The direct-conversion FSK digital paging receiver concept has also been successfully used at very low carrier frequencies (hundreds of kilohertz) with much lower data rates in implanted devices for biomedical applications [49].

3.5 ICs IN CELLULAR TELEPHONE TRANSCEIVERS

Mobile and handheld cellular telephones are the first widespread two-way radios for consumer use. They were preceded by cordless telephones for local area use. These wireless telephones must meet stringent demands for low weight and volume, long battery life, low cost, and reliable network access to be successful with consumers. In contrast, walkie-talkie transceivers were always aimed at specialized markets and did not face these pressures for miniaturization. The average consumer, for instance, will not voluntarily accept a transceiver of the size and weight that policemen or soldiers carry as part of their outfit. Further, to support a large numbers of users in a crowded radio spectrum, wireless telephones use more internal signal processing than other common transceivers and must be capable of connecting to the public switched telephone network [50]. Features such as digitally selected channels, direct-sequence spread spectrum, and diversity selection are now becoming common. The transceivers perforce must use highly integrated, low-power electronics. Thus it may be said that with the advent of the modern cellular telephone, the conventions of wireless design are being re-examined, and sometimes rewritten.

The first generation of cellular telephones carried voice signals by analog frequency modulation of a carrier. In the U.S. AMPS system, for example, the handset receives at a carrier selected from the 869- to 894-MHz band, while it transmits on a carrier in the 824- to 849-MHz band. These 25-MHz widebands are separated by 45 MHz between the uplink and downlink, enabling the user to talk and listen at the same time much as on the wired telephone (users are not too fond of the "over, over-and-out" protocol). An antenna duplexer suppresses coupling from the transmitter into the sensitive receiver, acting as the equivalent of a two-to-four wire hybrid transformer in a telephone. This duplexer is a passive three-port designed to pass energy in the receive frequency band from the antenna port to the receive port, while attenuating energy in the transmit band from the transmitter into the receiver. It is either made with high-dielectric ceramic resonators [51] or with a SAW filter and coaxial resonator in parallel [52]. The receive portion of a conventional handset resembles a broadcast FM receiver (Figure 3.9). If the local oscillator at the first mixer lies at a higher frequency than 894 MHz, the image is guaranteed to lie outside

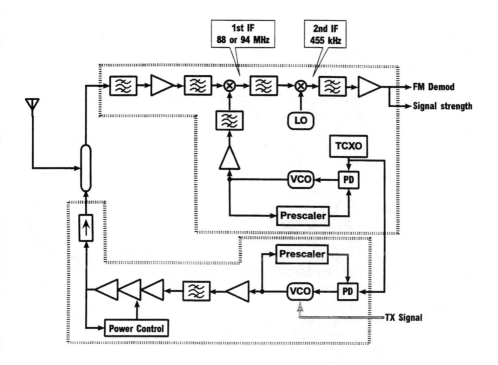

Figure 3.9 The transmit and receive front-end of an analog cellular telephone. Receiver uses conventional double-superheterodyne architecture, while transmitter is one-step upconversion.

the AMPS band. Furthermore, an LO offset of more than 45 MHz ensures that the image is attenuated by the receive band SAW filter, suffering at least 20-dB loss in each of the two filters in the handset [53]. A first IF of 90 MHz is therefore often used; this also avoids problems caused outside the handset by parasitic LO leakage through the antenna [54]. The desired channel is selected by locally synthesizing the first IF. This is followed by downconversion to a fixed second IF of 455 kHz, then demodulation by a frequency discriminator.

The first generation of small-scale ICs for portable communication devices offered building blocks for the intermediate-frequency chain, such as the mixer and LO conventional single or double-superheterodyne receivers [55], the IF amplifier chain and signal-strength indicator [55], or a stand-alone image-reject mixer [56]. Today, almost every major semiconductor company with an interest in the communications market offers building block ICs at this scale of integration.

Although ICs entered the IF portions, the RF front-end circuits continued to be made from discrete components. An RF amplifier and a first mixer may be mounted with the associated filters on a dense miniature board [57, 58]. Discrete bipolar transistors responded to needs for portable RF applications by offering, for instance, a low noise figure and f_T exceeding 5 GHz at less than 1-mA bias currents [59, 60]. Manufacturers of passive components, too, have steadily scaled down their package sizes for high-density board mounting.

There is little argument, though, that the RF front-end components must also be integrated to reduce power dissipation. In competing with passive solutions, the RF ICs must cross some important thresholds of low price and high performance [61]. However, they offer the prospect of an order-of-magnitude reduction in physical volume of the front-end electronics, and power savings will accrue by routing RF signals at a high impedance on-chip, while eliminating the low characteristic impedance interconnects between discrete packages. Various reasons have been advanced for why GaAs MMIC technology is now the right choice for cellular telephones [62–64]. They are summarized as follows: first, that owing to the semi-insulating nature of the GaAs substrate, reasonable size inductors may be integrated with transistors to make high-frequency monolithic tuned circuits, which allows for lower current operation at a given frequency than would be possible with RC broadbanding techniques; and, second, that MESFETs afford lower noise figures at a given bias current than a bipolar transistor in a comparable silicon bipolar technology. Most GaAs MMICs integrate front-end components for cellular applications at a small scale. For instance, a chip may integrate a tuned RF low-noise amplifier in the 900-MHz band, or a mixer and a local oscillator [64–72].

The typical RF amplifier may consist of only one or two FETs, with LC matching circuits on the input and the output ports for stand-alone operation in a 50Ω environment (Figure 3.10). A powerful and popular method to match the capacitive FET input is to insert a series feedback inductor, L, in the FET source, which at high frequencies contributes a *resistance* g_mL/C_{GS} at the input port [73]. This method is preferred to resistive feedback found in wideband amplifiers for impedance matching because, unlike feedback resistors, the inductor does not degrade the noise figure. MMICs from Matsushita favor the use of dual-gate MESFETs with RC matching instead of inductors [74], possibly because spiral inductors consume too large a chip area. GaAs IC designers must closely watch their chip area to remain competitive in price. The various low-noise amplifiers operate in the 0.9- to 2-GHz bands, with gains of 15 to 20 dB and noise figures of around 3 dB. These submicron MESFET ICs drain anywhere from 3 to 5 mA.

The received RF is typically downconverted by the local oscillator modulating the conductance of the mixer FET through a cascode FET (Figure 3.10). The mixers yield conversion gains greater than 10 dB, noise figures of 10 to 12 dB, and third-order input-referred intercept points of –5 to 0 dBm. Dual-gate MESFETs are naturally suited for mixer use, and offer a similar performance [74]. A recent 0.7-μm GaAs MMIC offers an LNA mixer pair draining 3 mA from 3V, with the LNA producing a 13-dB gain, a 3.6-dB noise figure, and an input-referred IP3 of –11 dBm [75].

Slicing across the system a little differently, another MMIC implements the downconversion and upconversion mixers for the receiver and transmitter, respec-

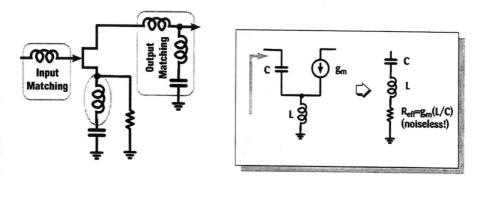

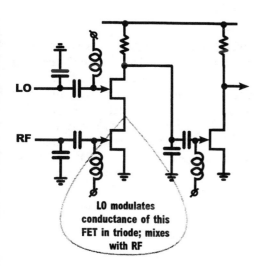

Figure 3.10 Typical stand-alone GaAs MMICs implementing low-noise amplifier (above) and downconversion mixer (below). Concept of series-feedback inductor for noise matching shown in box.

tively, integrating their shared LO on the same substrate [76]. The four-FET switch mixer (Figure 3.11) is very linear, but it suffers from two disadvantages: it requires a large LO drive to turn the switches on and off [77]; and, unlike the bipolar transistor Gilbert-cell analog multiplier, the switch mixer is lossy, requiring additional signal amplification from the following stages.

The current generation of analog cellular telephones transmits power levels of more than 1W (30 dBm). RF power amplifiers are usually packaged in a separate module with some form of integral heat sink. A preamplifier (or, in radio terminology, exciter) in the transmitter section boosts the modulated carrier level close to 0 dBm

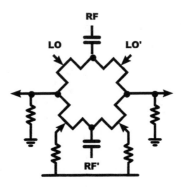

Figure 3.11 A four-FET commutating switch mixer. The balanced LO signal switches the FETs, which chop the RF signal to produce the sum and difference frequencies.

to drive the power amplifier input. The power amplifier module itself usually consists of a cascade of two or three FETs, tapering up to a single large-size FET that will deliver the required signal current into the antenna load. Furthermore, as the power amplifier is the largest single source of battery drain, it must have a high *conversion efficiency*. Much as in baseband power amplifiers, an efficient RF power amplifier is biased close to cutoff to reduce the dc standing current, and then driven by the input signal in Class A-B or Class B mode. Narrowband filters at the amplifier output remove harmonic distortion caused by nonlinear operation at RF [78]. These filters may be merged into the passive matching networks required for optimum power transfer from the amplifier to the load. As designers of RF power amplifiers have observed [79, 80], much of their work consists of synthesis and iteration of the interstage, input, and output *matching networks*. It is easier to design efficient power amplifiers for constant-envelope modulations, such as analog FM or digital FSK, where distortion may be tolerated because the useful information is all contained in the zero crossings. Further, it is argued that owing to the lower parasitic capacitance of a GaAs MESFET relative to silicon devices, a GaAs power amplifier at 1W power levels affords a higher efficiency (~ 60%) compared to silicon bipolars or FETs (~ 45%) [47]. A monolithic power amplifier consisting of a four-stage cascade of MESFETs with on-chip lumped LCR input and interstage matching networks, delivers 1W at 900 MHz at 63% efficiency from a 5.5V supply [80]. The distributed-element output matching network at 900 MHz would be exorbitantly large on an IC, and is therefore printed on an off-chip alumina substrate. The high efficiency is attributed to an improved method of suppressing the second harmonic at the amplifier output. This multicomponent module approach to mate ICs with matching networks is widely used in GaAs power amplifiers [81]. Another 900-MHz power module uses two discrete MESFETs, wirebonded to a hybrid IC containing a combination of distributed circuits and chip capacitors and resistors in the matching network to attain 65% efficiency when delivering over 1.3W from 4.7V supply [82]. The output FET is 12-mm wide, with a 1-μm channel length. A separate negative supply is required in many of these MESFET power amplifiers to bias the gate. The circuit was recently modified [83] to produce the same output power equally efficiently, now with a single 3.5V supply and FETs of 0.5-μm channel length. The desired low-voltage operation, much more suitable for one battery, is the result of an improved device structure. The matching networks in the RF signal path are fabricated on separate passive-only GaAs ICs, which the authors claim are three times cheaper than GaAs ICs with FETs, while the bias networks reside on a miniature PC board. The various components are wirebonded to one another, and mounted on an AlN substrate with ten times higher thermal conductivity than alumina. Power amplifiers may also be built with silicon NMOSFETs: among the major semiconductor vendors, Hitachi has pursued this option. A MOSFET gracefully accepts large voltage swings on the gate without possibility of Schottky conduction, as in a MESFET gate. With an offset-gate FET structure and 0.8-μm channel length, a MOSFET power amp delivers 2W at 1.5 GHz with 55% efficiency from a single 6V supply [84]. The power amplifier module embeds these FETs into matching networks [85].

The foregoing discussion is not meant to imply that cellular telephones mainly use GaAs MMICs in the RF front-end. Silicon bipolar and BiCMOS technologies have made tremendous strides in improving f_T, and today silicon MMICs offer comparable performance to what is available in GaAs. A recent 1-GHz BiCMOS LNA-mixer combination [86] uses familiar circuits to baseband designers, such as a Gilbert-multiplier type mixer, and the chip yields a comparable performance to the MMICs described above. The RF signal path consists of bipolar circuits only, while the FETs are used as switches to select various power-down modes. There are no on-chip inductors to tune the low-noise amplifier, which is wideband; instead, the LNA output is routed off-chip into a passive bandpass filter, then returned to the chip for downconversion. This is part of a complete chipset from Philips for a digital cellular telephone handset [87].

The first LO in a cellular telephone receiver is programmed to the incoming RF channel with a PLL synthesizer, while the second LO at IF is fixed in frequency. The transmit LO oscillates in yet another frequency band. Both the receiver and the transmitter therefore require 900-MHz voltage-controlled oscillators. These are most often implemented with bipolar transistors, whose very small flicker noise means low phase-noise sidebands in the VCO. A stripline resonator sets the nominal frequency of a Colpitts oscillator, and this is voltage-controlled by a varactor diode in parallel [54, 57, 88]. Phase noise levels of −110 to −120 dBc/Hz are attained at a 50-kHz offset from the oscillation frequency. The VCO application has created a brisk demand for varactors with large voltage coefficient and low loss [60]. The VCO and its buffer may be connected in series to reuse the bias current between the two stages [57, 88] (Figure 3.12). The nominal oscillation frequency may be slaved in a frequency multiplying PLL to a 12- or 15-MHz crystal oscillator.

The Motorola MicroTac, first introduced in 1989, set an industry standard for a miniature cellular handset. Small handsets [89] continue to use conventional architecture, but attain a small size with discrete filters in miniature packages [52, 53, 58, 90, 91] and with lower power electronics, which reduces battery weight. The electronics entail carefully designed standby modes, better software for the onboard microprocessors, and, among other items, lower power PLL prescalers— digital counters that must operate all the time for frequency synthesis and that often posed a significant power drain [92, 93]. In the transmit mode, greater efficiency is sought in the power amplifier to prolong battery life, as well as smart control circuits to maintain maximum efficiency across the range of output power levels [89]. It was generally agreed that by 1992, the handset volume should scale down to less than 150 cc, and that its weight should be less than 230 gm [20, 94]. NTT demonstrated the prototype of such a telephone in 1991 [95], while noting that the volume should be no smaller than this target, both for the sake of ergonomics and to ensure adequate heat removal from the package so that the internal temperature rises no more than 15°C. These small telephones might be powered by a 6V NiCd battery with 400-mA · h capacity.

With the emergence of the IS-54 standard, there is some convergence in the United States on a digital cellular technology. Upward compatibility is sought with the existing analog cellular bands, which is why handsets conforming to IS-54 are referred to as dual-mode cellular. A recent 12-GHz f_T silicon bipolar transceiver IC

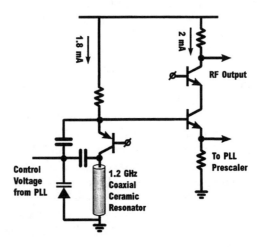

Figure 3.12 Colpitts oscillator with coaxial resonator. Varactor controls frequency with control voltage. Series-connected buffer distributes RF oscillation to mixer and to PLL prescaler.

from AT&T uses a conventional double-superheterodyne architecture with an 80-MHz first IF, and after IF amplification the signal path bifurcates into a conventional 455-kHz second IF FM demodulator or a digital I-Q PSK detector [96]. An interesting alternative is to downconvert the second IF, without preceding AGC, in a delta-sigma A/D converter for subsequent digital baseband signal processing [97].

3.6 DIGITAL CELLULAR AND CORDLESS TELEPHONES

Analog cellular telephones use the frequency spectrum inefficiently. The modulation schemes consume a large bandwidth and every cellular telephone transmits at constant power all the time it is in use, thereby appearing as an interferer to other users at nearby frequencies. As a result, the spectrum allotted to cellular phones in large metropolitan areas nears exhaustion.

Digital cellular telephones are one solution to better utilize the scarce spectrum. They use more efficient modulation schemes, such as minimum FSK or PSK, and multiple users may share time slots on the same part of the spectrum. The complex modulation formats used in these telephones and the greater capabilities required to withstand nearby blocking signals are prompting large-scale integration of the RF and IF electronics. Examples of digital telephony standards are the European GSM, North American Digital Cellular, and the emerging Japanese Personal Handy Phone (PHP) [98]. The salient characteristics of these various systems, as well as key digital cordless standards, DECT and CT-2, are summarized in Table 3.1, as well as Table 2.3 in Chapter 2.

The typical handset involves an RF/IF front-end, followed by a baseband digital signal processor (Figure 3.13). The first significant set of RF-ICs for GSM handsets appeared in 1990. Notable among them is a receiver and transmitter 7.5-GHz f_T silicon bipolar chipset from Siemens (Figure 3.14) [99, 100]. The double-superheterodyne receiver uses a selectable first IF of anywhere from 45 to 90 MHz, which places the image channel in the stopband of the 25-MHz-wide RF preselect bandpass filter. A 71-MHz IF, for instance, guarantees that the image lies outside the GSM band. The desired 200-kHz channel is selected by a SAW filter at the first IF. The receiver IC provides an on-chip buffer to drive this filter. Following an RF amplifier with an off-chip LC tuned load, the input signal is downconverted by a Gilbert-cell mixer. There is AGC with 70-dB range and a signal-strength indicator at the first IF, and the selected channel is then downconverted to baseband in a quadrature demodulator to acquire both amplitude and phase for GMSK vector demodulation. The on-chip gain may be as large as 80 dB at the highest VGA setting, but as it is distributed in different frequency bands there is little on-chip crosstalk, and the monolithic receiver operates stably. The receiver uses a fully balanced signal path and an input noise figure of 7 dB. It drains 27 mA from 5V in active mode and 10 mA in standby mode.

Table 3.1
Summary of Proposed Digital Cellular Telephone Standards

Specification	GSM Europe	NADC North America	J-PHP Japan	CT-2 Europe, Asia	DECT Europe
Downlink frequency band	935–960 MHz	869–894 MHz	1.9 GHz	864–868 MHz	1.88–1.9 GHz
Uplink frequency band	890–915 MHz	824–849 MHz	1.9 GHz	864–868 MHz	1.88–1.9 GHz
Multiple access method	TDMA	TDMA	TDMA/TDD	FDMA/TDD	TDMA/TDD
Modulation	GMSK	π/4-DQPSK	π/4-QPSK	B-FSK	GMSK
Speech data rate	13 Kbps	8 Kbps	32 Kbps	32 Kbps	32 Kbps
Handset output power	3.7 mW → 1W	2.2 mW → 6W	10 mW	1 mW → 10 mW	250 mW
Modulation rate	271 Kbps	49 Kbps	384 Kbps	72 Kbps	32 Kbps
Channel spacing	200 kHz	30 kHz	300 kHz	100 kHz	1.762 MHz
Burst length	156 bits	324 bits			424 bits

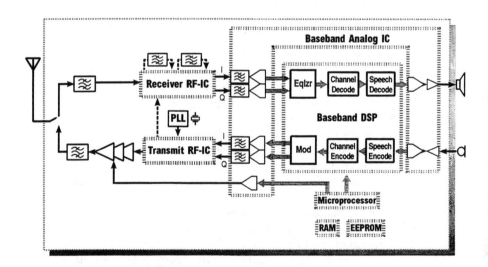

Figure 3.13 Block diagram of digital cellular telephone. Chip partitions reflect current state-of-the-art technology.

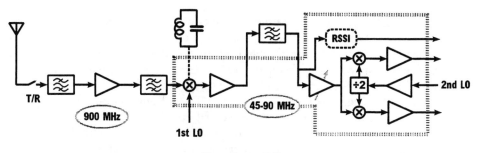

Receiver IC

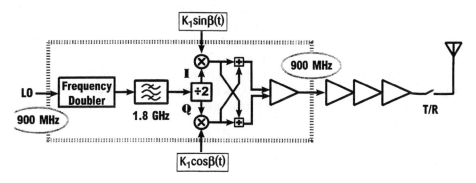

Transmitter IC

Figure 3.14 Siemens receiver and transmitter ICs for GSM handsets. Receiver is double-superheterodyne architecture. Requires one off-chip tuned circuit, SAW resonator at first IF, and low-noise amplifier. Transmits doubles LO frequency and then divides by 2 to accurately obtain quadrature phases.

In the transmitter IC, a precision quadrature upconversion mixer produces a single-sideband, suppressed-carrier QPSK output from a baseband vector input. An RF differential input amplifier selects one sideband from the upconverted outputs in the two arms of the mixer. As the other sideband would occupy a nearby channel's spectrum, it must be adequately suppressed (relative to the level of the wanted sideband). Imperfect cancellation of the unwanted sideband arises from gain mismatch in the I and Q channels of the upconversion mixer, and in deviations from quadrature in the two LO outputs. The goal in this transmitter was quadrature phase errors of less than 2 deg, which with adequate gain matching implies unwanted sideband suppression of at least 40 dB relative to the wanted sideband. The stringent GSM requirements on low LO phase noise are only met with a 900-MHz off-chip coaxial-resonator-based, varactor tuned Colpitts oscillator [101], which inherently provides a single-phase output. There are several methods to accurately derive quadrature phases. Siemens uses the double-frequency method. An on-chip nonlinear element doubles the oscillator frequency to 1.8 GHz, following which an on-chip bandpass filter removes harmonics. Then, the oscillation is divided by 2 by positive- and negative-edge triggered flip-flops, which will yield quadrature outputs with a phase accuracy set by how close the duty cycle of double-frequency oscillation is to 50%. Combined with an output stage to drive an off-chip power amplifier module, the IC drains 40 mA. Using a similar superheterodyne architecture with 71-MHz IF, AT&T Microelectronics has recently combined the GSM receiver and transmitter blocks, including the frequency synthesizer, onto one chip [102].

In an effort toward even greater miniaturization at Alcatel, the transmitter and receiver sections are both integrated onto one 9-GHz f_T silicon bipolar chip (Figure 3.15) [103, 104]. From the point of view of integration, the main advantage of the zero IF direct-conversion architecture is that it eliminates the IF passive high-frequency bandpass filters. The disadvantage, on the other hand, is that the downconverted signal has energy at dc, to which will add the receiver's offsets and low-frequency noise. An off-chip low-noise amplifier drives directly into quadrature (I-Q) mixers. The mixer dynamic range is wide enough so that a large blocking signal only 3-MHz away from the desired signal produces insignificant intermodulation distortion. The bipolar Gilbert-cell mixer is linearized with emitter degeneration

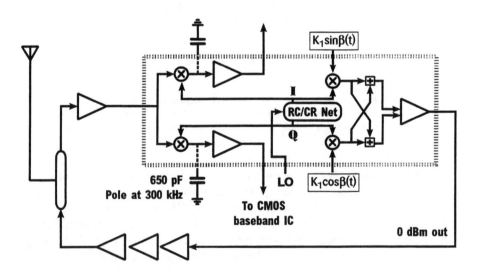

Figure 3.15 Alcatel direct-conversation single-chip transceiver. Channel selection done by switched-capacitor lowpass filters in companion CMOS mixed-signal baseband IC. On-chip, trimmed RC-CR network generates quadrature phases for mixers.

resistors, which also degrade its noise figure. The receiver section requires large (650 pF) off-chip capacitors at its output for antialias lowpass filters before the baseband signal is sent to a companion mixed-signal CMOS IC. This baseband CMOS IC contains a high-order, switched-capacitor lowpass filter for channel selection; a digital GMSK detector; and various DSP functions [105]. An RC and CR network with off-chip trimming shifts the phase of an external LO by precisely ±45 deg to produce the quadrature drive to the mixers. Upconversion mixers in the transmitter drive a power amplifier module. The chip drains 25 mA from 5V in the receive mode and 45 mA in the transmit mode, rather comparable figures to the previously described Siemens chipset. The transmitter suppresses the unwanted sideband and carrier by about 40 dB.

There is increasing interest in the 1.9-GHz band for digital cellular telephony. All the current approaches (Table 3.1) use time-division duplexed (TDD) receive and transmit frames, wherein users are assigned different timeslots. Two-way communications take place over the same frequency band. A host of ICs are appearing to serve the European DECT standard. The majority cater to a superheterodyne receive architecture with 110-MHz IF [106–109]; some of the literature [107, 109] also describes procedures for system design. The RF sections are usually integrated at small scales, embodying, say, the low-noise amplifier and first mixer on one chip, the quadrature modulator on another, and the exciter and power amplifier on a third and fourth chip. RF-ICs operating at 1.9 to 2.5 GHz exist in both GaAs [68, 110–112] and silicon [106, 113–115] technologies. Some of these stand-alone GaAs IC low-noise amplifiers achieve impressive gain and noise figure with 1- to 2-mA current drain from 3V [69, 116]. At the system level, though, functionality and overall dynamic range in these short-haul wireless links takes precedence over raw component performance. Thus, system-level input noise figures of 10 to 15 dB are acceptable [109], as are input-referred intercept points of -16 dBm and a 20-dB power amplifier control range [112]. The less than stringent system specifications lead to simplifications in the transmit path, such as direct VCO modulation by the baseband signal by opening the transmit PLL over the duration of the transmit-frame [107].

The power amplifier continues to be built almost exclusively as a separate GaAs IC, and at 1.9 GHz the on-chip wavelength is short enough that it is now possible to integrate distributed matching networks [68, 117, 118]. Efficiencies of 50% are attained when delivering almost 1W to the load from 3V. Either the transmitter or the receiver is active in TDD digital transceivers, so fast-switching power-down modes are designed into the various components. In particular, the switching trajectory must be shaped to suppress spurious emissions when the power amplifier is switched on and off [85]. Finally, the transmit and receive signals are directed to the antenna via a three-port passive circulator, or preferably through a low-loss, monolithic transmit/receive (T/R) RF switch [112, 119, 120]. A good FET switch must not appreciably distort the RF signal, or incur more than 1-dB insertion loss when on, yet when off it should offer at least 30-dB isolation. Microcell applications, such as the Japanese Handy Phone, require an average output power of only 10 mW (20 dBm), although the $\pi/4$-QPSK modulation requires the power amplifier to handle larger peak powers. An exciter amplifier, power amplifier, and T/R switch have been integrated together [121]. The exciter and power amplifier attain a 44% efficiency at 23-dBm output from a 4.8V supply, and the on-chip matching network uses LC lumped elements. Although matching networks will eliminate harmonic distortion, they are ineffective in suppressing near-carrier spurii produced by intermodulation distortion. Feedback linearization techniques have been proposed that predistort the digitally synthesized exciter input waveform to anticipate the power amplifier nonlinearities [122]. Use of these techniques make a highly nonlinear but efficient power amplifier appear linear. These remain at the experimental stage today, and with the move to microcells and low emitted power levels, they may soon not be necessary in handsets.

The building-block ICs described so far are important advances in realizing small, relatively low-power transceivers, but the ultimate goal remains to integrate the entire transceiver on to a single chip. To this end, some early breadboard-level experiments show that direct-conversion receiver architectures seem well-suited to the DECT application [123, 124]. A 16-GHz f_T silicon bipolar IC from Alcatel operating at 1.9 GHz contains a complete direct-conversion DECT transceiver [125]. The architecture resembles the previously described Alcatel direct-conversion GSM transceiver [103], requiring a separate low-noise amplifier in the receive path, a power amplifier module, and a baseband CMOS mixed-signal signal processor IC. From a 5V supply, the transceiver drains 50 mA in receive mode and 80 mA in transmit mode. One of the challenges in realizing this system was a fast-switching PLL frequency synthesizer with low phase noise and low spurious output levels, which was built here with an improved charge pump and loop filter.

Siemens has extended its dual-conversion GSM transmitter/receiver chip set [99, 126] to a generalized front-end for digital cellular telephones operating in any part of the RF spectrum from 800 MHz to 2.1 GHz [127]. The 25-GHz f_T silicon bipolar chipset operates at a supply as low as 2.7V, the transmitter chip draining 60 mA and the receiver 33 mA. Both have power-down modes. An external power amplifier module is required. A notable low-voltage circuit on this chip is a modified Gilbert-cell mixer, with resistors instead of current sources to the negative supply [31]. To accommodate a wide input dynamic range, the gain of the RF low-noise amplifier in the receiver may be switched from −5 dBm to +15 dBm, a powerful technique that has also been used in paging receivers [46]. This assumes that the receiver does not instantaneously require a very wide dynamic range, but either receives mostly strong signals or mostly weak ones. The downconversion mixer, with a 13-dB noise figure and −3-dBm input IP3, is followed by an IF variable-gain amplifier with an 80-dB digitally programmable range [126]. The on-chip RF oscillator requires an off-chip resonator and varactor, and its phase noise at 2-kHz offset is −88 dBc/Hz. A balanced signal path is used throughout. One good measure of the quality of on-chip gain matching in the I and Q paths in the direct upconverter, as well as of the quadrature accuracy of the upconversion clocks, is given by the relative levels of spurious emissions from the transmitter IC. The unwanted sideband is suppressed by 48 dB, while the RF carrier tone, emitted due to dc offsets in the two paths, is 37-dB down. A third-order modulation tone at −46 dB appears due to mixer nonlinearity. The absolute output noise level at a 25-MHz offset is −141 dBm/Hz.

3.7 DIRECT-CONVERSION TRANSCEIVERS

A receiver with zero IF is called a *direct-conversion* receiver. When the LO is synchronized in-phase with the incoming carrier frequency, this is also referred to as the *homodyne*. This architecture is sufficiently promising for single-chip transceivers to warrant a separate section to its study.

The desired channel is translated by the first mixer to a 0-Hz center frequency, and instead of adjacent channel rejection with a bandpass resonator, a more flexible and easier to implement lowpass filter is required—in effect, a bandpass filter centered at dc when the negative frequency axis is included. With zero IF, there is no image frequency. As early as 1924, radio pioneers had considered homodyne architectures for crude receivers requiring only a single vacuum tube, but it was in 1947 that a homodyne was first used to full effect, with a high-order, lowpass filter for channel selection in a measuring instrument for carrier-based telephony [32].

3.7.1 Direct-Conversion Single-Sideband Synthesizers

For reasons of spectral efficiency, the transmitted signal in digital communications is always single sideband with suppressed carrier. This is most often produced with the so-called *phasing method* [128]. The modulated signal is first synthesized in quadrature at baseband, *directly upconverted* into two paths by quadrature LO centered at the carrier frequency, and added or subtracted to select either the upper or lower sideband (Figure 3.16(a)).

The unwanted sideband is suppressed to an extent limited by the gain mismatch in the two upconversion paths, and by departures from quadrature in the two LO outputs (Figure 3.16(b)). Unequal dc offsets in the paths produce an output signal at the LO frequency. The unwanted sideband and LO leakage are spurious but unavoidable components of the transmitted spectrum. Although on the same IC the two upconversion paths are well-matched, a gain mismatch as small as 1% (0.1 dB) will limit suppression of the unwanted sideband to 45 dB. With this gain mismatch, a phase-error of up to 1 deg may be tolerated between the two LO outputs. These mismatches may be trimmed down at time of transceiver manufacture, or self-

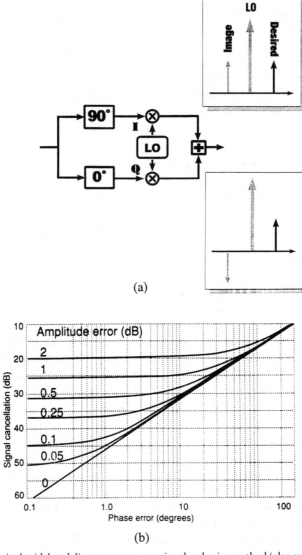

(a)

(b)

Figure 3.16 (a) A single-sideband direct upconverter using the phasing method (also called the quadrature modulator). (b) Extent of suppression of unwanted (image) sideband vs. gain mismatch and phase errors from quadrature in two arms of modulator.

calibrated with loopback modes that are activated during idle times to sense and suppress the unwanted spurii.

An off-chip resonator, often connected to an on-chip unbalanced oscillator circuit, may also become a cause of spurious RF leakage if it couples energy into the power amplifier or the antenna. Frequency-offset upconversion schemes have been proposed [129] to combat this coupling problem. Other spurious output tones may arise from parasitic remixing of the modulated output with the baseband signal and by intermodulation distortion in the output stage [96]. Balanced circuit topologies, on-chip LOs requiring no external resonators [130], and lowered transmit power levels in microcells are all expected to lessen the magnitude of the spurious leakage problem.

Cellular wireless systems operate best with tightly regulated power control, the handsets transmitting only the minimum power required for reliable reception by the base station, and vice versa. Whereas in the receiver, the front-end is the main source of noise, in a transmitter, the upconversion LO phase-noise appears as added noise on the emitted signal. Direct upconversion has the advantage over two-step schemes in that only *one* LO contributes noise.

3.7.2 Direct-Conversion Receivers for GSM Digital Cellular Telephones

Among the various RF-IC suppliers for the European GSM digital cellular telephone handset, only Alcatel at present uses a direct-conversion receiver architecture [103, 125, 131]. This results in a relatively small silicon bipolar RF front-end chip, and the remainder of the signal processing, including lowpass channel-select filtering, is at baseband in mixed-mode CMOS. Why are the others reluctant to use direct conversion in their receivers? Given the many decades of superheterodyne experience, the most likely reason is conservatism. But direct conversion also suffers from some unique problems.

A well-known problem is that spurious LO leakage from the receiver into the antenna becomes an in-band interferer to other nearby receivers tuned to the same band. Superheterodynes, with their frequency-offset LOs, do not suffer from this problem. However, experimental studies suggest that with standard shielding in the receiver, this problem is not so severe as to handicap the use of direct conversion [132].

A more serious problem is *dc offset* in the receiver. Offset arises from three sources [131]: transistor mismatch in the signal path; the LO signal leaking to the antenna because of poor reverse isolation through the mixer and RF amplifier, then reflecting off the antenna and self-downconverting to dc through the mixer (Figure 3.17); and a large near-channel interferer leaking into the LO port of the mixer, then self-downconverting to dc. Good circuit design may reduce these effects to a certain extent, but they cannot be eliminated.

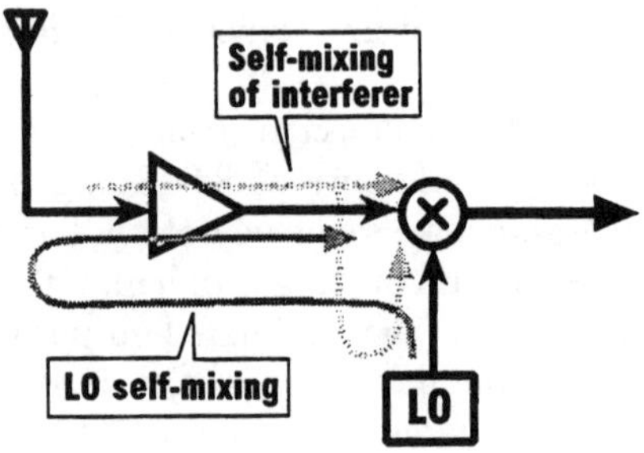

Figure 3.17 Sources of leakage in a direct-conversion receiver, and how through self-mixing they create dc offsets.

The spectrum of the GMSK modulation used in GSM has a peak at dc. Offsets will directly add to the spectral peak of the downconverted signal. These offsets are usually much larger than the rms front-end noise, and may therefore significantly degrade the SNR at the detector. To remove the offsets by ac coupling, the receiver will require impractically large capacitors if the signal-bearing spectrum around dc is not to be sacrificed. However, they may be compensated with DSP-based self-calibration [105, 133]. The baseband signal processor makes long-term measurements on the dc level in the receiver and subtracts off this level from the downconverted signal. The spectrum loss around dc is only a few hertz, and digital filtering does not distort the midchannel group delay in the receiver. Among the various RF-IC suppliers for the European GSM digital cellular system, Alcatel has demonstrated the feasibility of highly integrated direct-conversion receivers [103, 125, 131] where others use superheterodyne architectures. Alcatel's motivations are, predictably, that the channel filters may be lowpass and therefore active; a minimal silicon bipolar RF front-end is required; and that all baseband signal processing may be in CMOS.

3.7.3 Local Oscillators With Quadrature Outputs

A *carrier-frequency* LO with quadrature outputs is a key circuit component in direct-conversion transmitters and receivers. Usually, this oscillator is tuned by an LC circuit or an off-chip resonator and inherently produces a single-phase output. Quadrature phases are often derived by passing the oscillator output through a CR and an RC network, whose time constant is equal to the oscillation period. The two resultant outputs are then phase-shifted by +45 deg and −45 deg, respectively. Inaccuracies in the actual values of R and C will lead to errors in quadrature and are compensated by some form of on-chip trimming [103, 125].

An alternative method is to divide an oscillation at twice the carrier frequency with positive- and negative-edge triggered flip-flops [99]. The resulting two outputs at the desired frequency are in quadrature. Residual phase errors caused by unequal delays in the two flip-flops and by departures from 50% duty cycle in the double-frequency oscillation are found to be less than 1 deg. However, the double-frequency portions of the circuit may become a speed bottleneck.

Quadrature outputs may also be derived from a polyphase oscillator, such as a variable-frequency ring oscillator. In a four-delay stage ring oscillator, taps at diametrically opposite points will yield quadrature phases at all frequencies [134] (Figure 3.18). In oscillators with an odd number of unit delays, they may be synthesized from two taps by a voltage-controlled phase shifter in feedback around a quadrature-sensing circuit [135]. Mismatches in transistor characteristics limit the attainable phase accuracy. This type of resonatorless oscillator is practical when the specifications on phase noise are not too stringent (say up to −80 dBc/Hz at a 100-kHz offset from the carrier). The free-running oscillator must be slaved to a crystal reference in a frequency-locked loop, whose loop gain and bandwidth are specifically designed to suppress phase noise.

When circuit techniques cannot contribute further improvements to gain- and phase-matching in the two arms of a quadrature modulator or demodulator, digital

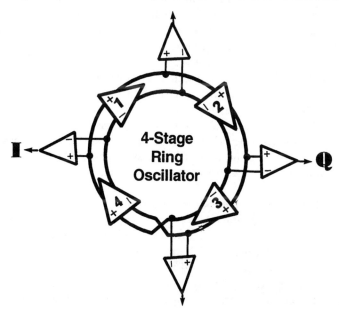

Figure 3.18 A four-phase voltage-controlled ring oscillator. Diametrically opposite taps give quadrature phase at any frequency. Unequal loadings and device mismatch set phase errors. VCO must be embedded in a PLL whose closed-loop bandwidth is tailored to suppress oscillator noise.

adaptive algorithms may be used at baseband to sense and compensate for these errors [136].

3.8 ICs FOR SPREAD-SPECTRUM WIRELESS TRANSCEIVERS

Spread-spectrum communications offer greater user capacity than narrowband techniques in a given piece of wireless spectrum. Much has been written on this subject [137]. Spread-spectrum techniques were developed during World War II as a form of secure communication with low probability of intercept and resilience to jamming [138]. Over the years, this technology has been further developed and refined for military communications. With the wireless revolution at hand and the IC technology now available to implement complex transceivers, spread spectrum has awakened commercial interest [139]. Spread-spectrum communications make it easy for users to access the wireless channel. Whereas conventional narrowband wireless communications require a careful discipline, which the FCC or other government agencies enforce by issuing licenses so as to prevent use of the same frequency by nearby operators, for spread-spectrum use, the FCC has allocated certain *unlicensed* bands in the United States, referred to as the instrumentation, scientific, and medical (ISM) bands, wherein the user is only required to spread spectrum by a minimum amount, and not to exceed an upper limit on transmitted power. The extent of spectrum spreading may be measured by the amount of *processing gain* in the receiver required to despread and detect the signal [140].

As analog cellular telephones begin to saturate the available RF allocations, spread-spectrum techniques are being deployed in cordless telephones and wireless modems [141]. These devices usually operate in the two lower ISM bands: 902 to 928 MHz, and 2.4 to 2.48 GHz. A ubiquitous wireless environment is envisioned, in which mobile users, wherever they may be, can access data and communications services through an intricate network of base stations [2]. The greatest hardware challenge in realizing this scenario is the development of a low-power miniature handset.

There are two different methods to spread the spectrum of a signal: by *direct-sequence* modulation or by *frequency hopping* [140]. Direct sequence is conceptually the simpler, as well as the more straightforward to implement. Each data bit (either +1 or −1) at the transmitter multiplies a prescribed sequence of bits, or *chips* as they are called. The chip sequence is selected for very low autocorrelation, and is often referred to as a pseudo-noise (PN) sequence. Each user is either assigned a unique PN sequence with very low cross-correlation with other user sequences, or a time-shifted version of some long PN sequence. The receiver, after an initial acquisition search to align itself with the start of the PN sequence, correlates the incoming sequence with the pattern it knows to be its own. On detection of a correlation peak, the sign of the peak signals the source data bit. The longer the PN sequence for each bit, the greater the spreading factor, and the more reliable the detection. Here, the processing gain is the sequence length per bit. The spectrum of the transmitted data, composed from a concatenation of PN sequences, is noise-like; thus, the term *spread-spectrum*. Several baseband and IF modem ICs have already appeared to support BPSK and QPSK direct-sequence spread-spectrum transceivers [142–144].

Whereas direct-sequence modulation spreads the spectrum by randomizing the waveform, frequency-hopping spreads spectrum by *hopping* the carrier according to a prescribed sequence across the entire band. The effect in the frequency domain is similar to direct sequence, except for one fundamental difference: with direct sequence, spreading across a wide bandwidth requires each data bit to be mapped into a long PN sequence, and the resultant high "chip-rate" requires high-speed signal processing at the receiver front-end. On the other hand, if the carrier frequency is hopped with a wideband synthesizer, the output spectrum may span an arbitrarily wide frequency range, even at low bit rates. Therefore, if a fast and agile frequency synthesizer is readily available, the receiver front-end operates at the actual data rate frequency-hopped spread spectrum, rather than at the considerably higher chip-rate in a direct-sequence receiver. The former will very likely lead to a low-power solution.

As a notable recent example of a miniature spread-spectrum transceiver, Plessey has introduced a 700-Kbps frequency-hopped device operating in the 2.4-GHz ISM band. The transceiver is entirely contained on a 2-in × 3-in PCMCIA card for insertion into notebook computers [109, 145]. Data FSK modulates the carrier frequency, and a variable-modulus PLL synthesizer slowly hops the carrier to spread the spectrum across the 80-MHz band. All the active devices in the transceiver are on three ICs, consisting of a GaAs RF front-end, a silicon bipolar IF receiver, and a CMOS IC for the hopping-frequency synthesis. As in any spread-spectrum, two-way communication system, transmission and reception is time-division duplexed on the same frequency band. The receiver architecture is a conventional double superheterodyne. In addition, the transceiver requires 50 passive components, including 6 rather bulky filters, and when transmitting 100-mW RF power it dissipates more than 1W.

By the norms of GaAs MMICs, the single-chip IC front-end in this transceiver is highly integrated (Figure 3.19(a)) [146]. It includes the power amplifier and drivers for 2-GHz passive filters in the transmit and receive paths. Interesting features are the dc series connection of the single-ended two-stage low-noise amplifier, which shares the same bias current in both stages through a bypass inductor; the Clapp VCO; and the four-FET switched mixers in the transmit and receive paths (Figure 3.19(b)). In its first version, the IC includes more than 20 on-chip spiral inductors, perhaps the largest number on any MMIC at the time of this design. A second bipolar IC processes the 350-MHz IF signal. In receive mode, the GaAs front-end IC drains 30 mA from a 5V supply. The receiver selects one of two external antennas through an on-chip RF switch to attain spatial diversity. This work has spurred a

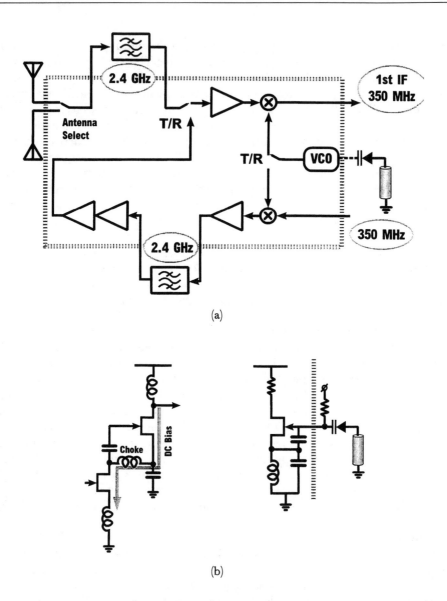

(a)

(b)

Figure 3.19 (a) Front-end GaAs MMIC transceiver for spread-spectrum communications in 2.4-GHz band. Receiver is double-superheterodyne. Note use of off-chip filters. (b) Low-noise amplifier in transceiver IF (left) reuses bias current in two stages. Clapp voltage-controlled local oscillator (right) combines on-chip LC network with off-chip resonators and varactors.

flurry of similar GaAs ICs at the same level of integration [147–149], all operating at ±5V supplies, with similar functionality and performance. One of them [149] uses a very high IF of 915 MHz, so that spurious products at the transmitter output lie well in the stopbands of the output filter, and where the IF signal is processed by a 915-MHz cellular-telephone-type IC.

3.9 ICs TO ENABLE FUTURE TRANSCEIVERS

What will portable wireless communicators of the future look like? We may draw some conclusions from the foregoing summary of RF-IC developments. The crowded spectrum means that future wireless communications will predominantly use spread-spectrum techniques. Coupled to this is a need for low-power dissipation, which will force *architectural innovations* and *higher levels of integration* in the electronics. As an example of such an advanced transceiver, the author, with his colleagues and their graduate students at UCLA, is investigating the architecture and circuit design of a frequency-hopped, binary frequency-shift keyed, zero-IF, all-CMOS two-chip transceiver capable of delivering up to 160 Kbps (the base ISDN rate) in the 900-MHz ISM band [39, 150]. The transceiver architecture is inspired by the modern paging receiver, a very low energy wireless device widely used today. The transceiver implementation (Figure 3.20) will freely mix analog and digital circuits, which makes CMOS the IC technology of choice. Further, to avoid the routing of high-frequency signals off- and on-chip and thereby save the power the buffers would use to drive stray capacitance and off-chip low-impedance lines, it is preferred to integrate all the blocks, *including* the RF front-end. Finally, it is most desirable to use an unmodified, standard production CMOS process.

Low-power operation requires the entire system to operate on a 3V supply. This supply voltage cannot be any lower because in many places the analog circuits contain stacked cascade transistors and the FET threshold voltages are about 0.8V. RF amplifiers are normally tuned with inductor loads, which on silicon ICs are almost always off-chip passive components [151]. However, this is itself wasteful of power, because in addition to the current supplied into the inductor, the circuit must also drive the various parasitic capacitances in off-chip routing the RF signal.

Spiral inductors on silicon substrates suffer from a large capacitance to the conducting silicon substrate. However, after many years of the belief that no useful inductors could be made on silicon, it was found that spiral inductors as large as 10 nH with self-resonance beyond 2 GHz could be fabricated with the standard interconnect metallization. These inductors were used to build passive filters, a tuned amplifier, and an LC voltage-controlled oscillator on a silicon bipolar IC [152–154]. This work has rekindled interest in the design and modeling of small-value spiral inductors on silicon (usually bipolar) ICs, with values in the range of 1 to 10 nH [155, 156]. In some instances, higher inductor Qs are obtained with thicker

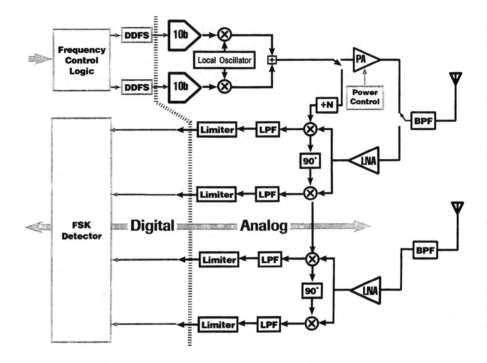

Figure 3.20 The UCLA frequency-hopped spread-spectrum transceiver. This uses direct conversion in both transmit and receive paths, and binary-FSK modulation. In the transmitter, the spread-spectrum waveform is synthesized at baseband and upconverted to RF. During receive mode, the DDFS is under control of a synchronization loop hops with the incoming data, and demodulates it to baseband. The entire mixed analog-digital, baseband/RF system is to be integrated on a 1-μm CMOS IC.

metallization, or higher resonant frequencies with thicker oxides. In search of larger value inductors for low-power, high-gain amplifiers, we have developed a fabrication method whereby the silicon substrate is selectively removed under the inductor, substantially reducing the capacitance to the substrate and thus extending self-resonance to a higher frequency (Figure 3.21(a)) [157]. The self-resonant frequency is now almost as high as it is on a semi-insulating GaAs substrate. Using two such inductors as loads in a balanced circuit, a 900-MHz RF amplifier has been built in 1-μm CMOS with a 30-dB gain draining only 3 mA (Figure 3.21(b)). The amplifier IP3 is about 0 dBm owing to the wide linear range of the MOS differential pair.

The front-end mixer in a direct-conversion receiver must be highly linear to suppress unwanted intermodulation produced by interferers. Also, as the local oscillator frequency is centered at RF in this direct-conversion receiver, inadequate reverse isolation or shielding may cause this frequency to leak through the antenna and to

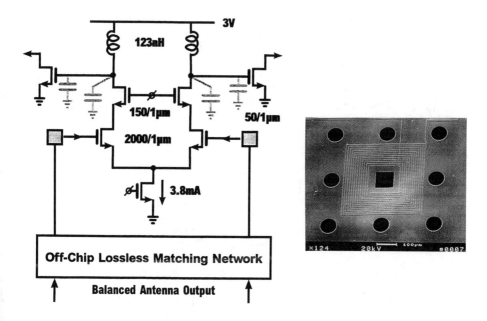

Figure 3.21 Balanced input CMOS RF amplifier. A cascode stage driving an on-chip inductor substantially enhances voltage gain. Outputs common-source stages are for measurement only. LNA will directly drive mixer. A microphotograph of a 100-nH spiral inductor suspended over a pit on substrate.

interfere with other nearby receivers tuned to the same RF. We use an unusual mixer—one that happens to be particularly well-suited to MOS implementation—to circumvent both these problems. The desired signal is downconverted to baseband by subsampling the incoming RF [158]. The mixer circuit is a track-and-hold, with such a wide track-mode bandwidth that it can follow the RF waveform, and a short enough aperture to acquire the *instantaneous* value of the RF waveform on receipt of the sampling clock edge (Figure 3.22). The received RF in our system is a 26-MHz wide spread-spectrum centered on a 915-MHz carrier, so by subsampling this waveform at a clock rate of 52 MHz or higher, the spectrum is translated to baseband without aliasing. However, the mixer also acquires wideband noise accompanying the signal from dc to up to the gigahertz track bandwidth, and aliases the rms noise into a bandwidth at half the sample rate, thereby raising the baseband noise spectral density by the ratio of the track bandwidth to the sample rate. Subsampling mixers therefore have a higher noise figure than conventional mixers. However, as such a mixer tends to be more linear than an analog multiplier, its dynamic range is also large, provided a high-gain, low-noise RF amplifier precedes it. The overall front-end spurious-free dynamic range is jointly set by the individual specifications of the RF amplifier and mixer. In this 1-μm CMOS prototype, the mixer circuit

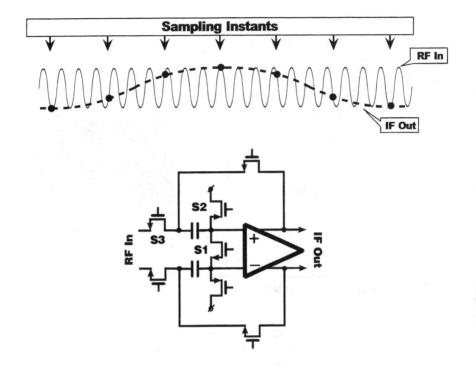

Figure 3.22 A downconversion mixer based on direct sampling of RF. During track mode, switches are closed and bandwidth of passive circuit is about 1 GHz. On receipt of sample clock, switches open within a very short aperture time, capturing instantaneous value of RF waveform. Op amp removes switch charge injection and buffers to subsequent circuits.

draws 4 mA from 3V to acquire samples of a 900-MHz modulated waveform at a 50-MHz rate, which it then translates to baseband. The linearity, as measured by a +26-dBm third-order intercept, exceeds that of most continuous-time 900-MHz monolithic mixers, and for the fundamental reasons given above, the noise figure is 18 dB.

A low-power, agile frequency synthesizer with high spectral purity is required in a frequency-hopped transceiver. Conventional synthesizers based on a PLL with a variable-modulus divider in feedback suffer from limited agility because the loop bandwidth may limit the speed of frequency switching, and it is difficult to design a VCO that both covers a wide range of frequency spreading and produces a pure spectrum. The DDFS-DAC combination is an alternative solution to this problem, which so far has only been used by the military in spread-spectrum communication systems. A direct-digital frequency synthesizer (DDFS) [159] consists of an accumulator and ROM, which produces a sequence of digital words representing samples of a sinewave at a frequency set by an input control word. A D/A converter converts the digital samples into a discrete-time analog sinewave, and a subsequent analog filter may smooth this into a continuous-time sinewave. Round-off errors in the DDFS and nonlinearity in the DAC and filter will limit the attainable spectral purity of the synthesized sinewave [159]. In practice, DAC imperfections may contribute the largest nonlinearity in the system. A characteristic feature of digitally-synthesized sinewaves is that the major imperfections do not necessarily produce harmonic tones, but instead spurious tones at rational multiples of the fundamental. These spurious tones often cluster around the fundamental and cannot be filtered. The DAC can add spurious tones of its own. We use a highly efficient DDFS combined with a high-speed, low-power charge-redistribution DAC to produce an FSK-modulated, frequency-hopped spread spectrum at baseband, which a fixed-frequency local oscillator upconverts to RF. A CMOS implementation of a DDFS-DAC at 3V dissipates only 40 mW when clocked at 50 MHz (of which the quadrature ROM-accumulator accounts for 35 mW), with worst-case spurious levels of −57 dBc or less across the entire spreading range (Figure 3.23) [160].

The 915-MHz LO is a four-stage MOS ring oscillator locked in a PLL to a lower frequency crystal reference. The quadrature outputs at 915 MHz for the image-reject mixers are tapped at diametrically opposite points [130, 134] to a phase accuracy of better than 1 deg. As the LO operates at a fixed frequency, it is locked to a low-frequency crystal reference in a PLL solely optimized to reduce phase noise. At a 100-kHz offset, a phase noise level of −75 dBc/Hz is measured on a prototype.

The power amplifier is a binary-weighted array of FETs biased near threshold, which is digitally selected to deliver power levels from −15 dBm to +15 dBm through a matching network to the antenna with a 40% conversion efficiency (Figure 3.24) [161].

The transceiver achieves spatial diversity with two individual receive channels, which are powered up continuously and connected to two different types of miniature antennas. Data decisions are made on an equally weighted sum of their baseband outputs. A high-order switched-capacitor elliptic lowpass filter selects the desired channel, which the onboard DDFS-based synthesizer course has dehopped to dc in the course of downconversion. The capacitively-coupled output is then amplitude-limited to produce a one-bit stream fed to an optimal FSK digital detector [39]. The detector quantizes the zero crossings in time by oversampling the limited baseband data and seeks correlations between one-bit representations of the two possible data values. A logarithmic amplifier connected to taps on the limiting amplifier chain measures the received signal strength [162]. Operating from a 3V supply, the 1-μm CMOS transceiver is expected to drain 70 mA in receive mode, and 100 mA in transmit mode (the power amplifier is on-chip). Considering that the transceiver communicates on 26-MHz-wide spectrum, this is indeed a low-power dissipation.

This research project is now exploring issues related to the single-chip integration of this transceiver. Such a highly-integrated CMOS "VLSI radio" would repre-

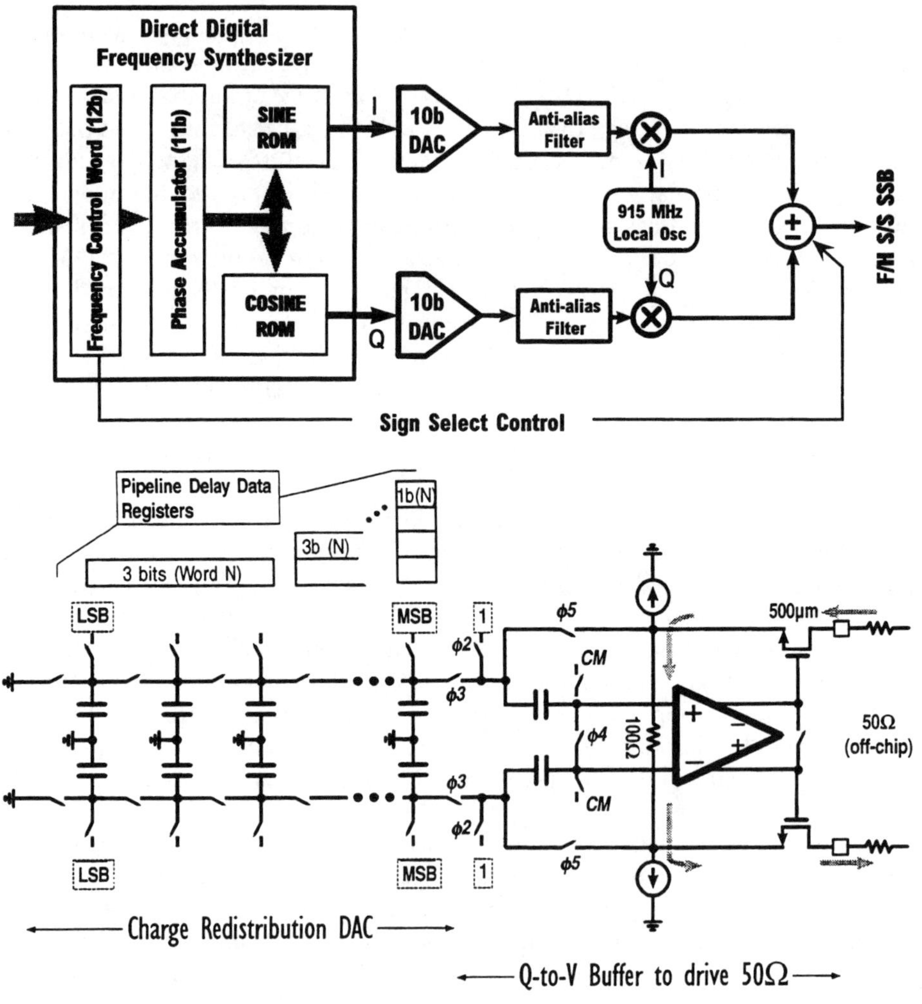

Figure 3.23 An agile frequency synthesizer. The DDFS produces digital samples of a sine and cosine at a frequency set by an input word. DACs convert these to analog domain, and after quadrature upconversion, either the upper or lower sideband is selected. Thus, 0- to 13-MHz baseband frequency synthesis results in output spanning 902 to 928 MHz. A first-order antialias filter suffices if DDFS/DAC operate at 80 MHz. DAC uses passive pipelined architecture.

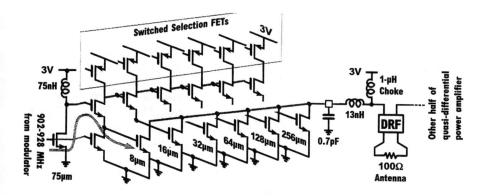

Figure 3.24 A digitally controlled power amplifier. The binary-weighted array of output NMOSFETs drive the antenna load differentially, to deliver up to +15 dBm from a 3V supply. A large on-chip inductor used as an exciter load gives swings above power supply. PMOSFETs selectively activate source follower buffers, and thus total delivered power.

sent a major step forward in the evolution of the radio integrated circuits described in this chapter.

3.10 CONCLUSIONS

This survey has covered some of the key existing and emerging communication applications that have prompted advances in RF-ICs. The emphasis was on widespread *portable consumer* applications. This has excluded coverage of integrated television tuners, for instance, which were historically the principal motivation for the development of UHF RF-ICs by the consumer electronics industry in the mid-1970s [163]. The fact that about 18 million tuners are produced every year means that there has been a steady stream of innovations in the building blocks and architectures for this application [164–178]. Then there are emerging areas, such as personal GPS receivers, where RF integration and low power will be the key for acceptance by consumers [179, 180]. Here, too, RF-IC designers are responding with integrated front-ends [181–184], although the greater challenges to realize a system with an acceptable precision may lie in the baseband processing.

The lucrative global wireless market is attracting a great deal of attention from circuit designers across the industry. The RF range of interest spans 400 MHz to 2.5 GHz. There is a multipronged industrywide assault to provide the right solutions. The solution must be low cost and low power, and it must give high functionality. MMIC designers bring to RF-ICs a microwave style derived from small-scale circuits that operated at much higher frequencies. On the other hand, as the frequency range lies within the capability of modern silicon IC technologies, there is a response from this community. Bipolar and BiCMOS RF-ICs are often extensions of baseband-style circuits to high-frequency, often with more functionality than GaAs MMICs offer. A low-transistor f_T is no longer a handicap, and experience rapidly accumulates on how to solve some of the unique problems of silicon substrates, such as high-frequency losses in the substrate [185] and on-chip coupling problems between circuit blocks. In the not too distant future, as microprocessors and memories drive the linewidths of silicon IC processes to deep submicron, RF-ICs might be designed as untuned wideband circuits, in the style of ICs for video applications today.

Small-scale GaAs MMICs might sometimes perform better as stand-alone components, but embedded in a system such as a digital cellular telephone or a spread-spectrum transceiver, they require a large overhead in support circuits, and thereby lose an edge in certain specifications, such as power dissipation. Silicon RF-ICs offer higher levels of integration, but even within the silicon IC design community there are differing points of view. While integrated silicon bipolar transceivers are now to be found in the next generation of products, CMOS RF solutions, albeit at an exploratory stage, now look very promising. They combine elements from well-established techniques from voiceband and video ICs, and are unfettered in partitioning tasks between analog and digital signal processing, while using off-chip passive components when necessary. Digital control components and baseband signal processors are readily integrated with, indeed merged into, the RF and IF sections. This array of competing technologies offers communication system designers more creative opportunity, and the best wireless transceiver solutions may well emerge from system design evolving together with architecture, circuits, antennas, and power allocation plans.

What are the prospects for this multitude of approaches to low-power wireless communications? If we project on the experience gained from baseband and video telecommunications circuits, the future wireless transceiver may be only one BiCMOS or CMOS large-scale, mixed analog-digital IC, requiring one passive filter and a crystal resonator. Baseband digital signal processing may make up for imperfections in the front-end RF sections. As it will likely be used in a microcell environment at low transmit power levels, the transceiver may be encapsulated in a plastic package. There is no fundamental reason why it should need more than a single 1.5V battery as the power source. After decades of thinking of it only as science-fiction, the electronics world is prodded forth by the communications explosion into making the two-way wristwatch-size radio a reality. A large community of researchers, not trained in the classical radio art, is busy re-examining the conventions, and at times challenging the received wisdom. We are poised at a turning point in the evolution of radio-circuit engineering.

Acknowledgment

My own involvement in RF-IC design is considerably enriched by collaborations with the many talented colleagues and graduate students participating in the Low-

Power Wireless Transceiver Program in the UCLA Electrical Engineering Department. Professors Henry Samueli, Greg Pottie, and Yahya Rahmat-Samii have made an exemplary teamwork possible on this project. The program is principally supported by the U.S. Defense Advanced Research Projects Agency, with key additional support, both intellectual and financial, from a consortium of semiconductor companies under the State of California MICRO Program.

The recent financial stringencies at UCLA have not so far eroded the marvelously rich engineering collection and online facilities at the UCLA Science and Engineering Library. It would be impossible to write this chapter without ready access to these resources. Ms. Karen Andrews and Ms. Aggi Raeder, two of the expert librarians there, enthusiastically helped me in using the latest tools to peruse the literature and amazed me with the ease with which they could track down obscure references from all corners of the world.

I thank Professor Bob Meyer for a very thorough reading of the draft manuscript and for his thoughtful comments on it, and Messrs Taiwa Okanobu and Mark McDonald who made useful suggestions. I am very grateful to Dr. Ran-Hong Yan, the Guest Editor of the *Proceedings of the IEEE* who solicited this manuscript for its original publication, for his extraordinary ability to obtain speedy reviews and patiently absorb delays on my part in bringing this work to completion.

References

[1] G. H. Heilmeier, "Personal Communications: Quo Vadis," *Int'l Solid State Circuits Conf*, San Francisco, CA, 1992., pp. 24–26.

[2] D. C. Cox, "Universal Digital Portable Radio Communications," *Proc. of IEEE*, Vol. 75, No. 4, 1987, pp. 436–476.

[3] S. Sheng, A. Chandrakasan, and R. W. Brodersen, "A Portable Multimedia Terminal," *IEEE Communications, Magazine*, Vol. 30, No. 12, Dec. 1992, pp. 64–75.

[4] B. A. Beatty, "The IBM Personal Communicator—Design Considerations," *Southeastcon*, Miami, FL, 1994, pp. 432–435.

[5] A. Matsuzawa, "Low-Voltage and Low-Power Circuit Design for Mixed Analog/Digital Systems in Portable Equipment," *IEEE J. of Solid-State Circuits*, Vol. 29, No. 4, 1994, pp. 470–480.

[6] E. A. Vittoz, "Low-Power Design: Ways to Approach the Limits," *Int'l Solid State Circuits Conf.*, San Francisco, CA, 1994, pp. 14–18.

[7] F. E. Terman, *Radio Engineers' Handbook*, New York, NY: McGraw-Hill, 1943.

[8] W. L. Everitt, ed., *Fundamentals of Radio and Electronics*, 2nd ed., Englewood Cliffs, NJ: Prentice-Hall, 1958.

[9] K. K. Clarke, and D. T. Hess, *Communication Circuits: Analysis and Design*, Reading, MA: Addison-Wesley, 1971.

[10] H. L. Krauss, C. W. Bostian, and F. H. Raab, *Solid State Radio Engineering*, New York: John Wiley & Sons, 1980.

[11] W. H. Hayward, *Introduction to Radio Frequency Design*, Englewood Cliffs, NJ: Prentice-Hall, 1982.

[12] U. L. Rohde, and T. T. N. Bucher, *Communications Receivers: Principles & Design*, New York, NY: McGraw-Hill, 1988.

[13] D. O Pederson, and K. Mayaram, *Analog Integrated Circuits for Communications: Principles, Simulation, and Design,* Boston, MA: Kluwer Academic, 1991.

[14] M. Ibuka, "How Sony Developed Electronics for the World Market," *IEEE Trans. on Engineering Management,* Vol. EM-22, No. 1, Winter 1975, pp. 15–19.

[15] A. Laundrie, and K. Negus, "GPS Receiver Chip Targets Low-Power Commercial Markets," *Microwaves & RF,* Aug. 1992, pp. 135–138.

[16] N. Camilleri, D. Lovelace, J. Costa, and D. Ngo, "New Development Trends for Silicon RF Device Technologies," *Microwave & Millimeter-Wave Monolithic Circuits Symp.,* San Diego, CA, 1994, pp. 5–8.

[17] W. Kanow, and I. Siewert, "Integrated Circuits for Hi-Fi Radios and Tuners," *Electronic Components and Applications,* Vol. 4, No. 1 Nov. 1981, pp. 11–27.

[18] R. M. Warner, and J. N. Fordemwalt, ed., *Integrated Circuits: Design Principles and Fabrication.* New York, NY: McGraw-Hill, 1965.

[19] C. S. Meyer, D. K. Lynn, and D. J. Hamilton, ed., *Analysis and Design of Integrated Circuits.* New York, NY: McGraw-Hill, 1968.

[20] Y. Hishida, and M. Kazama, "Mobile Communication System: Its Semiconductor Technology," *Hitachi Review.* Vol. 42, No. 3, June 1993, pp. 95–100.

[21] B. Crouch, ed., *Dick Tracy: America's Most Famous Detective.* Secaucus, NJ: Citadel Press, 1987.

[22] L. Blaser, and T. Taira, "An AM/FM Radio Subsystem IC," *IEEE Trans. on Consumer Electronics,* Vol. CE-23, No. 2, 1977, pp. 129–136.

[23] W. Peil, and R. J. McFayden, "A Single Chip AM/FM Integrated Circuit Radio," *IEEE Trans. on Consumer Electronics,* Vol. CE-23, No. 3, 1977, pp. 424–431.

[24] O. L. Richards, "A Complete AM/FM Signal Processing System," *IEEE Trans. on Consumer Electronics,* Vol. CE-24, No. 1, 1978, pp. 34–38.

[25] W. G. Kasperkovitz, "An Integrated FM Receiver," *Microelectronics Reliability,* Vol. 21, No. 2, 1981, pp. 183–189.

[26] A. A. Abidi, "Noise in Active Resonators and the Available Dynamic Range," *IEEE Trans. on Circuits and Systems,* Vol. 39, No. 4, 1992, pp. 296–299.

[27] W. G. Kasperkovitz, "FM Receivers for Mono and Stereo on a Single Chip," *Philips Technical Review,* Vol. 41, No. 6, 1983, pp. 169–182.

[28] W.H.A. van Dooremolen, and M. Hufschmidt, "A Complete FM Radio on a Chip," *Electronic Components and Applications.* Vol. 5, No. 3, 1983, pp. 159–170.

[29] J. Brilka, and G. Sieboerger, "A Single-Chip FM Radio," *Int'l Conf. on Consumer Electronics,* 1984, pp. 156–157.

[30] T. Okanobu, T. Tsuchiya, K. Abe, and Y. Ueki, "A Complete Single-Chip AM/FM Radio Integrated Circuit," *IEEE Trans. on Consumer Electronics,* Vol. CE-28, No. 3, 1982, pp. 393–408.

[31] T. Okanobu, H. Tomiyama, and H. Arimoto, "Advanced Low-Voltage Single Chip Radio IC," *IEEE Trans. on Consumer Electronics,* Vol. 38, No. 3, 1992, pp. 465–475.

[32] D. G. Tucker, "The History of the Homodyne and Synchrodyne," *J. of the British Inst. of Radio Engineers,* Vol. 14, No. 4, 1954, pp. 143–154.

[33] D. Mitchell, and K. G. van Wynen, "A 150-mc Personal Radio Signaling System," *Bell System Technical J.,* Vol. XI, No. 5, 1961, pp. 1239–1257.

[34] A. E. Kerwien, and L. H. Steiff, "Design of a 150-megacycle Pocket Receiver for the BELLBOY Personal Signaling System," *Bell System Technical J.,* Vol. XLII, No. 3, 1963, pp. 527–565.

[35] K. Nagata, M. Akahori, T. Mori, and S. Umetsu, "Digital Display Radio Paging System," *NEC Research & Development.,* No. 68, Jan. 1983, pp. 16–23.

[36] I.A.W. Vance, "An Integrated Circuit VHF Radio Receiver," *The Radio and Electronic Engineer,* Vol. 50, No. 4, 1980, pp. 158–164.

[37] I.A.W. Vance, "Fully Integrated Radio Paging Receiver," *IEEE Proc.,* Vol. 129, Part F, No. 1, 1982, pp. 2–6.

[38] R. C. French, "A High Technology VHF Radio Paging Receiver," *Int'l Conf. on Mobile Radio Systems & Techniques*, York, UK, 1984, pp. 11–15.

[39] J. Min, A. Rofougaran, H. Samueli, and A. A. Abidi, "An All-CMOS Architecture for a Low-Power Frequency-Hopped 900 MHz Spread-Spectrum Transceiver," *Custom IC Conf.*, San Diego, CA, 1994, pp. 379–382.

[40] K. Holtvoeth, "A Fully-Integrated Low-Power FSK Receiver for VHF and UHF Paging," *IEEE Trans. on Consumer Electronics*, Vol. 35, No. 3, 1989, pp. 707–713.

[41] S. Drude, and T. Rudolph, "New Chip-set Reduces Power Consumption of Radio Pagers," *Electronic Components and Applications*, Vol. 9, No. 2, 1989, pp. 112–124.

[42] J. F. Wilson, R. Youell, T. H. Richards, G. Luff, and R. Pilaski, "A Single-Chip VHF and UHF Receiver for Radio Paging," *IEEE J. of Solid State Circuits*, Vol. 26, No. 12, 1991, pp. 1944–1950.

[43] K. Nagata, D. Ishii, T. Mori, T. Oyagi, and T. Seo, "Slim Digital Pagers," *NEC Research & Development.*, No. 74, July 1984, pp. 55–64.

[44] K. Yamasaki, S. Yoshizawa, Y. Minami, T. Asai, Y. Nakano, and M. Kuroda, "Compact Size Numeric Display Pager with New Receiving System," *NEC Research & Development*, Vol. 33, No. 1, Jan. 1992, pp. 73–81.

[45] A. Burt, "Direct Conversion Receivers Come of Age in the Paging World," *GEC Review.* Vol. 7, No. 3, 1992, pp. 156–160.

[46] S. Tanaka, A. Nakajima, J. Nakagawa, and A. Nakagoshi, "High-Frequency, Low-Voltage Circuit Technology for VHF Paging Receiver," *IEICE Trans. on Fundamentals of Electronics, Communications and Computer Sciences*, Vol. E76-A, No. 2, 1993, pp. 156–163.

[47] B. F. Cambou, "MMIC Consumer Application and Production," *Microwave & Millimeter-Wave Monolithic Circuits Symp.*, Albuquerque, NM, 1992, pp. 1–3.

[48] H. Komatsu, R. Norose, and S. J. Symonds, "The Wristwatch as a Personal Communications Device," in *Pacific Telecommunications Conf.*, Honolulu, 1993, pp. 242–244.

[49] M. Pardoen, R. Pache, and E. Dijkstra, "Direct-Conversion Receiver Provides CP-FSK Operation," *Microwaves & RF*. March 1994, pp. 151–155.

[50] G. J. Lomer, "Telephoning on the Move—Dick Tracy to Captain Kirk," *IEEE Proc.*, Vol. 134, Part F, No. 1, 1987, pp. 1–8.

[51] K. Wakino, "Recent Developments of Dielectric Resonator Materials and Filters in Japan," *Ferroelectrics*, Vol. 91, 1989, pp. 69–86.

[52] M. Hikita, T. Tabuchi, and A. Sumioka, "Miniaturized SAW Devices for Radio Communication Transceivers," *IEEE Trans. on Vehicular Technology*, Vol. 38, No. 1, 1989, pp. 2–8.

[53] K. Eda, "Ultra-Small SAW Filter Works in Mobile Communications," *J. of Electronic Engineering.* May 1992, pp. 54–57.

[54] Y. Tamura, T. Maru, N. Hirasawa, H. Okuno, M. Komoda, and M. Hotsumi, "Hand-Held Portable Equipment for Cellular Mobile Telephone," *NED Research and Development*, No. 87, Oct. 1987, pp. 34–43.

[55] D. Anderson, and R. J. Zavrel, "RF ICs for Portable Communications Equipment," *Electronic Components and Applications*, Vol. 7, No. 1, 1985, pp. 37–44.

[56] I. A. Koullias, S. L. Forgues, and P. C. Davis, "A 100 MHz IF Amplifier/Quadrature Demodulator for GSM Cellular Radio Mobile Terminals," *Bipolar Circuits and Technology Mtg.*, Minneapolis, MN, 1990, pp. 248–251.

[57] Y. Mori, H. Yabuki, M. Ohba, M. Sagawa, M. Makimoto, and I. Shibazaki, "Miniaturized RF-Circuit Modules for Land Mobile Communication Equipment," *IEEE Vehicular Technology Conf.*, Orlando, FL, 1990, pp. 65–70.

[58] M.. Hikita, A. Yuhara, and K. Oda, "New Low-Loss, Miniature SAW Filters," *Hitachi Review*, Vol. 42, No. 3, June 1993, pp. 119–124.

[59] L. Scharf, H. Frank, and T. Pollakowski, "High f_T and Low Noise," *Siemens Components.* Vol. XXV, No. 4, 1990, pp. 119–122.

[60] S. Yamada, and K. Oguri, "Low-Power Front-End Oriented Discrete Semiconductor Devices," *Hitachi Review*, Vol. 42, No. 3, June 1993, pp. 101–106.

[61] C. Huang, "Where Do Integrated Circuits Replace Discretes?," *Applied Microwaves & Wireless*, Summer 1994, pp. 5–6.

[62] S. P. MacCabe, "GaAs MMICs for Cellular Telephones, GPS and DBS Markets," *WESCON*, San Francisco, CA, 1989, pp. 279–281.

[63] M. Rocchi, "GaAs & Si MMIC Building Blocks: A Moot Point Revisited," *Microelectronic Engineering*, Vol. 15, 1991, pp. 685–692.

[64] Y. Aoki, K. Kobayashi, and W. Kennan, "Monolithic Microwave ICs (MMICs) for Telecommunication," *Fujitsu Scientific & Technical J.*, Vol. 30, No. 1, June 1994, pp. 32–39.

[65] Y. Imai, M. Tokumitsu, A. Minakawa, T. Sugeta, and M. Aikawa, "Very Low-Current and Small Size GaAs MMICs for L-Band Front-End Applications," *GaAs IC Symposium*, Monterey, CA, 1989, pp. 71–74.

[66] Y. Imai, M. Tokumitsu, and A. Minakawa, "Design and Performance of Low-Current GaAs MMICs for L-Band Front-End Applications," *IEEE Trans. On Microwave Theory & Techniques*, Vol. 39, No. 2, 1991, pp. 209–215.

[67] K. R. Cioffi, "Ultra-low dc Power Consumptions in Monolithic L-Band Components," *IEEE Trans. on Microwave Theory & Techniques*, Vol. 40, No. 12, 1992, pp. 2467–2472.

[68] C. Kermarrec, M. Takano, F. McGrath, P. O'Sullivan, G. Dawe, E. Heaney, and G. St. Onge, "High Performance, Low Cost GaAs MMICs for Personal Phone Applications," *Int'l Symp. on GaAs and Related Compounds*, Karuizawa, Japan, 1992.

[69] E. Heaney, F. McGrath, P. O'Sullivan, and C. Kermarrec, "Ultra Low Power Low Noise Amplifiers for Wireless Communications," *GaAs IC Symp.*, San Jose, CA, 1993, pp. 49–51.

[70] S. Hara, K. Osato, A. Yamada, T. Tsukao, and T. Yoshimasu, "Miniaturized Low Noise Variable MMIC Amplifiers with Low Power Consumption for L-Band Portable Communication Applications," *Microwave and Millimeter-Wave Monolithic Circuits Symp.*, Atlanta, GA, 1993, pp. 67–70.

[71] T. Ohgihara, S. Kusunoki, M. Wada, and Y. Murakami, "GaAs JFET Front-End MMICs for L-Band Personal Communications," *Microwave and Millimeter-Wave Monolithic Circuits Symp.*, Atlanta, GA, 1993, pp. 9–12.

[72] C. Kusano, E. Hase, and K. Sakamoto, "Low-Power, Low-Noise GaAs MMIC," *Hitachi Review*, Vol. 42, No. 3, June 1993, pp. 107–112.

[73] R. E. Lehmann, and D. D. Heston, "X-Band Monolithic Series Feedback LNA," *IEEE Trans. on Microwave Theory & Techniques*, Vol. MTT-33, No. 12, 1985, pp. 1560–1566.

[74] H. Sakai, A Tezuka, Y. Mori, M. Sagawa, T. Katoh, J. Itoh, and K. Fujimoto, "Low-Power GaAs ICs for Mobile Communication Equipment," *Int'l Symp. on GaAs and Related Compounds*, Karuizawa, Japan, 1992.

[75] V. Nair, R. Vaitkus, D. Scheitlin, J. Kline, and H. Swanson, "Low Current GaAs Integrated Down Converter for Portable Communication Applications," *GaAs IC Symp.*, San Jose, CA, 1993, pp. 41–44.

[76] C. Woo, A Podell, R. Benton, D. Fisher, and J. Wachsman, "A Fully Integrated Transceiver Chip for the 900 MHz Communication Bands," *GaAs IC Symp.*, Miami Beach, FL, 1992, pp. 143–146.

[77] P. E. Chadwick, "High Performance Integrated Circuit Mixers," *In Radio Receivers & Associated Systems*, London, UK, 1981, pp. 1–9.

[78] D. M. Snider, "A Theoretical Analysis and Experimental Confirmation of the Optimally Loaded and Overdriven RF Power Amplifier," *IEEE Trans. on Electron Devices*, Vol. ED-14, No. 12, 1967, pp. 851–857.

[79] J. J. Komiak, "Design and Performance of an Octave Band 11 Watt Power Amplifier MMIC," *IEEE Trans. on Microwave Theory & Techniques*, Vol. 38, No. 12, 1990, pp. 2001–2006.

[80] T. Takagi, Y. Ikeda, K. Seino, G. Toyoshima, and A. Inoue, "A UHF Band 1.3W Monolithic

Amplifier with Efficiency of 63%," *Microwave & Millimeter-Wave Monolithic Circuits Symp.*, Albuquerque, NM, 1992, pp. 35–38.

[81] M. Gat, D. S. Day, S. Chan, C. Hua, and J. R. Basset, "A 3.0 Watt High Efficiency C-Band Power MMIC," *GaAs IC Symp.*, Monterey, CA, 1991, pp. 331–334.

[82] O. Ishikawa, M. Maeda, K. Nishii, M. Yanagihara, T. Yokoyama, Y. Ikeda, and Y. Ota, "Cellular Telecommunication GaAs Power Modules," *Applied Microwave & Wireless*, Vol. 4, No. 3, Fall 1992, pp. 83–88.

[83] M. Maeda, M. Nishijima, H. Takehara, C. Adachi, H. Fujimoto, and O. Ishikawa, "A 3.5V, 1.3W GaAs Power Multi-chip IC for Cellular Phones," *IEEE J. Of Solid-State Circuits*, Vol. 29, No. 10, 1994, pp. 1250–1256.

[84] I. Yoshida, M. Katsueda, S. Ohtaka, Y. Maruyama, and T. Okabe, "Highly Efficient 1.5 GHz Si Power MOSFET for Digital Cellular Front End," *Int'l Symp. on Power Semiconductor Devices*, Tokyo, 1992, pp. 156–157.

[85] T. Okabe, and K. Kobayashi, "Radio-Frequency MOS Power Module," *Hitachi Review*, Vol. 42, No. 3, June 1993, pp. 113–118.

[86] R. G. Meyer, and W. D. Mack, "A 1-GHz BiCMOS RF Front-End IC," *IEEE J. of Solid-State Circuits*, Vol. 29, No. 3, 1994, pp. 350–355.

[87] M. N. Sera, "Chip Set Addresses North American Digital Cellular Market," *RF Design*, Vol. 17, No. 3, March 1994, pp. 54–62.

[88] Y. Funada, "VCO Techniques Enable Compact, High-Frequency Mobilecom," *J. of Electronic Engineering*, Dec. 1993, pp. 40–43.

[89] Y. Tamura, A. Yonehata, S. Miyazaki, F. Kobayasi, Y. Fukuda, and J. Kamishiro, "Development of Advanced Mobile Telephone P3 (Personal Pocket Phone)," *NEC Research and Development*, No. 98, July 1990, pp. 60–70.

[90] H. Ohmori, "SAW Filters Hold Key to Creating Small, Portable Equipment," *J. of Electronic Engineering*, May 1992, pp. 63–67.

[91] G. Kainz, "Microwave Ceramics Boosts Miniaturization," *Siemens Components*, Vol. 27, No. 5, Sep-Oct 1992, pp. 11–14.

[92] Y. Kado, M. Suzuki, K. Koike, and Y. Omura, "A 1-GHz/0.9-mW CMOS/SIMOX Divide-by-128/129 Dual-Modulus Prescaler using a Divide-by-2/3 Synchronous Counter," *IEEE J. of Solid-State Circuits*, Vol. 28, No. 4, 1993, pp. 513–517.

[93] T. Seneff, L. McKay, K. Sakamoto, and N. Tracht, "A Sub-1 mA 1.5-GHz Silicon Bipolar Dual Modulus Prescaler," *IEEE J. of Solid-State Circuits*, Vol. 29, No. 10, 1994, pp. 1206–1211.

[94] M. Morishima, "Present and Future of Mobile Communications in Japan," *Microwave & Millimeter-Wave Monolithic Circuits Conf.*, Atlanta, GA, 1993, pp. 3–6.

[95] I. Shimizu, S. Urabe, K. Hirade, K. Nagata, and S. Yuki, "A New Pocket-size Cellular Telephone for NTT High-Capacity Land Mobile Communication System," *Vehicular Technology Conf.*, St. Louis, MO, 1991, pp. 114–119.

[96] I. A. Koullias, J. H. Havens, I. G. Post, and P. E. Bronner, "A 900 MHz Transceiver Chip Set for Dual-Mode Cellular Radio Terminals," *Int'l Solid State Circuits Conference*, San Francisco, CA, 1993, pp. 140–141.

[97] L. Longo, R. Halim, B.-R. Horng, K. Hsu, and D. Shamlou, "A Cellular Analog Front End with a 98 dB IF Receiver," *Int'l Solid State Circuits Conf.*, San Francisco, CA, 1994, pp. 36–37.

[98] D. J. Goodman, "Second Generation Wireless Information Networks," *IEEE Trans. on Vehicular Technology*, Vol. 40, No. 2, 1991, pp. 366–374.

[99] J. Fenk, W. Birth, R. G. Irvine, P. Sehrig, and K. R. Schon, "An RF Front-End for Digital Mobile Radio," *Bipolar Circuits and Technology Mtg.*, Minneapolis, MN, 1990, pp. 244–247.

[100] E. Badura, "New Integrated Circuits for GSM Mobile Radio," *Siemens Components*, Vol. XXVI, No. 2, 1991, pp. 72–76.

[101] S. Beyer and G. Lipperer, Low-Current Oscillator Design for 900 MHz GSM Applications, in *Microwave Engineering Europe*. Oct. 1991, pp. 35–41.

[102] R. E. Jenkins, "Transceiver Chip Simplifies GSM Cellular Design," *RF Design*, June 1994, pp. 26–35.

[103] J. Sevenhans, A. Vanwelsenaers, J. Wenin, and J. Baro, "An Integrated Si Bipolar RF Transceiver for a Zero IF 900 MHz GSM Digital Radio Front-end of a Hand Portable Phone," *Custom IC Conf.*, San Diego, CA, 1991, pp. 7.7/1–4.

[104] J. Sevenhans, and A. Vanwelsenaers, "A Silicon Transceiver for a 900 MHz GSM Hand Set," *Microwave Engineering Europe*, Dec./Jan. 1992/93, pp. 59–63.

[105] D. Haspeslagh, J. Ceuterick, L. Kiss, and J. Wenin, "BBTRX: A Baseband Transceiver for a Zero IF GSM Hand Portable Station," *Custom IC Conf.*, San Diego, CA, 1992, pp. 10.7.1–10.7.4.

[106] M. McDonald, "A DECT Transceiver Chip Set," *Custom IC Conf.*, San Diego, CA, 1992, pp. 10.6.1–10.6.4.

[107] B. Madsen, and D. E. Fague, "Radios for the Future: Designing for DECT," *RF Design*, April 1993, pp. 48–53.

[108] M. Eccles, "Gigahertz Systems on a Chip," *Electronics World+Wireless World*, Aug. 1993, pp. 662–663.

[109] I. White, "UHF Technology for the Cordless Revolution," *Electronics World+Wireless World*, July/Aug. 1993, pp. 542–546/657–661.

[110] A. Boveda, G. L. Bonato, and O. Ripolles, "GaAs Monolithic 1.5-2.5 GHz Image Rejection Receiver," *Int'l Symp. on Circuits & Systems*, San Diego, CA, 1992, pp. 232–235.

[111] A. Boveda, F. Ortigoso, and J. I. Alonso, "A 0.7-3 GHz GaAs QPSK/QAM Direct Modulator," *Int'l Solid State Circuits Conference*, San Francisco, CA, 1993, pp. 142–143.

[112] M. Williams, F. Bonn, C. Gong, and T. Quach, "GaAs RF ICs Target 2.4-GHz Frequency Band," *Microwaves & RF*, July 1994, pp. 111–118.

[113] M. D. McDonald, "A 2.5 GHz BiCMOS Image Reject Front-End," *Int'l Solid State Circuits Conf.*, San Francisco, CA, 1993, pp. 144–145.

[114] J. L. Wang, T. Tsuchiya, H. Takeuchi, K. Shirotori, S. Miyazaki, and T. Nakata, "A Miniature Ultra Low Power 2.5 GHz Downconverter IC for Wireless Communications," *NEC Research & Development*, Vol. 35, No.1, Jan. 1994, pp. 46–50.

[115] T. Tsukahara, M. Ishikawa, and M. Muraguchi, "A 2V 2 GHz Direct-Conversion Quadrature Modulator," *Int'l Solid State Circuits Conf.*, San Francisco, CA, 1994, pp. 40–41.

[116] M. Nakatsugawa, Y. Yamaguchi, and M. Muraguchi, "An L-Band Ultra Low Power Consumption Monolithic Low Noise Amplifier," *GaAs IC Symp.*, San Jose, CA, 1993, pp. 45–48.

[117] D. Ngo, B. Beckwith, P. O 'Neil, and N. Camilleri, "Low Voltage GaAs Power Amplifiers for Personal Communications at 1.9 GHz," *MTT-S Int'l Microwave Symp.*, Atlanta, GA, 1993, pp. 1461–1464.

[118] T. Kunihisa, T. Yokoyama, H. Fujimoto, K. Ishida, H. Takehara, and O. Ishikawa, "High Efficiency, Low Adjacent Channel Leakage GaAs Power MMIC for Digital Cordless Telephone," *Microwave & Millimeter-Wave Monolithic Circuits Symp.*, San Diego, CA, 1994, pp. 55–58.

[119] T. Tokumitsu, I. Toyoda, and M. Aikawa, "Low Voltage, High Power T/R Switch MMIC using LC Resonators," *Microwave and Millimeter-Wave Monolithic Circuits Symp.*, Atlanta, GA, 1993, pp. 27–30.

[120] K. Miyatsuji, S. Nagata, N. Yoshikawa, K. Miyanaga, Y. Ohishi, and D. Ueda, "A GaAs High-Power RF Single-Pole Double-Throw Switch IC for Digital Mobile Communication System," *Int'l Solid State Circuits Conf.*, San Francisco, CA, 1994, pp. 34–35.

[121] P. O'Sullivan, G. St. Onge, E. Heaney, F. McGrath, and C. Kermarrec, "High Performance Integrated PA, T/R Switch for 1.9 GHz Personal Communications Handsets," *GaAS IC Symp.*, San Jose, CA, 1993, pp. 33–35.

[122] A. Bateman, D. M. Haines, and R. J. Wilkinson, "Linear Transceiver Architectures," *IEEE Vehicular Technology Conf.*, Philadelphia, PA, 1988, pp. 478–484.

[123] G. Schultes, A. L. Scholtz, E. Bonek, and P. Veith, "A New Incoherent Direct Conversion Receiver," *Vehicle Technology Conf.*, Orlando, FL, 1990, pp. 668–674.

[124] G. Schultes, E. Bonek, P. Weger, and W. Herzog, "Basic Performance of a Direct Conversion DECT Receiver," *Electronics Letters*, Vol. 26, No. 21, 1990, pp. 1746–1748.

[125] J. Sevenhans, D. Haspesalgh, A. Delarbre, L. Kiss, Z. Chang, and J. F. Kukielka, "An Analog Radio Front-end Chip Set for a 1.9 GHz Mobile Radio Telephone Application," *Int'l Solid State Circuits Conf.*, San Francisco, CA, 1994, pp. 44–45.

[126] V. Thomas, J. Fenk, and S. Beyer, "A One-Chip 2 GHz Single Superhet Receiver for 2 Mb/s FSK Radio Communication," *Int'l solid State Circuits Conf.*, San Francisco, CA, 1994, pp. 42–43.

[127] W. Veit, J. Fenk, S. Ganser, K. Hadjizada, S. Heinen, H. Herrmann, and P. Sehrig, "A. 2.7V 800 MHz -2.1 GHz Transceiver Chipset for Mobile Radio Applications in 25 GHz fT Si-Bipolar," *Bipolar Circuits & Technology Mtg.*, Minneapolis, MN, 1994, pp. 175–178.

[128] D. K. Weaver, "A Third Method of Generation and Detection of Single-Sideband Signals," *Proc. of the IRE*, Vol. 44, No. 12, 1956, pp. 1703–1705.

[129] K. Negus, B. Koupal, J. Wholey, K. Carter, D. Millicker, C. Snapp, and N. Marion, "Highly Integrated Transmitter RFIC with Monolithic Narrowband Tuning for Digital Cellular Handsets," *Int'l Solid State Circuits Conf.*, San Francisco, CA, 1994, pp. 38–39.

[130] A. A. Abidi, "Radio-Frequency Integrated Circuits for Portable Communications," *Custom IC Conf.*, San Diego, CA, 1994, p. 151–158.

[131] J. Wenin, "ICs for Digital Cellular Communication," *European Solid State Circuits Conf.*, Ulm, Germany, 1994, pp. 1–10.

[132] H. Tsurumi and T. Maeda, "Design Study on a Direct Conversion Receiver Front-End for 280 MHz, 900 MHz, and 2.6 GHz Band Radio Communication Systems," *IEEE Vehicular Technology Conf.*, St. Louis, MO, 1991, pp. 457–462.

[133] A. Bateman and D. M. Haines, "Direct Conversion Transceiver Design for Compact Low-Cost Portable Mobile Radio Terminals," *IEEE Vehicular Technology Conf.*, San Francisco, CA, 1989, pp. 57–62.

[134] A. W. Buchwald and K. W. Martin, "High-Speed Voltage-Controlled Oscillator with Quadrature Outputs," *Electronics Letters*, Vol. 27, No. 4, 1991, pp. 309–310.

[135] S. K. Enam and A. A. Abidi, "NMOS ICs for Clock and Data Regeneration in Gb/s Optical-Fiber Receivers," *IEEE J. of Solid State Circuits*, Vol. 27, No. 12, 1992, pp. 1763–1774.

[136] J. K. Cavers and M. W. Liao, "Adaptive Compensation for Imbalance and Offset Losses in Direct Conversion Transceiver," *IEEE Trans. on Vehicular Technology*, Vol. 42, No. 4, 1993, pp. 581–588.

[137] J. T. Taylor and J. K. Omura, "Spread Spectrum Technology: A Solution to the Personal Communications Services Frequency Allocation Dilemma," *IEEE Communications Magazine*, Vol. 29, No. 2, Feb. 1991, pp. 48–51.

[138] R. A. Scholtz, "The Origins of Spread-Spectrum Communications," *IEEE Trans. on Communications*, Vol. COM-30, No. 5, 1982, pp. 822–852.

[139] D. T. Magill, F. D. Natali, and G. P. Edwards, "Spread-Spectrum Technology for Commercial Applications," *Proc. of IEEE*, Vol. 82, No. 4, 1994, pp. 572–584.

[140] R. C. Dixon, *Spread Spectrum Systems*, New York, NY: John Wiley & Sons, 1976.

[141] M. Leonard, "Wireless Data Links Broaden LAN Options," *Electronic Design*. March 19, 1992, pp. 51–58.

[142] D. T. Magill, "A Fully-Integrated, Digital, Direct-Sequence, Spread Spectrum Modem ASIC," *Int'l Symp. on Personal, Indoor and Mobile Radio Communications*, Boston, MA, 1992, pp. 42–46.

[143] R. Jain, H. Samueli, P. T. Yang, and C. Chien, "Computer-Aided Design of a BPSK Spread-Spectrum Chip Set," *IEEE J. of Solid State Circuits*, Vol. 27, No. 1, 1992, pp. 44–58.

[144] C. Chien, P. Yang, E. Cohen, R. Jain, and H. Samueli, "A 12.7 Mchip/s All-Digital BPSK Direct Sequence Spread-Spectrum IF Transceiver in 1.2mm CMOS," *Int'l Solid State Circuits Conf.*, San Francisco, CA, 1994, pp. 30–31.

[145] M. Leonard, "PCMCIA-Sized Radio Links Portable WLAN Terminals," *Electronic Design*. Aug. 5 1993, pp. 45–50.

[146] L. M. Devlin, B. J. Buck, J. C. Clifton, A. W. Dearn, and A. P. Long, "A 2.4 GHz Single Chip Transceiver," *Microwave & Millimeter-Wave Monolithic Circuits Symp.*, Atlanta, GA, 1993, pp. 23–26.

[147] T. Apel, E. Creviston, S. Ludvik, L. Quist, and B. Tuch, "A GaAs MMIC Transceiver for 2.45 GHz Wireless Commercial Products," in *Microwave & Millimeter-Wave Monolithic Circuits Symp.*, San Diego, CA, 1994, pp. 15–18.

[148] B. Khabbaz, A. Douglas, J. DeAngelis, L. Hongsmatip, V. Pellicia, W. Fahey, and G. Dawe, "A High Performance 2.4 GHz Transceiver Chip Set for High Volume Commercial Applications," in *Microwave & Millimeter-Wave Monolithic Circuits Symp.*, San Diego, CA, 1994, pp. 11–14.

[149] M. S. Wang, M. Carriere, P. O'Sullivan, and B. Maoz, "A Single-Chip MMIC Transceiver for 2.4 GHz Spread Spectrum Communication," *Microwave & Millimeter-Wave Monolithic Circuits Symp.*, San Diego, CA, 1994, pp. 19–22.

[150] J. Min, A. Rofougaran, V. Lin, M. Jensen, H. Samueli, A. A. Abidi, G. Pottie, and Y. Rahmat-Samii, "A Low-Power Handheld Frequency-Hopped Spread Spectrum Transceiver Hardware Architecture," *Virginia Tech's Third Symp. on Wireless Personal Communications*, Blacksburg, VA, 1993, pp. 10/1–8.

[151] M. Sakakura and S. Skiest, "Ultra-Miniature Chip Inductors Serve at High Frequency," *J. of Electronic Engineering*. Dec. 1993, pp. 48–51.

[152] N. M. Nguyen and R. G. Meyer, "Si IC-Compatible Inductors and LC Passive Filters," *IEEE J. of Solid State Circuits*, Vol. 25, No. 4, 1990, pp. 1028–1031.

[153] N. M. Nguyen and R. G. Meyer, "A Silicon Bipolar Monolithic RF Bandpass Amplifier," *IEEE J. of Solid State Circuits*, Vol. 27, No. 1, 1992, pp. 123–127.

[154] N. M. Nguyen and R. G. Meyer, "A 1.8-GHz Monolithic LC Voltage-Controlled Oscillator," *IEEE J. of Solid State Circuits*, Vol. 27, No. 3, 1992, pp. 444–450.

[155] D. Lovelace, N. Camilleri, and G. Kannell, "Silicon MMIC Inductor Modeling for High Volume, Low Cost Applications," *Microwave J.*, Vol. 37, No. 8, Aug. 1994, pp. 60–71.

[156] K. B. Ashby, W. C. Finley, J. J. Bastek, S. Moinian, and L. A. Koullias, "High Q Inductors for Wireless Applications in a Complementary Silicon Bipolar Process," *Bipolar Circuits & Technology Mtg.*, Minneapolis, MN, 1994, pp. 179–182.

[157] J. Y.-C. Chang, A. A. Abidi, and M. Gaitan, "Large Suspended Inductors on Silicon and their use in a 2-mm CMOS RF Amplifier," *IEEE Electron Device Letters*, Vol. 14, No. 5248, 1993, pp. 246.

[158] P. Y. Chan, A. Rofougaran, K. A. Ahmed, and A. A. Abidi, "A Highly Linear 1-GHz CMOS Downconversion Mixer," *European Solid-State Circuits Conf.*, Sevilla, Spain, 1993, pp. 210–213.

[159] H. T. Nicholas and H. Samueli, "A 150-MHz Direct Digital Frequency Synthesizer in 1.25 mm CMOS with -90 dBc Spurious Response," *IEEE J. of Solid State Circuits*, Vol. 26, 1991, pp. 1959–1969.

[160] G. Chang, A. Rofougaran, M. K. Ku, A. A. Abidi, and H. Samueli, "A Low-Power CMOS Digitally Synthesized 0-13 MHz Agile Sinewave Generator," *Int'l Solid State Circuits Conf.*, San Francisco, CA, 1994, pp. 32–33.

[161] M. Rofougaran, A. Rofougaran, C. Olgaard, and A. A. Abidi, "A 900 MHz CMOS RF Power Amplifier with Programmable Output," *Symp. on VLSI Circuits*, Honolulu, HI, 1994, pp. 133–134.

[162] K. Kimura, "A CMOS Logarithmic Amplifier with Unbalanced Source-Coupled Pairs," *IEEE J. of Solid State Circuits*, Vol. 28, No. 1, 1993, pp. 78–83.

[163] S. Watanabe, "Technology Transfer for High Frequency Devices for Consumer Electronics," *IEEE Trans. on Microwave Theory & Techniques*, Vol. 40, No. 12, 1992, pp. 2461–2466.

[164] U. Ablassmeier, W. Kellner, and H. Kniepkamp, "GaAs FET Upconverter for TV Tuner," *IEEE Trans. on Electron Devices*, Vol. ED-27, No. 6, 1980, pp. 1156–1159.

[165] K. Torii, S. Fujimori, S. Komatsu, S. Shimizu, K. Yoshihasa, J. Nagai, and Y. Yamagata, "Monolithic Integrated VHF TV Tuner," *IEEE Trans. on Consumer Electronics*, Vol. CE-26, 1980, pp. 180–187.

[166] J.-E. Muller, U. Ablassmeier, J. Schelle, W. Kellner, and H. Kniepkamp, "A Double-Conversion Broad Band TV-Turner with GaAs ICs," in *GaAs IC Symposium*, 1984, pp. 97–100.

[167] P. Dautriche, B. Y. Lao, C. Villalon, V. Pauker, M. Bostelmann, and M. Binet, "GaAs Monolithic Circuits for TV Tuners," in *GaAs IC Symp*, Monterey, CA, 1985, pp. 165–168.

[168] J. Fenk and R. Tauber, "TV VHF/Hyperband Tuner ICs," *IEEE Trans. on Consumer Electronics*, Vol. CE-32, No. 4, 1986, pp. 723–733.

[169] H. van Glabbeek, N. Baars, H. Schreurs, and W. Zwijsen, "VHF, Hyperband, and UHF Mixer/ Oscillator Sections of a TV Tuner on One IC," *IEEE Trans. on Consumer Electronics*, Vol. CE-33, No. 4, 1987, p. 619–622.

[170] T. Nakatsuka, S. Sakashita, K. Goda, and S. Nambu, "A GaAs RF Amplifier IC for UHF TV Tuners," *IEEE Trans. on Consumer Electronics*, Vol. 34, No. 2, pp. 366–371, 1988.

[171] H. Mizukami, K. Sakuta, H. Hatashita, T. Nagashima, and K. Shinkawa, "A High Quality GaAs IC Tuner for TV/VCR Receivers," *IEEE Trans. on Consumer Electronics*, Vol. 34, No. 3, 1988, pp. 649–658.

[172] T. Ducourant, P. Phillippe, P. Dautriche, V. Pauker, C. Villalon, M. Pertus, and J.-P. Damour, "A3-Chip GaAs Double-Conversion TV Tuner with 70 dB Image Rejection," *Microwave & Millimeter-Wave Monolithic Circuits Symp.*, pp. 87–90, Long Beach, CA, 1989.

[173] H. Yagita, A. Terao, M. Tsuneoka, A. Watanabe, T. Tambo, and S. Nambu, "Low Noise and Low Distortion GaAs Mixer-Oscillator IC for Broadcasting Satellite TV Tuner," *GaAs IC Symp.*, San Diego, CA, 1989, pp. 75–78.

[174] P. Philippe, and M. Pertus, "A 2 GHz Enhancement-Mode GaAs Down-Converter IC for Satellite TV Tuner," *Microwave Theory & Techniques Symp.*, Boston, MA, 1991, pp. 73–76.

[175] H. Mizukami, H. Ikedo, K. Ideno, T. Nagashima, and S. Yamada, "Low Supply Voltage GaAs ICs for a TV Tuner," *Int'l Symp. on GaAs and Related Compounds*, Karuizawa, Japan, 1992.

[176] K. Ideno, H. Mitzukami, T. Nagashima, and K. Sakamoto, "A Wide Band GaAs IC for a DBS/ TV Receiver," *Int'l Symp. on GaAs and Related Compounds*, Karuizawa, Japan, 1992.

[177] K. Washio, T. Okabe, K. Norisue, and T. Nagashima, "An All-Band TV Tuner IC with 10-GHz 100-V Mixed Analog/Digital Si Bipolar Technology," *IEEE J. of Solid-State Circuits*, Vol. 27, No. 9, 1992, pp. 1264–1269.

[178] N. Scheinberg, R. Michels, V. Fedoroff, and D. Stoffman, "A GaAs Upconverter Integrated Circuit for a Double Conversion Cable TV Set-Top Tuner," *IEEE J. of Solid-State Circuits*, Vol. 29, No. 6, 1994, pp. 688–692.

[179] J. Hurn, *GPS: A Guide to the Next Utility*, Sunnyvale, CA: Trimble Navigation, 1989.

[180] W. O. Dussell and J. Medina, "Using Creative Surface Mount Manufacturing and Packaging as a Competitive Weapon," *Surface Mount Int'l Conf. & Exposition*, San Jose, CA, 1993, pp. 241–245.

[181] R. M. Herman, C. H. Mason, H. P. Warren, and R. A. Meier, "A GPS Receiver with Synthesized Local Oscillator," *Int'l Solid-State Circuits Conf.*, New York, NY, 1989, pp. 194–195.

[182] R. M. Herman, A. Chao, C. H. Mason, and J. R. Pulver, "An Integrated GPS Receiver with Synthesizer and Downconversion Functions," *Int'l Microwave Symp.*, Boston, 1991, pp. 883–886.

[183] R. Benton, M. Nijjar, C. Woo, A. Podell, G. Horvath, E. Wilson, and S. Mitchell, "GaAs MMICs for an Integrated GPS Front End," *GaAs IC Symp.*, Miami Beach, FL, 1992, pp. 123–126.

[184] K. J. Negus, R. A. Koupal, D. Millicker, and C. P. Snapp, "3.3V GPS Receiver MMIC Implemented

on a Mixed-Signal, Silicon Bipolar Array," *Microwave & Millimeter-Wave Monolithic Circuits Symp.*, Albuquerque, NM, 1992, pp. 209–212.

[185] N. Camilleri, J. Kirschgessner, J. Costa, D. Ngo, and D. Lovelace, "Bonding Pad Models for Silicon VLSI Technologies and their effects on the noise figure of RF NPNs," *Microwave & Millimeter-Wave Monolithic Circuits Symp.*, San Diego, CA, 1994, pp. 225–228.

CHAPTER 4
▼▼▼

DIGITAL MODULATION AND DEMODULATION

Lawrence Burns
3Com Corporation
Santa Clara, CA

4.1 WHY DIGITAL MODULATION?

Most modern wireless communication systems use digital modulation/demodulation techniques. Digital modulation offers a number of advantages over analog modulation, such as increased channel capacity and the ability to transmit and receive messages with greater accuracy than an analog communication system in the presence of noise and distortion [1–9].

In digital communication systems, a finite number of electrical waveforms, or *symbols*, are transmitted. Each symbol can represent one or more bits. The job of the receiver is then to *estimate* which symbol was originally sent by the transmitter. The trick, of course, is estimating which symbol was sent after noise and distortion has been added. It is largely unimportant what size (amplitude) or shape (how distorted) the received signal is so long as the receiver can clearly distinguish one symbol from another.

In Figure 4.1(a), an undistorted binary bit stream is shown. As the figure shows, the original signal is free of distortion, with sharp, well-defined edges. It is easy at this point to determine the length of a bit period and whether each bit is a "1" or a "0." As the signal passes through the channel, distortion changes its shape.

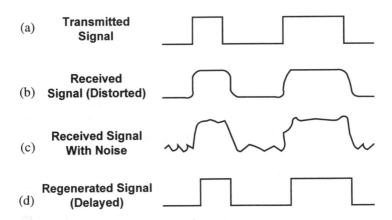

Figure 4.1 Waveforms in a typical digital communication system: (a) transmitted signal, (b) distorted receive signal, (c) distorted receive signal with noise, (d) regenerated signal (delayed).

Although it is somewhat less well-defined, it is still not hard to determine each bit period and its polarity. This is shown in Figure 4.1(b). Distortion in wireless communication systems can be caused by passing the signal through filters having insufficient bandwidth. This situation, where the effects of one bit affect succeeding bits, is called intersymbol interference (ISI). In addition to ISI, multiple versions of the original signal can be received at the same time. These delayed versions of the signal are produced when the transmitted signal reflects off of multiple objects on its way to the receiver. This effect is called *delay spread*. When the bit rate of the signal is high, delay spread can be on the order of one or more bit periods. Equalizers can be used to remove the ISI caused by filter effects and delay spread.

In addition to distortion, noise is added to the signal as it passes through the channel, as shown in Figure 4.1(c). Noise can come from the thermal noise within the receiver itself or can come from manmade or natural interference, such as motor noise or lightning. Other transmitters on adjacent channel frequencies can spill over into a receiver's band. This is called adjacent channel interference (ACI). In most cases, the amount of spectrum users are allowed to use is limited. This means users must reuse some or all of their channel frequencies. In general, transmitters operating at identical frequencies are placed some distance away from each other so that their transmissions do not interfere. When interference does occur, it is referred to as cochannel interference (CCI).

The addition of distortion and noise to the signal makes it harder to determine whether a 1 or a 0 was originally sent. If the amount of noise and distortion is small, then the receiver can easily decide what the original signal was. This is shown in Figure 4.1(d). In this case, the receiver was able to recreate the original bit stream without error or distortion. The ability of a digital communication systems to exactly regenerate the original signal is an advantage digital communication systems have over analog communication systems. It is this very feature that makes compact disc players have such high-fidelity audio, and the reason why laser disks have such sharp picture quality.

If the level of distortion and noise is too large, the receiver's estimate of whether a 1 or a 0 was sent will occasionally be wrong. When this happens, a *bit error* has occurred. A useful parameter that can be used to gauge the quality of a digital communication system is the probability of bit error, P_E. Because our ears have the ability to decipher what another person is saying, even in the presence of noise, most wireless voice systems can tolerate one bit error in every 1,000 bits received ($P_E = 10^{-3}$), before their performance is deemed unacceptable. Other applications, such as data transfer, demand a much lower P_E. Typical values are 1 error in every 100,000,000 to 1 error in every 10 million bits sent ($P_E = 10^{-8}$ or 10^{-10}) for systems such as Ethernet and token ring networks.

In analog communication systems, the level of performance tends to fall off somewhat gradually with increasing noise. In contrast, when the level of noise in a digital communication system reaches the point where the receiver can no longer tell one symbol from another, overall quality deteriorates rapidly. For example, a 1-dB decrease in signal-to-noise ratio (SNR) can cause a tenfold increase in P_E.

The rest of this chapter will discuss various phase and frequency modulations as well as other digital modulations, which are variants of these. We will discuss how to filter the signal so as to reduce the transmitted bandwidth, yet have no ISI at the sampling instants (Nyquist filtering). Finally, we will give examples of wireless systems that use these various digital modulations.

4.2 PHASE MODULATION

Digital phase modulation generally changes the phase of the carrier to a number of different phase angles. In the simplest case—binary phase shift keying (BPSK)—the carrier phase is either 0 deg or 180 deg. We can even use different phase angles to represent groups of bits. For example, we can use one of four different phase angles (i.e., 45 deg, 135 deg, −45 deg, and −135 deg) to represent two bits at a time. This is called quadriphase or quadrature phase shift keying (QPSK). This idea can be extended further by using eight phase angles to represent groups of three bits at a time (8-PSK), or by using 16 different phase angles to represent four bits at a time (16-PSK).

4.2.1 Binary Phase Shift Keying (BPSK)

The simplest form of digital phase modulation is BPSK. Binary phase shift keying is often used in applications such as garage door openers, direct-sequence spread-spectrum (DSSS) transceivers, and in pulse compression radars. In BPSK, the phase of the carrier is either 0 deg or 180 deg, depending on the input data. The phase domain of the carrier is shown in Figure 4.2. We can write the modulated carrier as

$$s_i(t) = A \cdot \cos[\omega_{RF}t + \varphi_i(t)] \qquad (4.1)$$

where the instantaneous phase angle, $\varphi_i(t)$, is given by

$$\varphi_i(t) = \pi \cdot i; \, i = 0, 1 \qquad (4.2)$$

and A is the amplitude of the carrier. In general, a logic "0" corresponds to a phase inversion of the carrier, while a logic 1 corresponds to no phase inversion, though this convention doesn't always hold. In general, a double-balanced mixer is used to modulate the carrier.

Binary phase shift keying is often referred to as an *antipodal* modulation, where each of the two binary waveforms is the negative of the other. In BPSK, the binary waveforms are the unshifted carrier or the inverted carrier. This can be more easily seen in time domain response, shown in Figure 4.3. As the figure shows, the binary bit stream inverts the phase of the carrier back and forth 0 deg or 180 deg. Each inversion of the carrier causes a sharp transition in the time domain response. As we'll see later, these transitions produce a very wide transmitted spectrum. In general, the carrier frequency is much higher than the modulation frequency, or data rate.

Figure 4.4 shows a block diagram for a typical BPSK modulator. An oscillator produces an unmodulated carrier that is fed to a double-balanced mixer. The non-return-to-zero (NRZ) data is mixed with the carrier to form the desired BPSK signal. As stated earlier, BPSK produces sharp transitions in the time domain, which result in a very wide transmitted spectrum. This is undesirable in most cases. While the transmitted signal could be directly filtered itself at its output, this is usually not done. A very narrow high-Q filter would be required and the center frequency of the filter would need to be changed for each channel frequency. A much better way

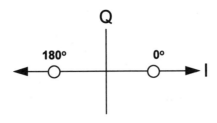

Figure 4.2 Phase domain of carrier for BPSK.

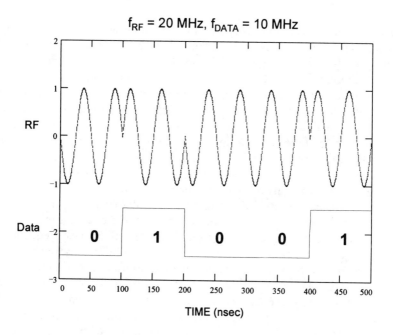

Figure 4.3 BPSK time domain waveforms.

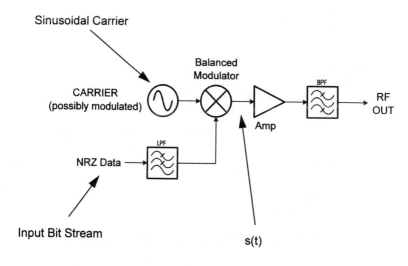

Figure 4.4 BPSK modulator.

to narrow the transmitted output spectrum is by filtering the baseband modulation signal itself. This can be done by inserting a lowpass filter between the NRZ data input and the double-balanced modulator. As we'll see later, some care must be taken when doing this in order to ISI and spectral regrowth.

In most cases, the signal coming out of the modulator is too weak to be sent to the antenna directly, so an amplifier is usually used to increase the transmitted signal level. Although quite clean near the fundamental frequency, most RF amplifiers produce harmonics of the signal at integer multiples of the fundamental transmitter frequency. For this reason, a final filter is used to eliminate these out-of-band signals. Often, this filter doubles as a "roofing filter" for the receiver, suppressing strong out-of-band received signals prior to downconversion.

As mentioned above, double-balanced modulators are often used in BPSK systems. A common form of such a modulator is the diode ring mixer, shown in Figure 4.5. In general, the LO waveform is at the RF carrier frequency and is overdriven, with a typical power level of +6 dBm to +7 dBm. As Figure 4.5 shows, this causes current to flow in and out to the mixer output at the same frequency as the LO. Diodes shown in solid lines are turned "on" and can be modeled as low-value resistors (almost short circuits). Diodes shown with dotted lines are "off" and can be modeled as small capacitors (almost open circuits). The modulation signal changes the sign of this output current in conjunction with the input data. The overall action of the mixer is equivalent to multiplying a square wave, $r(t)$, at the LO frequency with the modulation signal, $m(t)$.
This can be written as

$$r(t) = \frac{4}{\pi} \left[\cos(\omega_{RF}t) + \frac{1}{3} \cdot \cos(3\omega_{RF}t) + \frac{1}{5} \cdot \cos(5\omega_{RF}t) + \ldots \right] \quad (4.3)$$

$$m(t) \cdot r(t) = \frac{4}{\pi} \left[m(t) \cdot \cos(\omega_{RF}t) + \frac{1}{3}m(t) \cdot \cos(3\omega_{RF}t) + \frac{1}{5}m(t) \cdot \cos(5\omega_{RF}t) + \ldots \right] \quad (4.4)$$

Recall that an ideal square wave has energy only at the fundamental frequency and odd harmonics. To see where the frequency components of the modulated signal lie, we take the Fourier transform of both sides of the equation. Let $x(t)$ be a time domain signal. Its Fourier transform is given by

$$S(\omega) = \mathscr{F}\left[s(t)\right] = \int_{-\infty}^{\infty} s(t) \cdot e^{-j\omega t} dt \quad (4.5)$$

If we now apply the transform to the modulated waveform

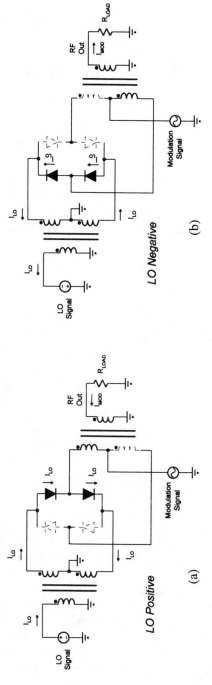

Figure 4.5 Diode-based BPSK modulator: (a) LO is positive, (b) LO is negative.

$$S_{MOD}(\omega) = \mathcal{F}\left[r(t) \cdot m(t)\right] = \int_{-\infty}^{\infty} r(t) \cdot m(t) \cdot e^{-j\omega t}dt \qquad (4.6)$$

$$S_{MOD}(\omega) = \int_{-\infty}^{\infty} r(t) \cdot \left\{\frac{4}{\pi}\left[\cos(\omega_{RF}t) + \frac{1}{3}\cos(3\omega_{RF}t) + \ldots\right]\right\} \cdot e^{-j\omega t}dt \qquad (4.7)$$

Performing the integration, gives

$$S_{MOD}(\omega) = M(\omega - \omega_{RF}) + M(\omega + \omega_{RF}) + M(\omega - 3\omega_{RF}) + M(\omega + 3\omega_{RF}) + \ldots (4.8)$$

where $M(\omega)$ is the Fourier transform of the baseband NRZ data signal. The action of the modulator causes the baseband data spectrum to be shifted to odd multiples of the LO frequency. Of course, the term we are interested in is the fundamental (and its negative frequency component), so after bandpass filtering we get

$$S'_{MOD}(\omega) = M(\omega - \omega_0) + M(\omega + \omega_0) \qquad (4.9)$$

where $S'_{MOD}(\omega)$ is the filtered version of the signal.

Other circuit configurations can be used to perform the BPSK mixing function. Figure 4.6 shows the schematic of a Gilbert-cell based double-balanced modulator. This circuit is very convenient for monolithic IC implementation. The active devices can either be bipolar transistors (BJT or HBT) or field-effect transistors (MOSFET or MESFET). In this circuit, the LO signal is applied to the upper transistor pairs. These pairs act as current switches and steer the current to one side or the other of the output. In the example shown, a center-tapped transformer is used, though this could be replaced by a transmission-line type transformer. In fact, each output of the mixer (collectors of Q1 through Q4) could be singly terminated as well. This results in a simpler interface, but suffers from the disadvantages of lower output power level and a strong dc component in the modulated signal. In some monolithic implementations, an active push-pull, Class AB output stage is used. More details on the design of this circuit are included in Chapter 5.

The Gilbert-cell-based modulator has a number of advantages. All of the active devices of the Gilbert-cell can be fabricated on a monolithic IC. In most cases, the transformer is off-chip, though it is not uncommon to put the transformer on-chip when GaAs or other semi-insulating substrate IC processes are used. The LO input to the upper pair need only be large enough to switch the transistors on and off. In general, 100 mV (about -10 dBm) is sufficient when bipolar devices are used.

The operation of the Gilbert-cell-based modulator is as follows. The modulation signal is applied to the lower transistor pair, formed by Q5 and Q6. In the case of unfiltered BPSK, this input can also be thought as a switching input. However, in almost all cases, the BPSK signal is filtered at baseband (via a lowpass filter) before

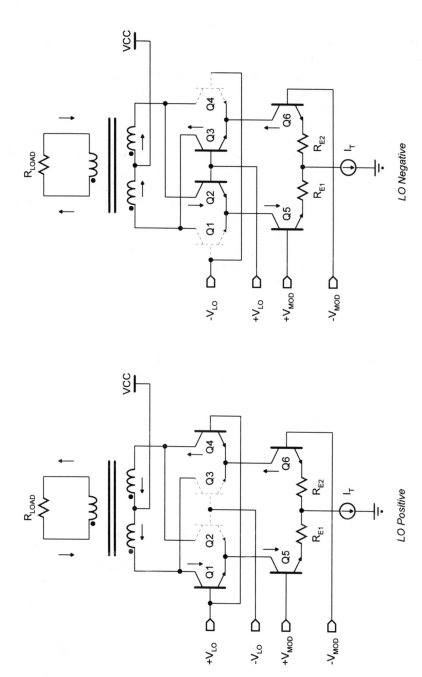

Figure 4.6 Gilbert-cell BPSK modulator (for monolithic IC applications).

being applied to the modulator. For this reason, the modulation input needs to be linear. This is accomplished by using emitter degeneration resistors in the lower pair. Unfortunately, the addition of degeneration resistors also increases the noise figure of the mixer to, typically, 10 to 15 dB. For this reason, the Gilbert-cell mixer is seldom used as the first mixer in a low-noise receiver. However, most modulators perform an upconverting function where noise figure is much less of a problem.

As stated earlier, the upper pair acts as current switches, configured in a cross-coupled fashion to change the polarity of the currents from the lower pair's collectors. Depending on which of the two LO inputs is greater, current is applied to the center-tapped output transformer either in-phase or out-of-phase.

A diagram of a typical coherent BPSK demodulator is shown in Figure 4.7. Basically, the action of the demodulator is the reverse of the modulator. The modulated signal is applied to a double-balanced mixer, which is mixed with an LO having the same frequency and phase as the modulated signal itself. Coherent demodulation, as opposed to noncoherent demodulation, implies that the exact phase and frequency of the modulated signal is known at the receiver. Circuits that produce this local carrier reference are called carrier recovery circuits. We can write the received, modulated signal as

$$r_i(t) = B \cdot \cos[\omega_{\mathrm{RF}}t + \varphi_i(t)] \qquad (4.10)$$

where B is the amplitude of the received signal and $\varphi_i(t)$ is the phase of the transmitted signal—either 0 deg or 180 deg. Multiplying the LO signal by the received signal gives

$$s_i(t) = B \cdot \cos[\omega_{\mathrm{RF}}t + \varphi_i(t)] \cdot \cos(\omega_{\mathrm{RF}}t) \qquad (4.11)$$

$$s_i(t) = \frac{B}{2} \cdot \{\cos[2\omega_{\mathrm{RF}}t + \varphi_i(t)] + \cos[\varphi_i(t)]\} \qquad (4.12)$$

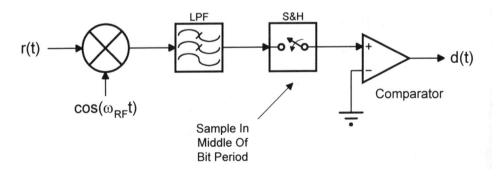

Figure 4.7 BPSK demodulator.

Lowpass filtering removes the component at twice the RF frequency, leaving

$$s_i(t) = \frac{B}{2} \cdot \cos[\varphi_i(t)] \tag{4.13}$$

If $\varphi_i(t)$ is 0 deg, then $s_i(t)$ is 1. If $\varphi_i(t)$ is 180 deg, then $s_i(t)$ is -1. Thus, the original binary NRZ signal is recovered by the demodulator. In Figure 4.7, the decision as to whether the received signal was 1 (or binary 1) or -1 (binary 0) can be performed by an analog comparator. In most cases, a special-purpose circuit, called a *bit (or data) slicer*, is used. The bit slicer tracks the dc level of the recovered data so as to sample the signal at precisely the middle of the data period, at the point of maximum SNR, providing for minimum P_E.

The NRZ data signal resembles a square wave. However, unlike a square wave, which alternates periodically between binary 1 and 0, the NRZ data signal alternates between binary 1 and 0 in a random fashion. This characteristic means that the NRZ signal's power spectral density is smooth (rather than consisting of discrete spectral lines), with a shape that resembles a sinx/x shape. The power spectral density of the NRZ baseband signal is

$$s(f) = 2(T_b A)^2 \frac{\sin^2(\pi f T_b)}{(\pi f T_b)^2} \tag{4.14}$$

where T_b is the bit period, A is the amplitude of the signal, and f is the frequency in hertz.

Recall that the modulated signal is produced by multiplying the baseband data signal by a square wave at the RF frequency. If we filter out all the terms except the fundamental, we are left with the original baseband data spectrum shifted to the RF frequency. This is given by

$$S_{TX}(f) = \frac{1}{2} \cdot [S(f - f_{RF}) + S(f + f_{RF})] \tag{4.15}$$

$$S_{TX}(f) = (AT_b)^2 \cdot \left\{ \frac{\sin^2[\pi(f - f_{RF})T_b]}{[\pi(f - f_{RF})T_b]^2} + \frac{\sin^2[\pi(f + f_{RF})T_b]}{[\pi(f + f_{RF})T_b]^2} \right\} \tag{4.16}$$

Figure 4.8 shows a plot of the power spectral density for an unfiltered BPSK signal. The data rate is 10 Mbps and the carrier frequency is 80 MHz. Only one side of the spectrum is shown for greater detail. Notice that the sharp transitions in the time domain lead to a relatively wide power spectral density that rolls off quite slowly. The first null occurs at a frequency equal to the (data rate)$^{-1}$ away from the carrier. The amplitude of the first lobe is only 13-dB down from its value

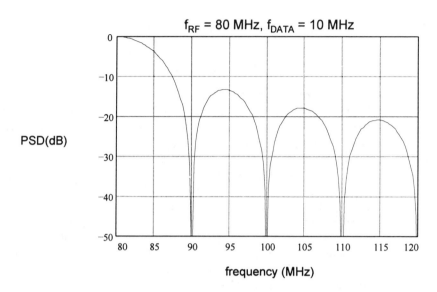

Figure 4.8 BPSK power spectral density.

at the carrier frequency. A good rule of thumb is that 99% of the transmitted power of an unfiltered BPSK signal lies within a bandwidth equal to the bit rate. As we will see later, filtering the signal can greatly reduce the transmitted signal bandwidth. However, care must be taken in order to avoid ISI and spectral regrowth. The theoretical spectral efficiency of BPSK is 1 bit/sec/Hz. In practice, a value of 1.4 bits/sec/Hz can be achieved using Nyquist filtering techniques (discussed in Section 4.2.5).

Another important parameter of any digital modulation is how the P_E varies with SNR. Figure 4.9 shows a BPSK received signal with and without noise. The demodulator decides whether the recovered signal was 1 or −1 by sampling the waveform at the center of the bit period. This is easy to do when there is no noise added to the signal, as shown in Figure 4.9(a). When noise has been added to the signal, as shown in Figure 4.9(b), it is easier to make an error. An error would occur if the noise level caused the signal to be above or below zero at the center of the bit period.

Before we can quantify how many bit errors occur for BPSK in the presence of noise, we must first have a means of quantifying the noise itself. In general, thermal noise is Gaussian in nature, having random amplitude values with a relatively flat frequency spectrum. Noise is added to the signal as it passes through the channel. We refer to such noise as additive white Gaussian noise (AWGN). A plot of AWGN is shown in Figure 4.10(a).

It is often convenient to assume that the noise power spectral density extends without limit in frequency in both directions. Of course, this can't really happen;

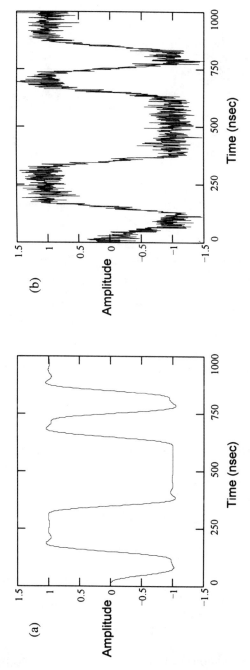

Figure 4.9 Demodulated BPSK waveforms: (a) without noise, (b) with noise.

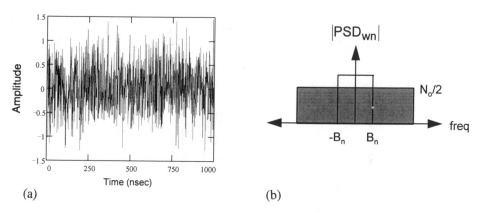

Figure 4.10 Additive white Gaussian noise: (a) noise time domain waveform, (b) noise power spectral density.

else, the noise would have infinite power. In reality, thermal noise does tend to fall off as the frequency approaches 10^{12} to 10^{14} Hz. For most of the frequencies we normally deal with in digital communications, assuming a flat power spectral density is a good approximation. A diagram of white noise versus frequency is shown in Figure 4.10(b). If we integrate the noise over a bandwidth ranging from $-B_n$ to $+B_n$ (where B_n is the one-sided noise bandwidth of the receiver), then we get the rms power of the noise. With an rms noise power equal to N, the noise power spectral density across the band must be equal $N_0/2$. This can be written mathematically by

$$N = \left(\frac{N_0}{2}\right) \cdot (2B_n) \tag{4.17}$$

Rearranging the equation, we can relate the noise power spectral density to the noise rms noise power by

$$N_0 = \frac{N}{B_n} \tag{4.18}$$

Figure 4.11 shows a plot of the probability of making a bit error versus a quantity called E_b/N_0 for BPSK. The quantity E_b/N_0 is the ratio of the *energy per bit* over the noise power spectral density. To see how E_b/N_0 is related to SNR, we can write

$$E_b = S \cdot T_b \tag{4.19}$$

where S is the average signal power of the carrier and T_b is the bit period. This gives

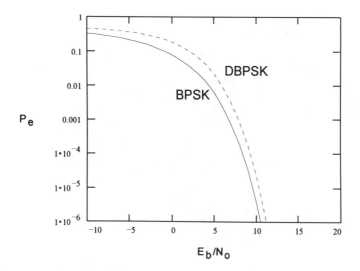

Figure 4.11 BPSK probability of bit error.

$$\frac{E_b}{N_0} = \frac{S \cdot T_b}{N/B_n} = \frac{S}{N} \cdot B_n T_b = \frac{S}{N} \cdot \frac{B_n}{f_b} \qquad (4.20)$$

where f_b is the bit rate of the data. As (4.20) shows, E_b/N_0 is equal to S/N, or SNR, *when the noise bandwidth of the receiver is equal to the data rate.* This is a theoretical result, since the noise bandwidth of the receiver needs to be somewhat wider than the bit rate to avoid ISI in the received waveform. A practical rule of thumb is that a BPSK signal can fit into a bandwidth equal to 1.4 times the bit rate by using special filtering techniques. This is significantly narrower than the spectrum of the BPSK signal shown in Figure 4.8. These filtering techniques will be discussed in more detail in Section 4.2.5.

E_b/N_0 is often used to relate probability of bit error because it is a dimensionless quantity. By using (4.20), we can relate this quantity back to SNR. Two curves are shown in Figure 4.11. The curve labeled BPSK is the P_e for coherently demodulated BPSK. The curve labeled DBPSK is for differentially encoded/decoded BPSK. In DBPSK, we compare the received signal with a delayed version of itself. This is shown in Figure 4.12, where both the DBPSK modulator and DBPSK demodulator are shown. Notice that the data must be differentially encoded prior to modulation. If the original NRZ data stream is

$$0, 1, 1, 0, 1, 1, 1, 0, 0 \ldots \qquad (4.21)$$

then the encoded NRZ data signal (assuming the initial bit was a 0) is

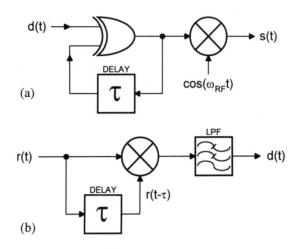

Figure 4.12 Differentially encoded/decoded BPSK (DBPSK): (a) DBPSK transmitter, (b) DBPSK receiver.

$$0, 1, 0, 0, 1, 0, 1, 1, 1, \ldots \qquad (4.22)$$

In DBPSK, we can write the received signal as before

$$r_i(t) = B \cdot \cos[\omega_{RF}t + \varphi_i(t)] \qquad (4.23)$$

The delayed version of the signal can be written

$$r_i(t - T_b) = B \cdot \cos[\omega_{RF}(t - T_b) + \varphi_i(t - T_b)] \qquad (4.24)$$

Multiplying the delayed and undelayed signals together yields

$$r_i(t - T_b) \cdot r_i(t) = B \cdot \cos[\omega_{RF}(t - T_b) + \varphi_i(t - T_b)] \cdot B \cdot \cos[\omega_{RF}(t) + \varphi_i(t)] \qquad (4.25)$$

After some algebra and lowpass filtering, we are left with

$$[r_i(t - T_b) \cdot r_i(t)]_{LPF} = \frac{B^2}{2} \cdot \cos[\varphi_i(t - T_b) - \varphi_i(t)] \qquad (4.26)$$

This result also assumes that there are an integral number of RF carrier periods in one bit period.

If the last angle and the present angle are both the same, then their difference is 0 and the filtered output is 1. If the present angle and the last angle are different, then their difference is 180 deg, and the filtered output is −1. Since a delayed version of the received signal, complete with added noise, is used as a reference for demodulation, we would expect that P_e performance for DBPSK would be worse than that of coherently demodulated BPSK. This is in fact the case, as Figure 4.11 shows.

Another way to view degradation due to noise is by picturing the transmitted signal as a vector. This is shown in Figure 4.13. Noise can be visualized as a random vector that is added to the tip of the received signal. In the case of BPSK, when the noise vector has greater amplitude than the signal vector—but in the opposite direction—a bit error has occurred. This way of viewing things becomes more useful when we talk about higher order modulations, such as QPSK.

4.2.2 Quadriphase (or Quadrature) Phase-Shift Keying (QPSK)

In QPSK, we introduce the concept of *symbols*. A symbol is an electrical waveform that can represent one or more bits. To begin with, let each of the icons in Figure 4.14 represent a group of 2 bits. In Figure 4.14(a), the dog represents the bit pattern 00. The cat represents 01, the elephant represents 10 and, finally, the bear represents the pattern 11. Let the transmitted bit stream be

$$0, 1, 0, 0, 1, 1, 1, 0, \ldots \qquad (4.27)$$

To send a stream of symbols corresponding to these bits, we first separate the bits into groups of two

$$0\ 1, 0\ 0, 1\ 1, 1\ 0 \qquad (4.28)$$

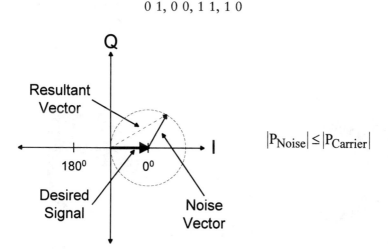

Figure 4.13 Resultant BPSK signal vector with noise.

Figure 4.14 Using symbols to represent groups of two bits each: (a) symbol assignment for groups of bits, (b) symbol sequence for the bit stream 01001110.

This grouping results in the symbol stream of cat, dog, bear, and, finally, elephant, as shown at the bottom of the figure.

Notice that by using symbols to represent groups of one or more bits, we can reduce the rate at which the symbols are sent. In other words, if our bit rate is 1 Mbps, then by using symbols to represent groups of 2 bits each, the symbol rate is one-half the bit rate, or 500 Kbps.

In the example above, where animals were used, it is easy to tell which symbol was sent since each looked distinctly different from the others. For example, it is hard to confuse the picture of the bear with that of the dog. But, what if we decided to use pictures of different types of dogs. Can most people tell the difference between the silhouette of a dalmatian from that of a labrador retriever? It might be difficult, especially if the pictures were fuzzy or out of focus. This illustrates the point that each symbol should also be distinctly different from all the others, so that we can still decide what symbol was sent in the presence of noise.

So, what kind of symbols can we use in digital communications? If we are doing phase modulation, then we might choose different phase angles. This is exactly what we do in QPSK. Figure 4.15 shows the phase domain of a QPSK signal. Each group of two bits is represented by a different phase angle, 90-deg apart from the next. Notice that each angle lies 45-deg off the in-phase, or *I*, and quadrature, or *Q*, axes. Again, as in the case of BPSK, we assume that a logic 0 corresponds to a −1.

We can write the QPSK signal as

$$s_i(t) = A \cdot \cos[\omega_{RF}t + \varphi_i(t)] \qquad (4.29)$$

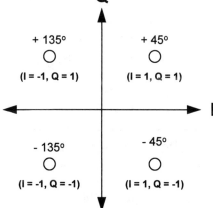

Figure 4.15 Phase domain of carrier for QPSK (gray-coded).

where A is the carrier amplitude, i is an integer, and $\varphi_i(t)$ is the instantaneous phase angle of the modulated signal with respect to the unmodulated carrier. The value $\varphi(t)$ can be written as

$$\varphi_i(t) = \frac{\pi}{4} \cdot (2i + 1) \tag{4.30}$$

where $i = 0, 1, 2, 3$.

As mentioned earlier, each symbol represents a group of two bits. Therefore, the symbol rate of QPSK is half the bit rate of BPSK. This means that the spectrum is only 1/2 as wide as that of BPSK for a given bit rate. QPSK is used in applications such as early telephone modems, satellite communications, the global positioning system (GPS) and in some forms of code division multiple access (CDMA).

One way to generate QPSK is by simply changing the phase angle of the carrier using a brute-force approach as shown in Figure 4.16. Techniques similar to this are used in active-array radars to steer the transmitted beam pattern. Unfortunately, this technique is not as useful in digital communication systems. Since the phase shifters operate at the carrier frequency, errors can be introduced into the transmitted phase via back-coupling from the powerful transmitted signal back into the phase shifting circuitry itself (phase pulling). Also, the phase shifters are open-loop devices,

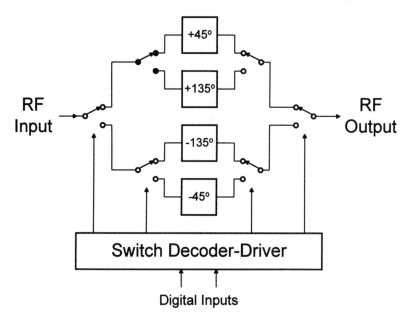

Figure 4.16 Brute-force QPSK modulator.

which tend to shift their phase with respect to one another over temperature. However, the most significant disadvantage of the brute-force phase shifter is that the phase transitions are abrupt. As we shall see, filtering the digital inputs to the QPSK modulator can greatly reduce the transmitted output spectrum.

A more common implementation used to perform QPSK modulation is shown in Figure 4.17. The serial data stream is demultiplexed into even and odd components via the serial-parallel converter (or demultiplexer). Each data component (now at one-half the bit rate) is applied to a set of double-balanced modulators. As will be discussed later, optional lowpass filters can be used to bandlimit the transmitted spectrum. The set of double-balanced mixers form what is known as a quadrature modulator. The quadrature modulator can vary both the phase and amplitude of the transmitted signal, though for QPSK, we only care about the phase. Finally, the outputs from each double-balanced mixer are added together and fed through an amplifier to a bandpass filter, which removes any harmonics of the modulated signal.

Figure 4.18 shows a simplified block diagram of the quadrature modulator. To understand the operation of the quadrature modulator, we can write the in-phase and quadrature components of the modulated signal in their exponential form

$$I \cdot \cos(\omega_{RF}t) = I \cdot \left[\frac{e^{j\omega_{RF}t} + e^{-j\omega_{RF}t}}{2} \right] \tag{4.31}$$

$$Q \cdot \sin(\omega_{RF}t) = Q \cdot \left[\frac{e^{j\omega_{RF}t} - e^{-j\omega_{RF}t}}{2j} \right] \tag{4.32}$$

The output of the quadrature modulator is then

$$s(t) = I \cdot \cos(\omega_{RF}t) + Q \cdot \sin(\omega_{RF}t) = I \cdot \left[\frac{e^{j\omega_{RF}t} + e^{-j\omega_{RF}t}}{2} \right] + Q \cdot \left[\frac{e^{j\omega_{RF}t} - e^{-j\omega_{RF}t}}{2j} \right]$$

$$\tag{4.33}$$

Rearranging the terms in the equation gives

$$s(t) = \frac{1}{2} \cdot [(I - jQ) \cdot e^{j\omega_{RF}t} + (I + jQ) \cdot e^{-j\omega_{RF}t}] \tag{4.34}$$

If we let

$$I - jQ = \sqrt{I^2 + Q^2} \cdot e^{j\varphi} \tag{4.35}$$

$$I + jQ = \sqrt{I^2 + Q^2} \cdot e^{-j\varphi} \tag{4.36}$$

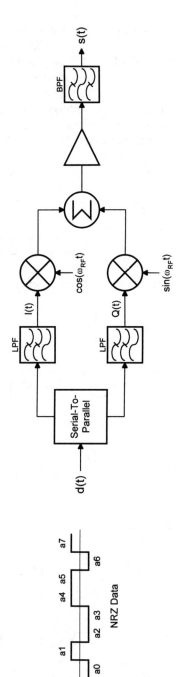

Figure 4.17 QPSK modulator.

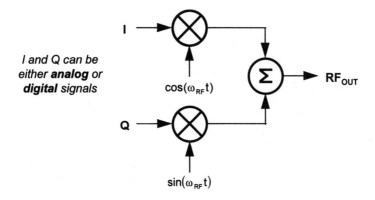

Figure 4.18 Quadrature modulator.

where

$$\varphi = \tan^{-1}(Q/I) \tag{4.37}$$

then we can write

$$s(t) = \sqrt{I^2 + Q^2} \cdot \left[\frac{e^{j\omega_{RF}t + \varphi} + e^{-j\omega_{RF}t - \varphi}}{2} \right] \tag{4.38}$$

$$s(t) = \sqrt{I^2 + Q^2} \cdot \cos(\omega_{RF}t + \varphi) \tag{4.39}$$

We see that the quadrature modulator can be used to modulate both the amplitude and the phase of the carrier. If I and Q are chosen to be ± 1, then the amplitude of the modulated signal is a constant $\sqrt{2}$. Notice also that having I and Q equal to ± 1, all phase angles lie along 45-deg lines off the axes. In general, we assume that the carrier amplitude is normalized to a value of 1, therefore we assume that both I and Q inputs can range between $\pm 1/(\sqrt{2})$, or ± 0.707. We can also use the quadrature generator to perform modulations that require both amplitude and phase variation. An example of such a modulation is quadrature amplitude modulation (QAM), in which each symbol has a different phase and/or amplitude from the other symbols.

A number of circuits can be used to generate the in-phase and quadrature components of the carrier, and some examples are shown in Figure 4.19. In Figure 4.19(a), two op amp allpass filters are used to generate two phase-shifted signals having constant amplitude. The allpass filters are designed such that there is a region in frequency where the relative phase shift at the output of the two filters is 90 deg,

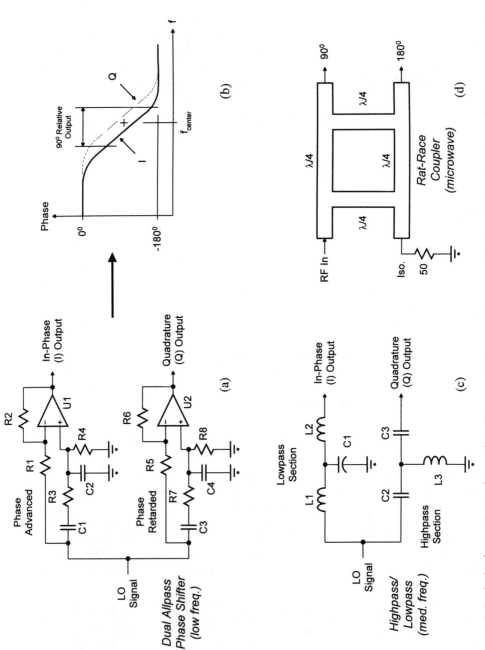

Figure 4.19 Methods of generating in-phase and quadrature LO references: (a) OPAMP allpass filter phase shifter *I/Q* reference, (b) output phase of OPAMP allpass filter, (c) lumped-element highpass/lowpass *I/Q* reference, (d) rat-race coupler *I/Q* reference.

as shown in Figure 4.19(b). With two second-order filters as shown, the phase difference can be maintained to an accuracy of a few degrees over a decade in frequency. This approach is limited to relatively low frequencies where the op amps work well.

Figure 4.19(c) shows a higher frequency version of the differential phase shifter. In this approach, a lowpass filter is used to provide a relative −45-deg phase shift and a highpass filter is used to provide a relative +45-deg phase shift at the center of the frequency band. The lowpass/highpass approach can be used to much higher frequencies than the op amp approach, but is limited to a smaller useful bandwidth due to the fact that the amplitude of each filter changes with frequency.

An approach that is used at microwave frequencies is shown in Figure 4.19(d). A rat-race coupler is used to provide a relative 90-deg phase shift at each output. In general, this approach works over about an octave in frequency. As the frequency is lowered, the size of the rat-race coupler gets physically large, limiting its usefulness to relatively high frequencies. The phase shifters of Figure 4.19(c, d) are discussed in more detail in Chapter 5.

The quadrature signals can also be generated by digital means. Figure 4.20 shows two methods by which this can be done. In Figure 4.20(a), an input at four times the desired LO frequency is applied to a divide-by-4 circuit constructed using two D-type flip-flops. The divider action ensures that each output of the circuit is a perfect square wave, shifted by one period of the input signal (or 90 deg). The digital circuit can be used from dc up to the highest frequency of operation of the divider. Nowadays, digital logic circuits can operate well above 1 GHz, even tens of gigahertz. In order to ensure the circuit tracks out all errors over temperature and process variations, it is essential that every node in the divider is equally loaded. It is not uncommon to use dummy gates to equally load each output.

The major disadvantage of the divide-by-4 circuit is that a signal at four times the desired LO frequency needs to be applied as an input. If it can be guaranteed that the input is perfectly symmetrical (i.e., has a 50% duty cycle), then the circuit can be connected as shown in Figure 4.20(b) where one of the flip-flops is driven out-of-phase from the other. This approach requires the input signal to be only twice the desired LO frequency.

We've discussed how to generate the required in-phase and quadrature versions of the LO signal, but how are the two double-balanced modulators and summing junction implemented? Figure 4.21 shows a schematic of a dual Gilbert-cell mixer. Recall from earlier discussions on BPSK modulation, that the Gilbert-cell mixer can be used as a double-balanced modulator. Since the outputs of the Gilbert-cell mixer are really currents, we can form the summing junction by simply "wire OR-ing" the outputs together. The dual Gilbert-cell implementation is often used in monolithic versions of the quadrature modulator.

As shown in Figure 4.17, the input to the quadrature modulator is split into two parallel bit streams, each at one-half the rate of the original serial data stream.

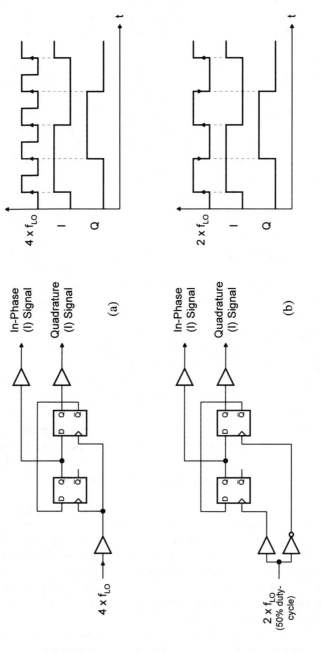

Figure 4.20 Digital *I/Q* references: (a) 4X LO reference (no duty cycle restrictions), (b) 2X LO reference (50% duty cycle required).

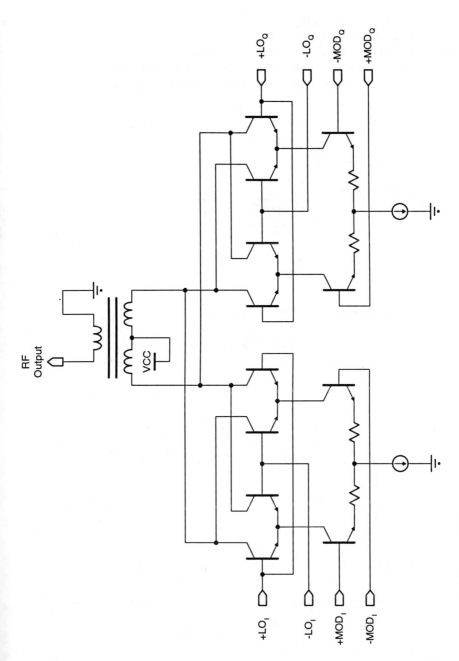

Figure 4.21 Dual Gilbert-cell QPSK modulator.

Figure 4.22 shows the block diagram for a digital demultiplexer that does this. The first two D-type flip-flops take snapshots of alternating bits of the serial data. The third flip-flop adds a 1-bit delay to the *I* data to realign the two bit streams.

Figure 4.23 shows the time domain response for QPSK. As the figure shows, the phase of the carrier can be changed 0 deg, ±90 deg, or 180 deg from its present

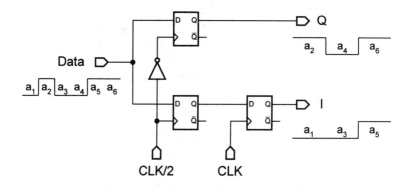

Figure 4.22 Data demultiplexer (serial-to-parallel) for QPSK.

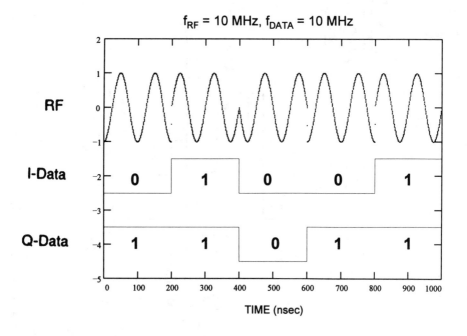

Figure 4.23 QPSK time domain waveforms.

value. The two bit streams in the bottom of the plot correspond to the demultiplexed *I* and *Q* data streams. In this example, the input data bits are unfiltered. This causes abrupt changes in phase in the modulated carrier, much in the same fashion as BPSK. Figure 4.24 shows the transmitted power spectral density of a typical QPSK signal at a carrier frequency of 80 MHz and a data rate of 10 Mbps, along with a BPSK spectrum having the same parameters. Not surprisingly, the QPSK spectrum is only half as wide as the BPSK spectrum. Notice however, that the spectrum still has sharp nulls and rolls off very slowly with relatively large sidelobes. In theory, the spectral efficiency of QPSK is 2 bits/sec/Hz, which is better than BPSK by a factor of two.

As we might expect, the QPSK demodulator is simply the reverse of the modula-tor. Figure 4.25 shows a block diagram of a typical QPSK demodulator, which is for all intents and purposes simply two BPSK demodulators. The received signal is multiplied by cosine and sine versions of the local carrier. The block labeled C/R contains the carrier recovery circuitry, which provides a local in-phase reference precisely aligned in phase to the transmitted carrier. Of course, the quadrature signal is generated by simply shifting the in-phase signal by 90 deg. The block labeled STR contains the symbol timing recovery circuitry, which extracts the data clock and samples the lowpass filtered outputs at precisely the optimum position within the bit period. Finally, a data multiplexer reinterleaves the data back into the serial bit stream.

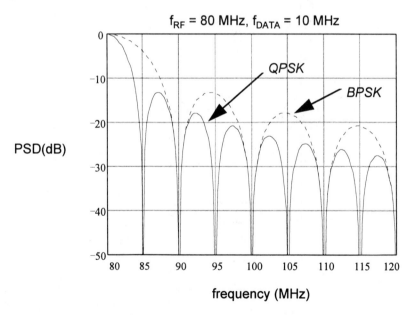

Figure 4.24 QPSK power spectral density.

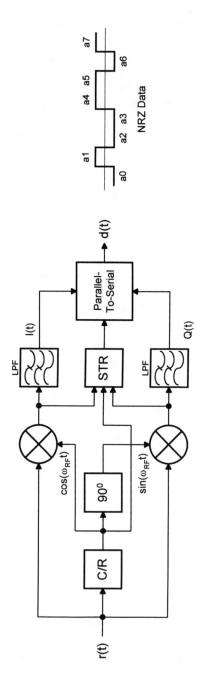

Figure 4.25 QPSK demodulator.

As in the case of BPSK, we can think of the recovered QPSK signal as a vector to which gets added a noise vector. This is shown in Figure 4.26. It is easy to see that any added noise need only be strong enough to knock the resultant vector into the next quadrant for a symbol error to result. In fact, the noise required to maintain a given symbol error rate for QPSK must be 3-dB lower than that required to maintain the same P_E in BPSK. However, when we demodulate the QPSK signal, we break it into two BPSK bit streams by multiplying by cosine and sine waveforms, as shown in the figure. This is further illustrated in Figure 4.27, where we have assumed, for now, that the carrier has no extra phase noise imparted onto it from the noisy received signal. Looking at the recovered signal in this form leads us to conclude that the bit error rate of QPSK is the same as for BPSK. This is in fact the case.

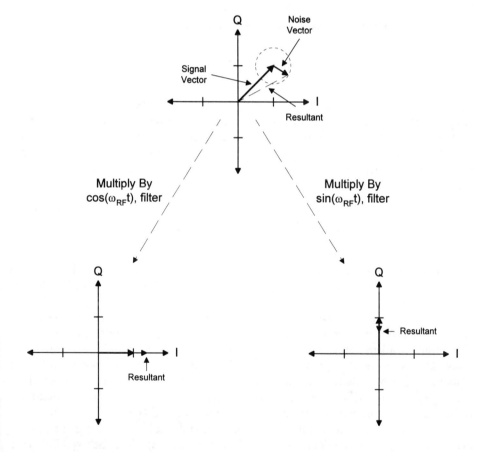

Figure 4.26 Resultant QPSK signal vector with noise (broken into *I* and *Q* components).

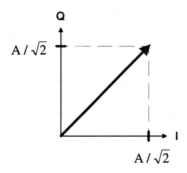

Figure 4.27 QPSK signal vector in the *I/Q* phase plane.

Recall, from our earlier discussions that we can write E_b/N_0 as

$$\frac{E_b}{N_0} = \frac{S}{N} \cdot \frac{B_n}{f_b} \tag{4.40}$$

where E_b is the energy per bit, N_0 is the noise power spectral density, B_n is the effective noise bandwidth of the receiver and f_b is the bit rate. In QPSK, we have created two data streams, each at one-half the rate of the original bit stream and each having maximum amplitude of $\pm(\sqrt{S})/\sqrt{2}$ (recall that S is the power of the received signal, therefore it is the square of its amplitude). Plugging this information into (4.40) gives

$$\frac{E_b}{N_0} = \frac{(\sqrt{S}/\sqrt{2})^2}{N} \cdot \frac{B_n}{(f_b/2)} = \frac{(S/2)}{N} \cdot \frac{B_n}{(f_b/2)} = \frac{S}{N} \cdot \frac{B_n}{f_b} \tag{4.41}$$

Therefore, we see that the BER of QPSK is the same as that of BPSK. Figure 4.28 shows a plot of the probability of bit error for QPSK versus E_b/N_0, which is identical to BPSK.

As we discussed above, a noise spike could very easily cause the resultant vector to fall into the wrong quadrant. If we code the data properly, a symbol error will result in a single bit error and not two bit errors. Figure 4.29 illustrates this point. In this figure, the symbols on the QPSK phase plane are shown for different assignments of bits. Because the noise is Gaussian by nature, it is much more likely that a noise spike will cause the resultant vector to fall in an adjacent quadrant, rather than the opposite quadrant. If the symbols are coded as shown in Figure 4.29(a), then falling into an adjacent quadrant could result in two bit errors. However, if we *gray code* the data, as shown in Figure 4.29(b), then each symbol error into an adjacent quadrant causes only one bit error. A gray code is a mapping of

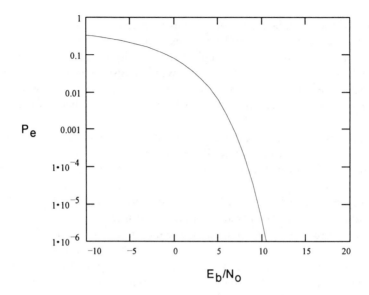

Figure 4.28 QPSK probability of bit error.

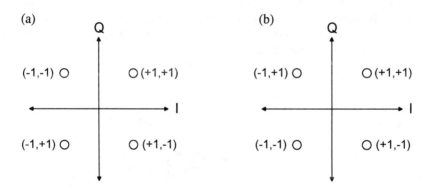

Figure 4.29 Assignment of symbols for QPSK: (a) nongray-coded QPSK, (b) gray-coded QPSK.

input bits to output bits, such that only one output bit at a time changes for sequential input bit changes.

Earlier, we saw that the power spectral density of QPSK was similar to that of BPSK, but only half as wide, due to the reduced symbol rate. In most wireless communication systems, such a spectrum would be too wide to meet FCC regulations and would also cause unwanted interference by spilling into adjacent channels. We could narrow the transmitted spectrum by placing a very narrow bandpass filter on the output of the modulator. However, this is usually not practical, since the Q of the filter would have to be very high. This is due to the fact that the transmitted carrier frequency is usually much higher than the data rate. Such a high-Q filter would be difficult to build, expensive, and would cause significant distortion on the transmitted waveform due to rapid phase variations at the band edges. In addition, if we have many channels, we would also need to design such a filter so as to be tunable.

A better way to limit the transmitted output spectrum is by filtering the baseband I and Q modulation signals before applying these to the double-balanced modulators, as shown in Figure 4.30. Figure 4.31 shows the resulting transmitted spectrum of filtered and unfiltered QPSK. As the figure shows, the action of the filter greatly reduces the bandwidth of the transmitted signal. Figure 4.32 shows the corresponding time domain signal for a baseband filtered QPSK signal. Notice that every time a 180-deg phase change is encountered, the amplitude dips to zero. However, 90-deg phase changes result in much lower amplitude variations.

In most wireless communication systems, the output of the modulator is too small to be directly applied to the antenna. Therefore, the modulator output signal is sent to a transmit power amplifier for further amplification. In order to get the highest efficiency possible from the transmit power amplifier, it is common to run the circuit in nonlinear, Class C operation. In Class C operation, the amplifier consumes little power when no signal is applied to it. When a signal is applied to the amplifier, the amplifier is driven into limiting (sometimes called compression or saturation). Most of the power applied to a hardlimited amplifier is directly transferred to the load, resulting in maximum efficiency.

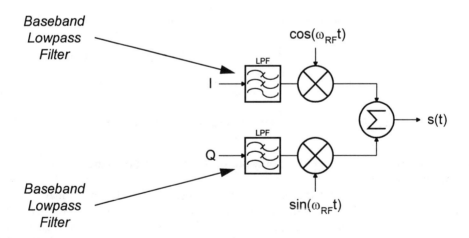

Figure 4.30 Using premodulation lowpass filters to limit QPSK transmitted bandwidth.

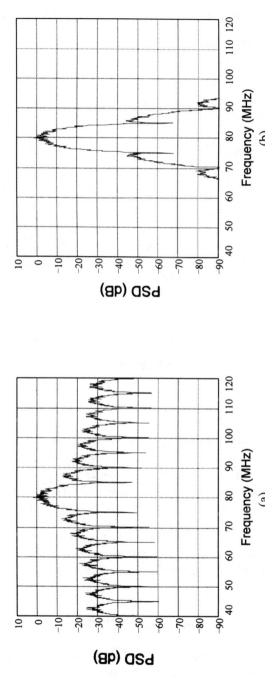

Figure 4.31 QPSK transmit power spectral density: (a) unfiltered (10-Mbps data rate, 80-MHz carrier), (b) baseband filtered (10-Mbps data rate, 80-MHz carrier, 2.5-MHz Gaussian lowpass premodulation filter).

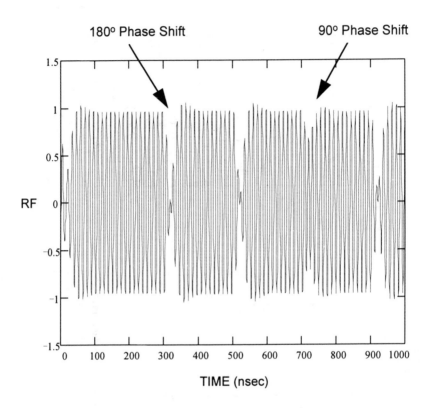

Figure 4.32 Time domain waveform of baseband-filtered QPSK transmit signal.

When the filtered QPSK signal is passed through a limited amplifier, much of the amplitude variations in the time domain envelope are reduced. However, it is these very amplitude variations that resulted in the narrow transmitted spectrum. Figure 4.33 shows the spectrum of a QPSK signal that has been first filtered and then limited. As the figure shows, the limiting action of the amplifier causes the spectrum to regrow. This is called *spectral regrowth*. In general, modulation schemes that have nonconstant envelopes suffer from spectral regrowth when passed through a hardlimited amplifier (i.e., an amplifier in saturation or compression). For this reason, these types of modulations generally require linear amplifiers in the transmit path.

4.2.3 Offset or Staggered Phase-Shift Keying (OQPSK)

As we have just seen, large amplitude variations in the QPSK waveform result whenever the phase is changed by 180 deg. We might expect that if we could make

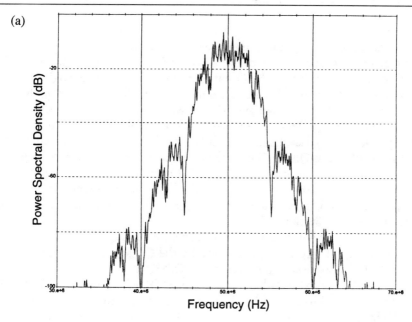

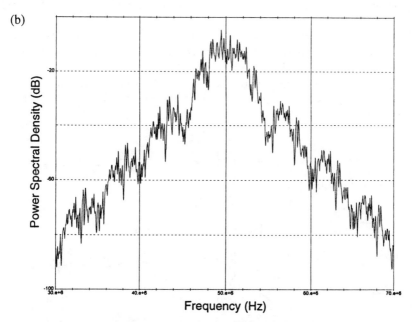

Figure 4.33 QPSK transmit output spectrum with baseband filtering: (a) without limiting, (b) with limiting (amplifier 10 dB into compression).

the phase transitions more gradual, then the QPSK waveform might be more tolerant to limiting. In offset, or staggered, QPSK (OQPSK), we do just that. OQPSK is identical to QPSK, except that the Q data stream is delayed by a single bit period (or 1/2 of a symbol period). This is easily done by modifying the data demultiplexer, as shown in Figure 4.34. As the figure shows, we simply remove the extra flip-flop from the I data stream. This has the same effect as delaying the Q data stream. Figure 4.35 shows the phase domain, I and Q data streams and corresponding phase angles for QPSK and OQPSK. The same two data streams are used in both cases, but notice how the action of the one bit period delay in QPSK causes no 180-deg phase transitions. OQPSK is sometimes used in satellite communications, but the real advantage of OQPSK is its use as a building block for other modulations, such as MSK (discussed in Section 4.3.1).

Figure 4.36 shows the time domain response of OQPSK. All of the phase transitions in OQPSK are 90 deg. Notice how the Q data stream is delayed by a bit period from its position in the example of QPSK.

Figure 4.37 shows a block diagram of a typical OQPSK modulator. The OQPSK modulator is identical to that of QPSK with the exception that an extra bit period delay is inserted into the Q data stream. In this figure, the demultiplexer is assumed to be identical to the one used in QPSK. Not surprisingly, the OQPSK demodulator is very similar to the QPSK demodulator. Figure 4.38 shows a block diagram of the demodulator, where an extra bit of delay is inserted in the I data stream to realign the data.

Due to the similarities with QPSK, both the power spectral density and the P_E of OQPSK are the same as for QPSK. However, when filtered and limited, OQPSK shows much smaller spectral regrowth than QPSK. Figure 4.39 shows a plot of a time domain response of a baseband-filtered OQPSK signal. As the plot shows, the amplitude fluctuations are greatly reduced. When this signal is passed through a

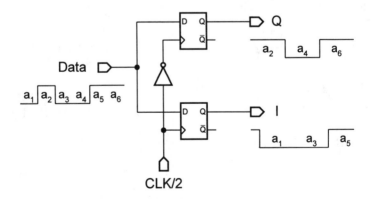

Figure 4.34 OQPSK data demultiplexer (serial-to-parallel).

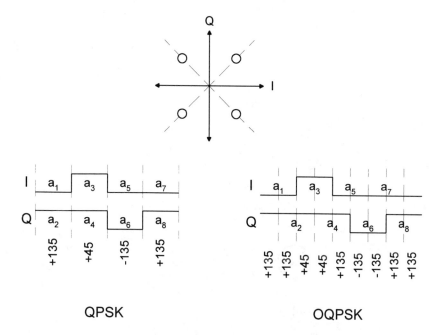

Figure 4.35 QPSK and OQPSK I/Q data waveforms with their corresponding phase angles.

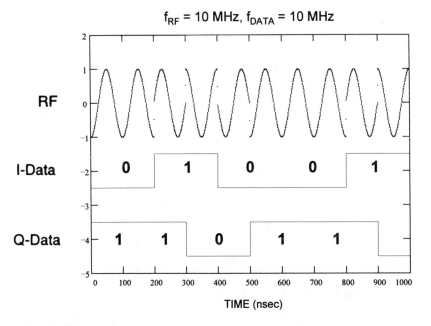

Figure 4.36 OQPSK time domain waveforms.

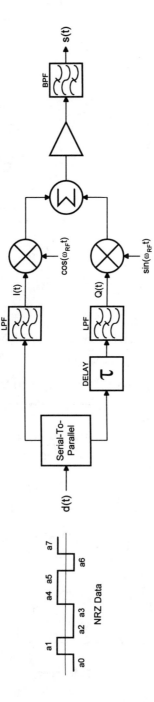

Figure 4.37 OQPSK modulator.

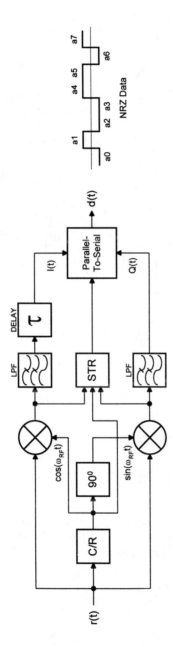

Figure 4.38 OQPSK demodulator.

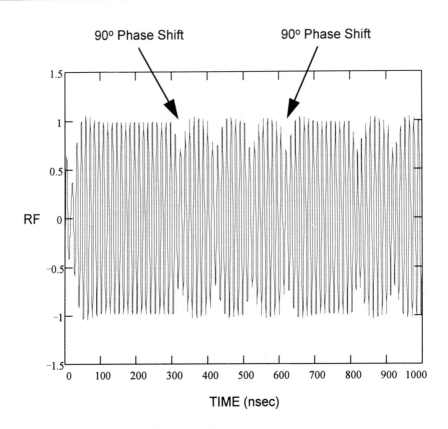

Figure 4.39 Time domain waveform of baseband-filtered OQPSK transmit signal.

limiting amplifier, its spectral regrowth is much lower than that of QPSK. Figure 4.40 shows a plot of this.

4.2.4 8-PSK and 16-PSK

The concept of using symbols can be extended further. By grouping three bits at a time into one symbol, we can transmit one of eight different phase angles, in reference to the unmodulated carrier. This type of phase modulation is known as 8-PSK. The phase domain of 8-PSK is shown in Figure 4.41. As the figure shows, each symbol lies on the unit circle in eight equally spaced phase angles, 45-deg apart. In order to maintain constant amplitude, the allowable values of I and Q are

$$I, Q = 0, \pm1, \pm\frac{\sqrt{2}}{2} \tag{4.42}$$

(a)

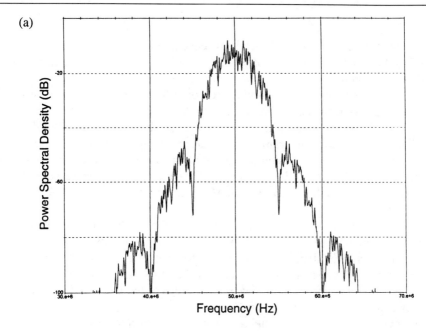

(b)

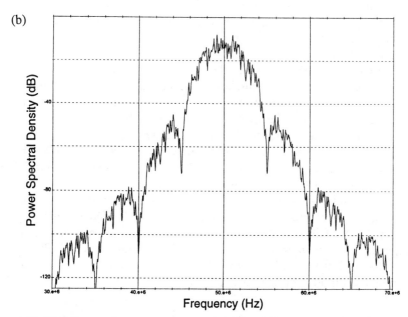

Figure 4.40 OQPSK transmit output spectrum with baseband filtering: (a) without limiting, (b) with limiting (amplifier 10 dB into compression).

Eight-PSK is sometimes used with special data encoding to form a modulation known as $\pi/4$-DQPSK (discussed in Section 4.4.2). 8-PSK is also used in point-to-point microwave communications.

Figure 4.42 shows the time domain response for an 8-PSK signal. The three upper traces correspond to the three demultiplexed bit streams taken from the original data. The modulated signal can vary in phase by any increment of 45 deg. Figure 4.43 shows a block diagram of an 8-PSK modulator. It is very similar to the QPSK modulator; however, the demultiplexer contains a data encoder that not only demultiplexes the original data in to groups of 3 bits, but also provides a set of outputs to drive the digital-to-analog-converters (DACs) with the allowable modulation levels given in (4.42).

Figure 4.44 shows a block diagram for an 8-PSK demodulator. The demodulator is basically the reverse of the 8-PSK modulator. Sine and cosine mixers are used to recover the I and Q signals corresponding to each phase angle. An analog-to-digital converter (ADC) is used to quantize the I and Q signals. The output of the ADC drives a decoder circuit that recovers the groups of three bits and reinterleaves them back into the original bit stream.

Once we exceed four phase angles, the bit error performance begins to deteriorate. Figure 4.45 shows the probability of symbol error versus E_b/N_0 for 2-PSK (i.e., BPSK), 4-PSK (i.e., QPSK), 8-PSK, and so on. Note that the figure shows symbol error probability, not bit error probability. Therefore, the P_E for BPSK is better than that of QPSK, given the same E_b/N_0. As the figure shows, more and more SNR is required to maintain the same P_E as the number of symbols increases.

Figure 4.46 shows a plot of the power spectral density for unfiltered 8-PSK. As in the case of both BPSK and QPSK, the sharp phase transitions result in the same classic $\sin x/x$ shape—with sharp nulls, large sidelobes and slow roll-off. However, the spectrum is only one-third as wide as that of BPSK, since groups of three bits are combined into one symbol, effectively lowering the symbol rate by 3.

Taking the concept one step further still, we take groups of four bits represented by a symbol. This is called 16-PSK and is shown in Figure 4.47. Each symbol is separated from the others by multiples of 22.5 deg. 16-PSK is used in some wireless local area network (WLAN) applications.

Eight-PSK and 16-PSK are generally considered as spectrally efficient modulations (having a spectral efficiency greater than 2 bits/sec/Hz). Quadrature amplitude modulation (QAM) falls into this class as well, though we do not cover it here. In QAM, we vary both the *phase* and *amplitude*. QAM modulations typically combine groups of 4 to 10 bits into individual symbols. QAM is considered one of the most spectrally efficient modulations in use today [5].

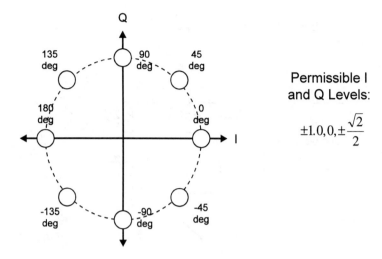

Permissible I and Q Levels:

$$\pm 1.0, 0, \pm \frac{\sqrt{2}}{2}$$

Figure 4.41 Phase domain of carrier for 8-PSK.

f_{data} = 10 Mbps, f_{TX} = 80 MHz

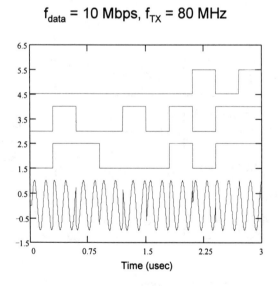

Figure 4.42 8-PSK time domain waveforms.

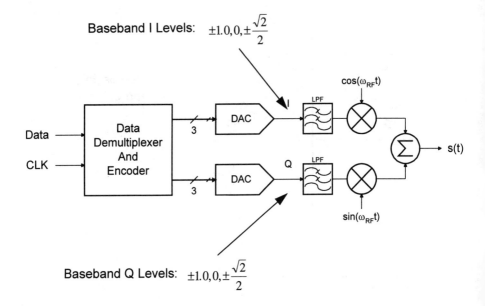

Figure 4.43 8-PSK modulator.

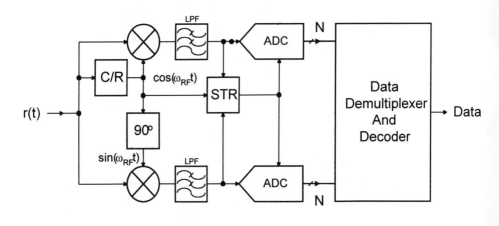

C/R = carrier recovery
STR = symbol timing recovery

Figure 4.44 8-PSK demodulator.

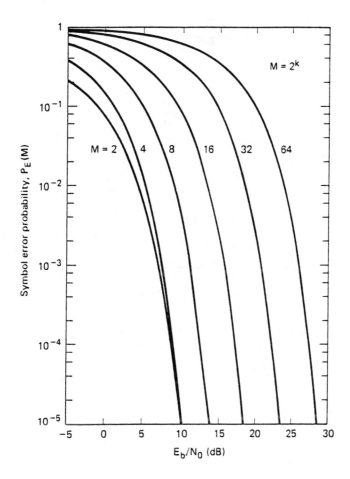

Figure 4.45 Probability of bit error for MPSK.

4.2.5 Nyquist Filtering as Applied to PSK Modulations

Earlier, we used (4.40), repeated here, to express the relation between E_b/N_0 and S/N

$$\frac{E_b}{N_0} = \frac{S}{N} \cdot \frac{B_n}{f_b} \tag{4.43}$$

where E_b is the energy per bit, N_0 is the noise power spectral density, S is the signal power, N is the noise power, B_n is the effective noise bandwidth of the receiver, and f_b is the bit rate.

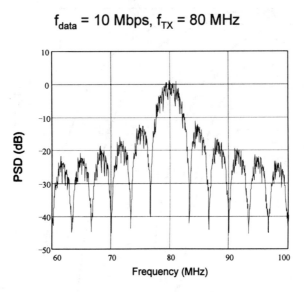

Figure 4.46 8-PSK transmit power spectral density (no baseband filtering).

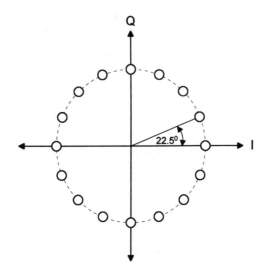

Figure 4.47 Phase domain of carrier for 16-PSK.

When the effective noise bandwidth is equal to the bit rate, E_b/N_0 equals S/N. However, we saw earlier that a modulation such as unfiltered BPSK occupies a much wider bandwidth than that equal to the bit rate. This means, in practice, the bandwidth of the receiver needs to be wider than the bit rate in order to avoid excessive ISI caused by passing a spectrally wide signal through too narrow a filter. If B_n gets larger, then (4.43) says that E_b/N_0 must increase in order to maintain the same P_E. This is because the wider bandwidth will allow more noise into the receiver, thus lowering the effective S/N (or SNR). Therefore, the narrower we can make the transmitted spectrum, the more sensitive we can make the receiver. However, too narrow a filter can cause excessive ISI, as we will see in the case of GMSK (Section 4.4.1).

A special class of filters, called Nyquist filters, solves the bandwidth problem without introducing excessive ISI into the waveform. Consider the frequency response of an ideal "brick-wall" lowpass filter shown in Figure 4.48(a). As the figure shows, the filter has unity gain within the passband and zero gain outside the passband. The impulse response can be found by simply taking the inverse Fourier transform of the filter's frequency response

$$h(t) = \int_{-\infty}^{\infty} H(f) \cdot e^{+j\omega t} df \tag{4.44}$$

$$h(t) = \int_{-f_b/2}^{f_b/2} 1 \cdot e^{+j\omega t} df = f_b \cdot \frac{\sin(\pi f_b t)}{(\pi f_b t)} \tag{4.45}$$

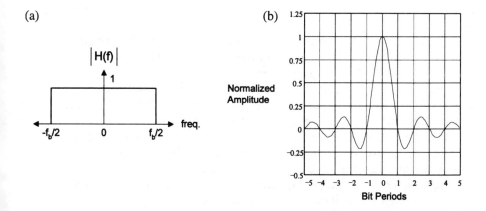

Figure 4.48 Frequency response and impulse response of an ideal brick-wall lowpass filter: (a) frequency response, (b) impulse response.

We see that the impulse response of the filter, shown in Figure 4.48(b), has the familiar $\sin x/x$ shape. A point of considerable interest is the fact the impulse response is equal to zero at the center of all bit periods except for $t = 0$. This can be easily seen by plugging $t = mT_b$ into the equation

$$h(mT_b) = f_b \cdot \frac{\sin(\pi f_b mT_b)}{(\pi f_b mT_b)} \tag{4.46}$$

Recall that $T_b = 1/f_b$. This gives

$$h(mT_b) = f_b \cdot \frac{\sin(m\pi)}{(m\pi)} \tag{4.47}$$

Thus,

$$h(0 \cdot T_b) = f_b \tag{4.48}$$

$$h(mT_b) = 0; \text{ for } m \neq 0 \tag{4.49}$$

Nyquist [10] proposed that the shape of each data pulse be such that the signal be bandlimited and yet there be no ISI due to preceding or following pulses at each sampling (or decision making) instant. He proposed three different methods for doing this. Filters, which produce pulse shapes with no ISI at each sampling instant, are referred to as Nyquist I type filters. Nyquist II filters, which will not be discussed here, employ filters whose pulse shape is such that pulses can be sent at a rate of $2f_b$ within a bandwidth of f_b and still be unambiguously interpreted. Nyquist III filters, also not discussed here, have pulse shapes such that the total area in all pulses not at $t = 0$ are zero.

Passing the sampled-data signal (actually ideal impulses) through an ideal brick-wall lowpass creates a pulse shape that meets Nyquist's first criterion. These types of filters are often called Nyquist filters.

Unfortunately, we can never actually realize the perfect brick-wall lowpass response. First of all, the impulse response for the brick-wall filter is noncausal. This is due to the fact that the impulse response is infinite in both directions in time. Secondly, the $\sin x/x$ shape tends to fall off rather slowly in time. We can, however, compromise slightly on the frequency response by allowing some of the energy in the filter to spill past the band edges of the filter.

Figure 4.49 shows one possible way in which this can be done. In this approach, we allow the edge of the band to taper off to zero following a cosine function. The filter response is equal to 1/2 at $\pm f_b/2$, but does not equal to zero until the frequency is equal to $\pm(1 + r)f_b/2$, where r is a constant. The quantity r is called the *roll-off factor*, and can range between 0 through 1. In addition, we allow the filter to begin

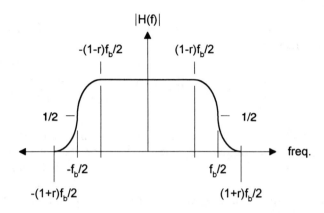

Figure 4.49 Frequency response for realizable raised cosine (Nyquist) lowpass filter.

to roll off starting at $\pm(1 - r) f_b/2$. Within the band $-(1 - r) f_b/2 < f < +(1 - r) f_b/2$, the filter has unity gain. By shaping the spectrum in this way, we can maintain the filter's characteristic of no ISI at the sampling instants of other bit periods, yet roll off the impulse response much more quickly in time. A filter with such a response is called a raised cosine filter. Notice that when $r = 0$, the filter response is equal to that of an ideal brick-wall lowpass filter. When $r = 1$, the frequency response begins to roll off at $f = 0$, with no flat-topped unity gain region.

To find the impulse response of the raised cosine filter, we take the inverse Fourier transform of the frequency response. The frequency response of the filter can be divided into five different regions

$$H(f) = H_1(f) + H_2(f) + H_3(f) + H_4(f) + H_5(f) \qquad (4.50)$$

Regions $H_1(f)$ and $H_5(f)$ lie outside the filter band and are equal to zero. Regions $H_2(f)$ and $H_4(f)$ are on the cosine-shaped portion of the band, while region $H_3(f)$ has a frequency response equal to unity. The frequency response of the filter can be written as

$$|H(f)| = 1; \text{ for } 0 \le |f| < \frac{f_b}{2}(1 - r) \qquad (4.51)$$

$$|H(f)| = \cos^2\left\{\frac{1}{4f_br} \cdot [2\pi f - \pi(1 - r) f_b]\right\}; \text{ for } -\frac{f_b}{2}(1 - r) \le |f| < +\frac{f_b}{2}(1 - r) \qquad (4.52)$$

$$|H(f)| = 0; \text{ for } |f| < \frac{f_b}{2}(1 + r) \qquad (4.53)$$

Letting $r = 0.5$, we can find the impulse response of the filter by taking the inverse Fourier transform. The result is

$$h(t) = \int_{-\infty}^{\infty} H(f) \cdot e^{+j\omega t} df = f_b \cdot \frac{\cos(\pi f_b t)}{1 - 4f_b^2 t^2} \cdot \frac{\sin(\pi f_b t)}{(\pi f_b t)} \qquad (4.54)$$

Figure 4.50 shows a plot of raised cosine filter impulse responses for $r = 0$, 0.5, 1.0 over ±3 bit periods. Notice that $r = 0$ is for the ideal brick-wall filter response, $r = 1.0$ is for the response where the bandwidth begins to taper off immediately, and $r = 0.5$ is in between these two extremes. By looking at the plot, we see that $r = 0.5$ results in a pulse shape that is almost zero by time it gets out to ±3 bit periods.

Figure 4.51 shows a block diagram of an FIR filter, which can be used to construct a raised cosine filter. The input to the filter is really an analog variable. Therefore each delay is a pure analog delay. Such an analog filter is often fabricated

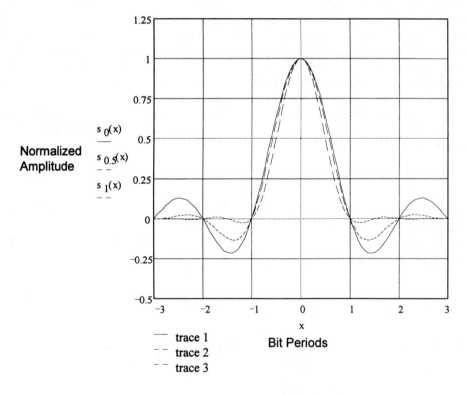

Figure 4.50 Raised cosine impulse response for different values of roll-off factor (r).

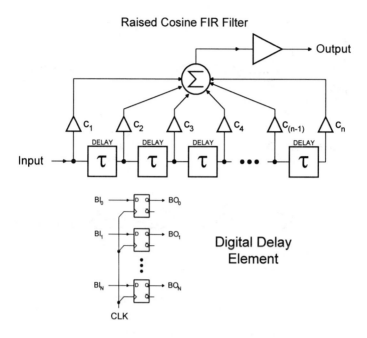

Figure 4.51 FIR filter realization of raised cosine filter.

in a surface acoustic wave (SAW) filter. The adders and tap-weight multipliers are implemented by the spacing and configuration of the metallization on the surface of the SAW substrate.

In many cases, the raised cosine filter is realized in digital form. The input to the filter is first digitized by an ADC and converted into a parallel combination of bits. Each adder and multiplier is implemented as a digital circuit. Each delay consists of a collection of parallel registers, as shown in the inset. The digital output of the filter is then converted back to analog form by a DAC, whose output is then filtered by an analog filter.

When the filter is implemented in digital form, a general purpose DSP processor is often used. The DSP processor performs the filter function by recursively using a single multiplier and accumulator to perform the tap-weight multiplication and summation. This approach performs the same function shown in Figure 4.51, but greatly reduces the amount of digital circuitry required.

Figure 4.52 shows an "eye" diagram of a raised cosine filter with $r = 0.5$. Notice that there is no ISI at the sampling instants. The overall pulse shape is sinusoidal in appearance, suggesting a narrow power spectral density. This is in fact the case. Figure 4.53 shows the power spectral densities for an unfiltered BPSK signal and a BPSK signal whose modulation signal has been passed through an

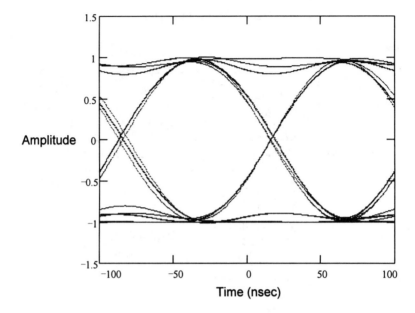

Figure 4.52 Eye diagram of output of raised cosine filter (no *x*/sinx equalization).

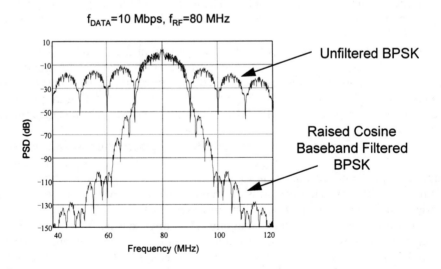

Figure 4.53 Transmit output spectrum of BPSK with and without raised cosine baseband filtering.

$r = 0.5$ raised cosine filter. As the figure shows, the spectrum of the filtered signal falls off dramatically.

The impulse response of a filter is the filter output when the input is an ideal impulse function (i.e., delta function). However, the output of a DAC driven by a digital signal is not a train of impulses, but a flat-topped succession of pulses that form a stairstep-looking response. This type of signal is often called a "boxcar" response due to its resemblance to a series of boxcars passing by on a freight train. Figure 4.54 shows a plot of the output of an ideal raised cosine filter and the output of an actual raised cosine filter. The boxcar output of the DAC causes the signal to prematurely roll off in frequency, instead of maintaining its flat-frequency response.

We can compensate the roll-off of the DAC within the raised cosine filter itself. This can be done by visualizing the filter output as a stream of impulses convolved with an ideal rectangular pulse. The result is the desired flat-topped boxcar output. This is shown in Figure 4.55. Recall that convolution in the time domain results in multiplication in the frequency domain. This implies

$$s_{\text{Box}}(t) = p(t) \ast s_{\text{RC}}(t) \tag{4.55}$$

where $s_{\text{Box}}(t)$ is the time domain boxcar output, $p(t)$ is the time domain pulse shape of the ideal rectangular pulse and $s_{\text{RC}}(t)$ is the sampled data impulse train of the raised cosine filter output. In the frequency domain, we can write

$$S_{\text{Box}}(f) = P(f) \cdot S_{\text{RC}}(f) \tag{4.56}$$

where

$$S_{\text{RC}}(f) = \frac{1}{T_{\text{samp}}} \cdot \sum_{n=-\infty}^{\infty} S\left(f - \frac{n}{T_{\text{samp}}}\right) \tag{4.57}$$

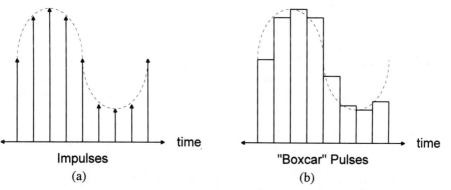

Impulses
(a)

"Boxcar" Pulses
(b)

Figure 4.54 Sampled data output of raised cosine filter: (a) ideal filter output (unit impulses), (b) actual filter output (sample-and-hold, or boxcar).

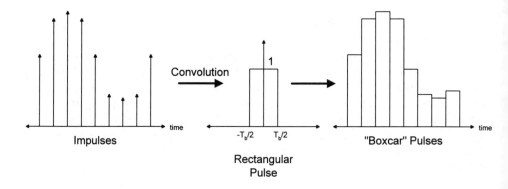

Figure 4.55 Modeling the boxcar output of the raised cosine filter.

Here, $S(f)$ is the spectrum of the raised cosine filter's impulse response and $S_{RC}(f)$ is the spectrum of the impulse train output of the raised cosine filter. The Fourier transform of the ideal rectangular pulse is given by

$$P(f) = T_{samp} \cdot \frac{\sin(\pi f T_{samp})}{(\pi f T_{samp})} \qquad (4.58)$$

This results in the following boxcar spectrum

$$S_{Box}(f) = \left[T_{samp} \cdot \frac{\sin(\pi f T_{samp})}{(\pi f T_{samp})} \right] \cdot \frac{1}{T_{samp}} \sum_{n=-\infty}^{\infty} S_{RC}\left(f - \frac{n}{T_{samp}} \right) \qquad (4.59)$$

$$S_{Box}(f) = \left[\frac{\sin(\pi f T_{samp})}{(\pi f T_{samp})} \right] \cdot \sum_{n=-\infty}^{\infty} S_{RC}\left(f - \frac{n}{T_{samp}} \right) \qquad (4.60)$$

The above equations show that the flat-topped boxcar output of the DAC is equivalent to having the ideal raised cosine filter output multiplied by a sin$x$/x shaped roll-off. At the Nyquist frequency of the filter (i.e., one-half the sample rate), the sin$x$/x term causes significant roll-off. This can be calculated by letting $f = f_{samp}/2$

$$S_{Box}(f_{samp}/2) = \left[\frac{\sin\left(\pi \dfrac{f_{samp}}{2} T_{samp} \right)}{\left(\pi \dfrac{f_{samp}}{2} T_{samp} \right)} \right] \cdot S_{RC}(f_{samp}/2) = \frac{2}{\pi} \cdot S_{RC}(f_{samp}/2) \qquad (4.61)$$

So, we see the boxcar effect reduces the output at the Nyquist frequency by about 3.9 dB.

This roll-off can be compensated for by "pre-emphasizing" the frequency response of the filter, before taking the inverse Fourier transform to find the impulse response. We do this by multiplying the frequency response by an *x/sinx* term as follows:

$$|H(f)| = \frac{(\pi f T_b)}{\sin(\pi f T_b)}; \text{ for } 0 \le |f| < \frac{f_b}{2}(1 - r) \qquad (4.62)$$

$$|H(f)| = \frac{(\pi f T_b)}{\sin(\pi f T_b)}\cos^2\left\{\frac{1}{4f_b r} \cdot [2\pi f - \pi(1 - r)f_b]\right\}; \text{ for } -\frac{f_b}{2}(1 - r) \le |f| < +\frac{f_b}{2}(1 - r)$$
$$(4.63)$$

$$|H(f)| = 0; \text{ for } |f| < \frac{f_b}{2}(1 + r) \qquad (4.64)$$

The desired impulse response is then computed from the above equations. Figure 4.56 shows a plot of an equalized eye diagram for a raised cosine filter. As the plot shows, the sections of the waveforms between the sampling instants tend to be much more pronounced.

4.3 FREQUENCY MODULATION

Up to now, we have discussed modulations where the phase of the carrier is abruptly changed to a new value in response to the input data signal. We have seen that this can result in a wide transmitted spectrum and can also result in spectral regrowth if the signal does not have a constant envelope. We have also seen that we can use multiple phase angles to represent groups of bits and thus reduce the effective symbol rate at the transmitter.

Instead of breaking the phase angle into small pieces, perhaps we could smoothly change the phase angle from one value to another. If we do this in such a way so as to remain on the unit circle, we will have a transmitted signal with constant amplitude. A constant amplitude signal would be impervious to the effects of filtering/hardlimiting and spectral regrowth.

Recall that frequency is the derivative of phase, or

$$\omega(t) = \frac{d\varphi(t)}{dt} \qquad (4.65)$$

Smooth, continuous phase modulation corresponds to frequency modulation. In this section, we will show how frequency can be used as a modulation parameter.

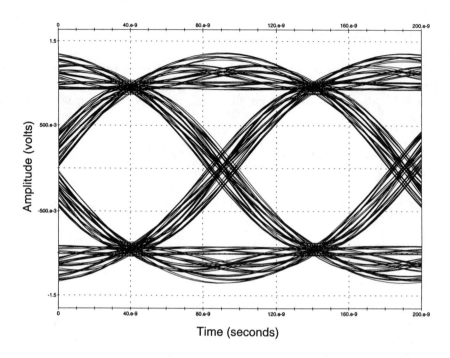

Figure 4.56 Raised cosine filter output with x/sinx equalization.

4.3.1 Minimum Shift Keying

Minimum shift keying (MSK) can be thought of as either a phase or a frequency modulation. Figure 4.57 shows the phase domain plot for MSK. In MSK, the carrier changes phase by ±90 deg over the course of a bit period. For example, suppose the carrier phase at the beginning of the data period is 0 deg. If the next data bit is a 1, then the phase smoothly advances by 90 deg by the end of the bit period. If the data bit remains a 1, then the phase advances another 90 deg to 180 deg. Instead, if the next bit were a 0, then the phase would smoothly retard by 90 deg to −90 deg. Again, if the data were to remain at 0, then phase would continue to retard itself by another 90 deg. Since the phase will continue to advance or retard itself over the course of each bit period, the derivative of the phase, or frequency, is varied. In this way, MSK can also be considered a frequency modulation. MSK is used in satellite communications, military tactical radio, and extremely low frequency (ELF) underwater communications.

Figure 4.58 shows how the phase of MSK can be plotted over time. This is known as a phase trellis. As the figure shows, the phase angle changes smoothly by 90 deg over each bit period. We can write the MSK signal as

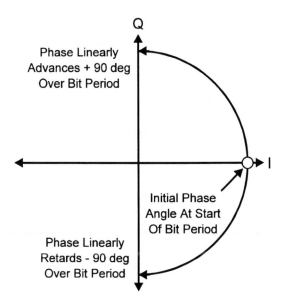

Figure 4.57 Phase domain of carrier for MSK.

$$s(t) = A \cdot \cos\left[\omega_{RF}t + \int_{-\infty}^{t} d(t) \cdot \frac{\pi}{2} dt\right] = A \cdot \cos[\omega_{RF}t + \frac{\pi}{2} \cdot (\pm 1) \cdot t + \varphi_0] \quad (4.66)$$

where $d(t)$ gives rise to the ± 1 term in the above equation.

To find the effective frequency difference, assume that $d(t)$ changes from one bit period to another. In one case, the frequency will be higher, and when the input changes, the frequency will be lower. We can write this as

$$\Delta f^+ = \frac{\Delta \omega}{2\pi} = \frac{\Delta \varphi(t)}{\Delta t} \cdot \frac{1}{2\pi} = \frac{\pi/2}{T_b} \cdot \frac{1}{2\pi} = \frac{f_b}{4} \quad (4.67)$$

$$\Delta f^- = \frac{\Delta \omega}{2\pi} = \frac{\Delta \varphi(t)}{\Delta t} \cdot \frac{1}{2\pi} = \frac{-\pi/2}{T_b} \cdot \frac{1}{2\pi} = -\frac{f_b}{4} \quad (4.68)$$

where $f_b = 1/T_b$ where f_b is the bit rate and T_b is the bit period. Since we are interested in the frequency difference, we subtract the two terms above to get

$$\Delta f^+ - \Delta f^- = \left(\frac{f_b}{4}\right) - \left(-\frac{f_b}{4}\right) = \frac{f_b}{2} \quad (4.69)$$

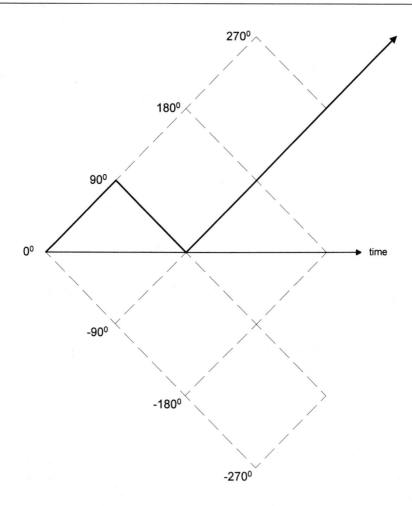

Figure 4.58 MSK phase trellis.

So, we see in MSK the frequency difference that corresponds to a 90-deg phase advance or lag over a bit period is equal to one-half the bit rate. In fact, the word minimum corresponds to the minimum frequency difference between the "mark" and "space" frequencies, which can be coherently demodulated.

We have stated earlier that the quadrature modulator is a sort of "universal" modulator. We might ask ourselves if there is a set of waveforms we can apply to the I and Q inputs that result in the desired constant envelope phase shift required for MSK. The answer is yes, and these waveforms are shown in Figure 4.59. As the figure shows, if the data stream does not change, then the I and Q modulation signals are simply two sinusoids 90 deg out of phase with one another. We can

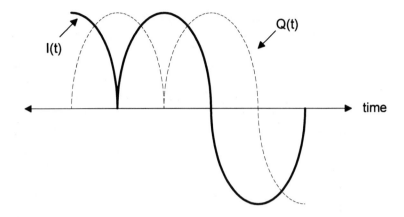

Figure 4.59 *I* and *Q* waveforms corresponding to MSK phase trellis shown in Figure 4.58.

think of this as a rectangular-to-polar conversion process. Simply look at any point on the unit circle and determine what values of *I* and *Q* are required for that phase angle. If the next bit is a 1, find a point on the unit circle that corresponds to 90-deg more of phase angle. This will dictate what the *I* and *Q* modulation signals should look like. As the data stream changes value, the *I* and *Q* signals will look like half-period sinusoids.

This is better illustrated in Figure 4.60. At the bottom of the plot, the *I* and *Q* values are shown as rectangular data pulses. After passing through special pulse-shaping filters, the *I* and *Q* signals become "half-sinusoidal" in nature. These are shown overlaid on the data streams. Notice that the corresponding output signal is indeed constant envelope, with its frequency changing from a slightly higher to a slightly lower value.

A block diagram of an MSK modulator is shown in Figure 4.61 [11]. Notice that the MSK modulator is identical to an OQPSK modulator, except for the special pulse-shaping filters. Figure 4.62 shows a detailed block diagram of each pulse-shaping filter, which is implemented as a counter-based finite impulse response (FIR) filter using read-only memories (ROMs). In this case, however, the filter has only one bit as an input, which greatly reduces complexity. The input to the filter simply reverses the polarity of the output signal at the end of each symbol period. Figure 4.63 shows a block diagram of the MSK demodulator. The MSK demodulator is identical to an OQPSK modulator. In general, the lowpass filters shown in the diagram are implemented as *matched filters*. Matched filters are special filters that are shaped to allow the signal itself to pass with little or no attenuation and reject all other waveforms (including noise).

The careful reader will notice that if a data stream of logic 1s or logic 0s are fed into the MSK modulator, the phase will not rotate about the unit circle. Instead,

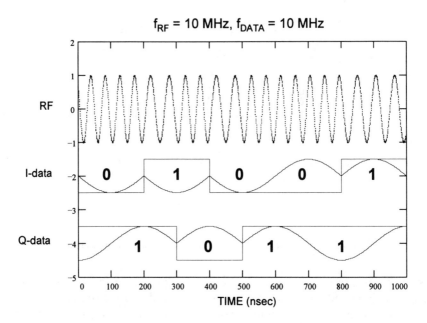

Figure 4.60 MSK time domain waveforms.

the phase bounces back and forth 90 deg. This really isn't a problem in most applications, since the incoming data stream is generally very close to a random bit stream anyway and so long as the demodulator "undoes" what the modulator does, the original bit stream is recovered. However, in applications that require true FM-like modulation, the *I* and *Q* bit streams can be differentially encoded.

As we have said earlier, MSK can be viewed as either frequency or phase modulation. This implies that we should be able to produce an MSK signal via standard FM circuitry. This is in fact the case, as Figure 4.64 shows. In the figure, the bit stream is fed directly into the modulation input of a VCO. We have already seen that the MSK signal has total frequency deviation equal to one-half the bit rate. Recall that the modulation index for an FM signal is given by

$$h \equiv \beta = \frac{\Delta f}{f_m} \tag{4.70}$$

where the variable, h, is more commonly used as the index of modulation for digital signals and β is the more familiar FM (analog) index of modulation. Δf is the total frequency difference and f_m is the modulation frequency. For MSK, we set the level of the modulation signal into the VCO to produce a modulation index, h, of 0.5.

Returning to Figure 4.64, the oscillator output is fed to a transmit power amplifier and filtered before being sent to the antenna. Because MSK can be viewed

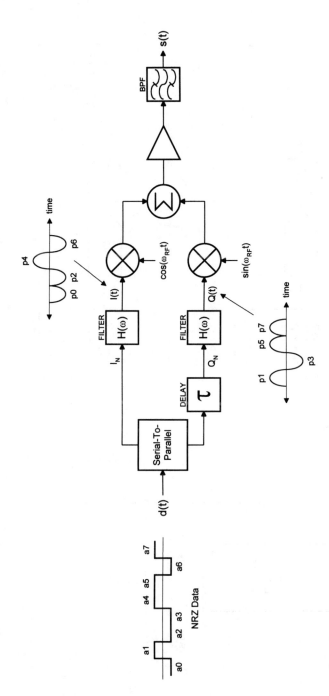

Figure 4.61 MSK modulator.

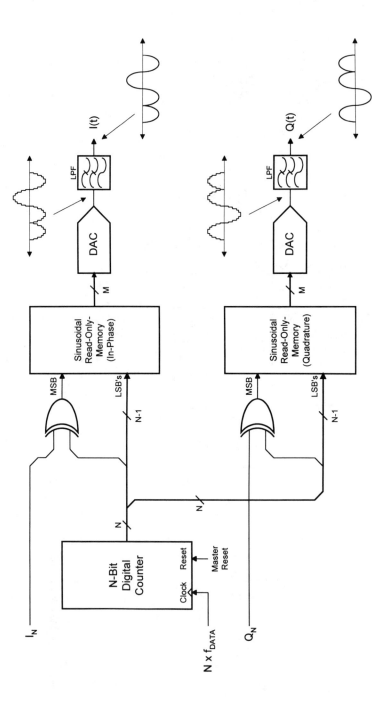

Figure 4.62 Implementation of MSK modulator using digital counters, ROMs, and D/A converters.

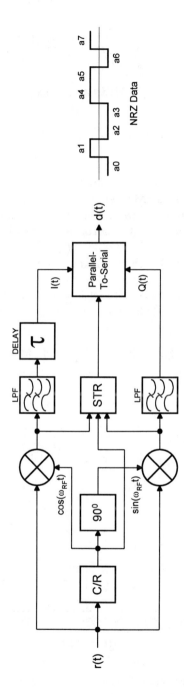

Figure 4.63 MSK demodulator.

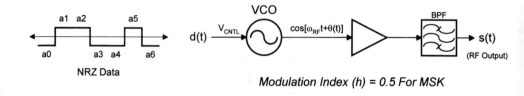

Figure 4.64 Using a voltage-controlled oscillator (VCO) as an MSK modulator.

as a type of frequency modulation, it is sometimes known as *fast frequency shift keying* (FFSK). Using the standard FM modulator is acceptable when MSK is noncoherently demodulated. However, it can be difficult to maintain the required index of modulation with an open-loop VCO approach due to variations of modulation sensitivity in the VCO. Treating MSK as a type of *I/Q* phase modulation ensures that the modulation index is fixed to 0.5, thus providing for fully coherent demodulation.

Figure 4.65 shows the time domain response of MSK. The modulated carrier is shown at the top of the plot. As stated earlier, the phase changes linearly 90 deg over the course of a bit period. This phase change is shown in the second trace.

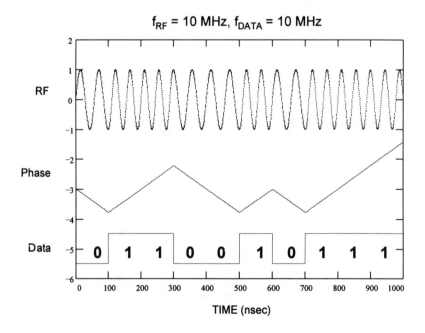

Figure 4.65 Time domain waveforms from VCO-based MSK modulator.

Notice that the relative slope is the same for a logic 0 as it is for a logic 1, only with a negative slope. The lower portion of the plot shows the input data stream.

A plot of the power spectral density of an MSK signal, compared to that of an unfiltered QPSK signal, is shown in Figure 4.66 for a carrier frequency of 80 MHz and a data rate of 10 Mbps. Notice how the MSK spectrum falls off much more rapidly than the QPSK spectrum. This is due to the fact that the MSK signal is continuous in-phase. Although the phase is continuous in MSK, the frequency (i.e., the slope of the phase) is not. It is this discontinuous frequency that results in a somewhat wider power spectral density with sharp nulls.

Figure 4.67 shows a plot of bit error probability versus E_b/N_0. Notice that the curve is the same as that of BPSK and QPSK. The reason for this is that MSK can be viewed as QPSK (actually OQPSK due to the one bit period offset) with special pulse-shaping filters, acting over two bit periods. The I and Q modulation signals are each half-sinusoidal pulses two bit periods long, which can assume either positive or negative values. This makes MSK another form of antipodal modulation. So long as we integrate over two bit periods in the demodulator, we can achieve the same P_E as for other antipodal modulations. In effect, we have knowledge of where the phase has been and where it is going if we look at the signal over two bit periods. This knowledge can be used to predict the value of the next signal, and hence lower the P_E of the signal.

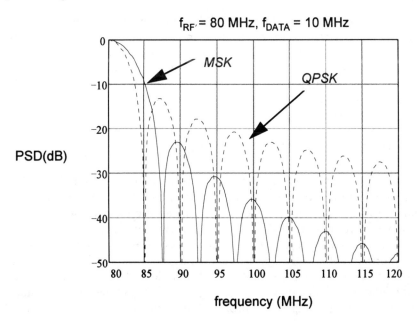

Figure 4.66 MSK versus QPSK power spectral density.

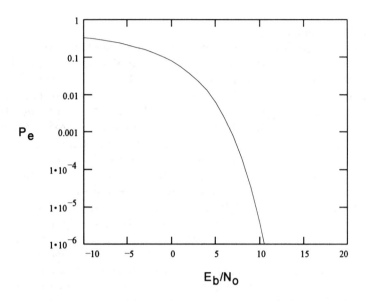

Figure 4.67 MSK probability of bit error.

Figure 4.68 shows a plot of an MSK signal in the time domain. As theory suggests, MSK is a true constant envelope modulation. No amplitude fluctuations can be seen in the waveform. Figure 4.69 shows a plot of the MSK signal before and after hardlimiting. As the figure shows, there is essentially no spectral regrowth. This feature of MSK makes it very attractive in applications where saturated power amplifiers are employed.

4.3.2 Frequency Shift Keying

MSK can be considered as either frequency or phase modulation. The only caveat is that we make the modulation index, h, equal to 0.5. Recall, that $h = 0.5$ is the minimum value of modulation index that we can use for coherent modulation. If we remove the restriction that $h = 0.5$, we can classify the type of modulation as a form of frequency shift keying (FSK). FSK has been used as a means of transmitting digital information for years. This is primarily due to the simplicity of the modulation and demodulation circuitry. Figure 4.70 shows block diagrams for a noncoherent FSK modulator and demodulator. As Figure 4.70(a) shows, FSK modulation is accomplished by applying the appropriate-level signal to the VCO's modulation input. The amplitude of the modulation signal is adjusted for the desired index of modulation. In general, the received FSK signal is passed through a limiting IF amplifier and fed to a noncoherent demodulator such as a frequency discriminator

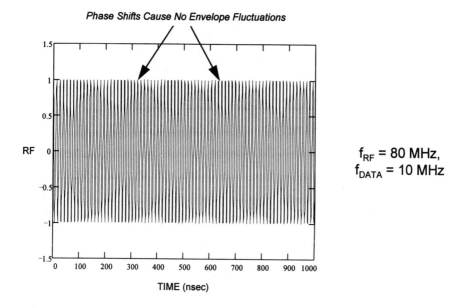

Figure 4.68 Time domain waveform of baseband-filtered MSK transmit signal (using 5-MHz Gaussian lowpass filter).

or quadrature detector. The image of the detected signal is filtered out and fed to a bit slicer or comparator to recover the original bit stream.

Figure 4.71 shows the time domain response for an FSK signal with $h = 1.0$. Notice that the modulated signal is constant amplitude, but the difference between the mark and space frequencies is larger due to the higher modulation index. With an $h = 1.0$, the total phase variation over one bit period is equal to 180 deg. This results in much larger—though still linear—phase variations over a bit period. The input bit stream is shown at the bottom of the figure.

In order to coherently demodulate the FSK signal, we need reference waveforms of the two possible transmitted frequencies. A block diagram of a coherent demodulator is shown in Figure 4.72. The received signal is applied to two double-balanced mixers. One mixer is driven by an LO at a frequency corresponding to a logical input of 0 on the modulator input. The other mixer is driven by the frequency that corresponds to a logic 1 on the modulator input. The corresponding outputs of the mixer are then integrated over a bit period and sampled. The outputs of the sample-and-hold circuits go to a decision circuit that decides which of the two signals was larger. Coherent FSK modulation is not often used in practice due to the difficulty in generating two reference frequencies close together at the receiver.

Figure 4.73 shows a plot of bit error probability of FSK versus E_b/N_o for both coherent and noncoherent signals. Not surprisingly, the noncoherent signal requires a higher SNR than the coherent signal in order to maintain the same P_E.

(a)

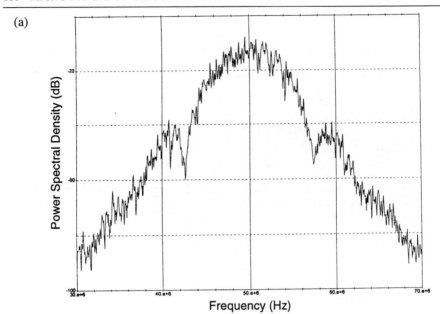

(b)

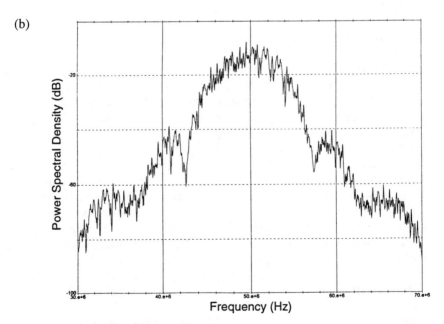

Figure 4.69 MSK transmit output spectrum with baseband filtering: (a) without limiting, (b) with limiting (amplifier 10 dB into compression).

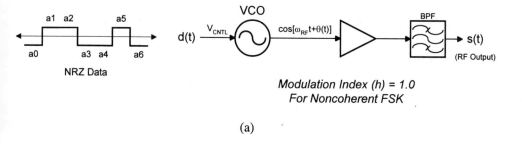

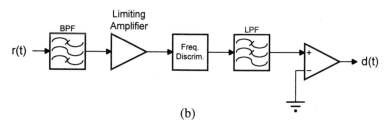

Figure 4.70 Noncoherent FSK modulator and demodulator: (a) FSK modulator, (b) FSK demodulator.

Figure 4.74 shows a plot of the power spectral density of an FSK signal with $h = 1.0$. Here, the carrier frequency is 80 MHz and the data rate is 10 Mbps. Notice that the spectrum is very similar to that of MSK, except that the main lobe is wider due to the larger modulation index. Also notice that FSK with $h = 1.0$ produces a spur at both the mark and space frequencies (though only one of the two spurs can be seen because of the single-sided nature of the plot). This spur has been used in some applications as a means to recover the carrier frequency.

Up to now, we have only been talking about using binary FSK modulation. As in the case of PSK, we can represent groups of bits by using different symbols. This is done in FSK by using more than two different frequencies. For example, we might choose to use four closely spaced frequencies as our symbol and thus reduce the data rate by one-half. Figure 4.75 shows a diagram of this. This is called 4-level FSK, or 4-FSK. As in the case of QPSK, the input bit stream is demultiplexed into two independent bit streams at half the original data rate. Each group of two bits, or symbols, is then represented by one of four different frequencies.

Figure 4.76 shows a block diagram of a 4-FSK modulator. An encoder is used to demultiplex the data and produce a corresponding digital word that is used to drive the DAC. The DAC output is then applied through a filter to the VCO. Figure

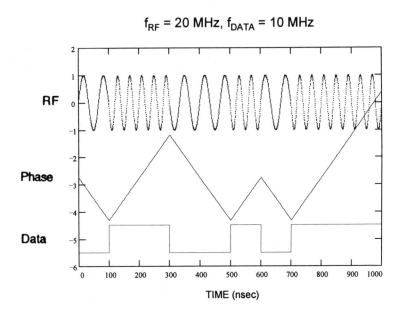

Figure 4.71 FSK time domain waveforms.

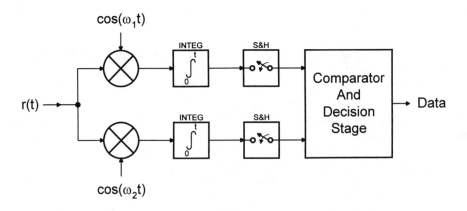

Figure 4.72 Coherent FSK demodulator.

4.77 shows a plot of a 4-FSK modulated signal with $h = 2.0$ having a carrier frequency of 80 MHz and a bit rate of 10 Mbps (corresponding to a symbol rate of 5 Msps). In this case, the maximum frequency deviation, Δf, is the frequency difference between the two outermost symbols. Notice the relatively flat-topped spectral response and the individual spurs at each of the corresponding symbol frequencies.

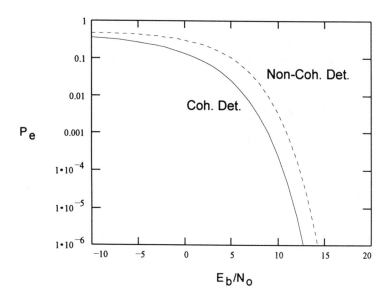

Figure 4.73 FSK probability of bit error.

Four-FSK can be demodulated by simply filtering and detecting each of the corresponding symbol frequencies. Figure 4.78 shows this brute-force approach. This approach is useful where the relative spacing of each symbol frequency is widely separated. This approach is commonly used in touch-tone phone decoders at audio frequencies. A disadvantage with this approach is that we essentially need as many different IF strips as we have symbol frequencies. If the IF center frequency is high and the spacing between each symbol frequency is small, it can be difficult to maintain adequate separation between each symbol frequency due to insufficient filter stopband performance.

A more modern, and higher speed, approach is shown in Figure 4.79. In this approach, a single frequency discriminator is used. The output of the frequency discriminator produces a multilevel analog signal. The four different levels corresponding to each of the symbols are detected by means of an ADC. In the past, the relatively high complexity of ADCs made this approach less attractive. However, with modern IC techniques, it is possible to fabricate the ADC in the same IC as the decoder, and other portions of the receiver.

Figure 4.80 (a) shows a plot of the symbol error probability of a multilevel FSK signal versus E_b/N_0 for coherent demodulation. In the Figure, M is an integer that represents the number of symbols. Figure 4.80(b) shows the same plot for noncoherent demodulation. Comparing the two plots reveals that there is very little difference in the symbol error probabilities when the number of symbols is large and E_b/N_0 is high. This is an advantage of multilevel FSK.

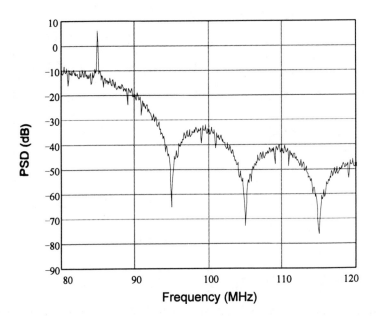

f_{RF} = 80 MHz, f_{DATA} = 10 MHz

Figure 4.74 FSK (h = 1.0) power spectral density.

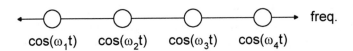

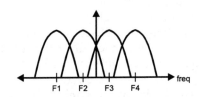

Figure 4.75 4-level FSK (4-FSK).

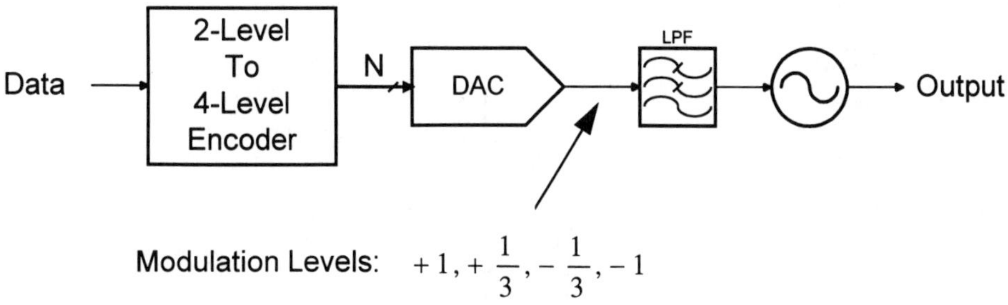

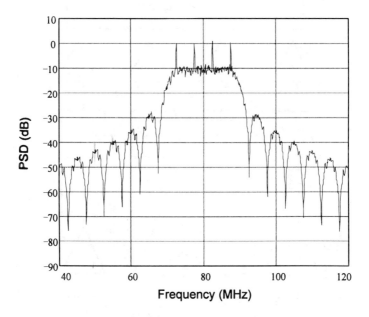

Modulation Levels: $+1, +\dfrac{1}{3}, -\dfrac{1}{3}, -1$

Figure 4.76 4-FSK modulator.

f_{RF} = 80 MHz, f_{DATA} = 10 MHz

Figure 4.77 4-FSK power spectral density ($h = 2.0$).

4.4 RELATED MODULATIONS

The basic principles of digital phase and frequency modulation have been covered in the preceding sections. A number of variations are possible on each of these modulations, which can produce a number of additional benefits.

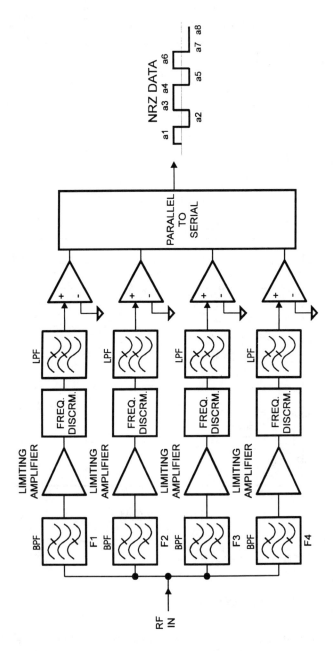

Figure 4.78 Brute force 4-level FSK demodulator (noncoherent).

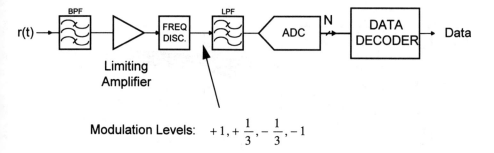

Figure 4.79 Modern 4-level FSK demodulator (noncoherent).

4.4.1 Gaussian-Filtered MSK (GMSK)

We have shown that baseband filtering of the modulation signal can result in a much narrower transmitted spectrum. However, as we saw in the case of QPSK, care must be taken when the signal is nonlinearly amplified to avoid spectral regrowth. The best way to avoid spectral regrowth is to use a constant envelope modulation, such as MSK. However, even then, the transmitted spectrum of MSK is still too wide for many applications.

One way to narrow the transmitted spectrum of MSK is by filtering the modulation signal before applying it to the VCO. This can be done by passing the modulation signal through a lowpass filter. It is important, however, that the lowpass filter have a well-behaved time domain response. A class of filters that has a well-behaved time domain response is Gaussian filters. The frequency response of a Gaussian lowpass filter is Gaussian in nature, following the following relation

$$H(\omega) = \tau \cdot \sqrt{2\pi} \cdot e^{-\frac{1}{2}(\tau\omega)^2} \tag{4.71}$$

This is the familiar $\exp(-x^2)$ shape of the Gaussian, or normal, probability density function (i.e., the classic "bell-shaped" curve). In the equation, ω is the frequency in rads/sec and τ is a constant.

A peculiar property of a filter with a Gaussian frequency response is that its impulse, or time domain, response is Gaussian as well. This can be seen by taking the inverse Fourier transform of the spectrum

$$h(t) = \frac{1}{2\pi}\int_{-\infty}^{\infty} H(\omega) \cdot e^{-j\omega t} d\omega = \frac{1}{2\pi}\int_{-\infty}^{\infty}\left[\tau \cdot \sqrt{2\pi} \cdot e^{-\frac{1}{2}(\tau\omega)^2}\right] \cdot e^{-j\omega t} d\omega \tag{4.72}$$

Solving the integral yields

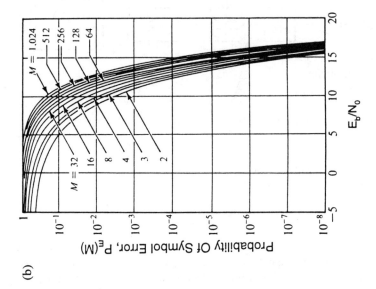

Figure 4.80 MFSK probability of bit error: (a) coherent MFSK, (b) noncoherent MFSK.

$$h(t) = e^{-\frac{1}{2}\left(\frac{t}{\tau}\right)^2} \tag{4.73}$$

Figure 4.81 shows plots of both the time domain and frequency response of a Gaussian lowpass filter. Notice that both curves have the same bell-shaped response.

The Gaussian impulse response is well-behaved, exhibiting no ringing or overshoot. However, the frequency response of the filter tends to fall off rather slowly. In general, the more gradual the frequency response of a filter, the better its time domain response (an exception to this rule are digital FIR filters). Gaussian filters have one of the most gradual frequency responses of any analog filter type.

Notice that the frequency response, given by (4.71) is purely real. This implies that the phase of the Gaussian filter is linear. Linear phase filters can be designed using digital FIR filter techniques or can be implemented with SAW filters. However, it is possible to approximate the response of a Gaussian filter by using standard reactive filters such as LC ladder filters. Figure 4.82 shows a schematic for a passive LC ladder filter with a table of normalized Gaussian filter coefficients for different filter orders. The table is for double-terminated filters.

For example, the design of a Gaussian lowpass filter with a cut-off (or −3 dB) frequency of 500 kHz with a 50-ohm system impedance is presented here [12]. The following equations are used to find the actual component values from the prototype values listed in the table:

$$L = \frac{R}{2\pi f_c} L_{\text{proto}} \tag{4.74}$$

$$C = \frac{1}{2\pi f_c R} C_{\text{proto}} \tag{4.75}$$

From the table, for a third-order filter, we find

$$C_{\text{proto},1} = 0.2624 \tag{4.76}$$

$$L_{\text{proto},2} = 0.8167 \tag{4.77}$$

$$C_{\text{proto},3} = 2.2262 \tag{4.78}$$

This gives the following component values:

$$C_1 = 1.67 \text{ nF} \tag{4.79}$$

$$L_2 = 13.0 \text{ uH} \tag{4.80}$$

$$C_3 = 14.2 \text{ nF} \tag{4.81}$$

Figure 4.81 Gaussian lowpass filter waveforms: (a) impulse response, (b) frequency response.

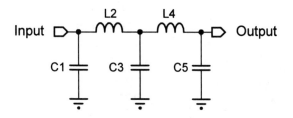

Table Of Normalized Values

N	$C1_{proto}$	$L2_{proto}$	$C3_{proto}$	$L4_{proto}$	$C5_{proto}$
2	0.4738	2.1850			
3	0.2624	0.8167	2.2262		
4	0.1772	0.5302	0.9321	2.2450	
5	0.1312	0.3896	0.6485	0.9782	2.2533

Actual component values given by:

$$L = \frac{R}{2\pi f_c} L_{proto} \qquad C = \frac{1}{2\pi f_c R} C_{proto}$$

Figure 4.82 Using lumped-element LC ladder filter to approximate Gaussian lowpass filter.

Figure 4.83 shows the completed, third-order Gaussian lowpass filter. Gaussian filters designed using reactive elements have a more gradual frequency response than other reactive filters, such as Bessel, Butterworth, and Chebyshev. However, Gaussian filters exhibit no overshoot or ringing in the time domain. This smooth, well-behaved impulse response results in very little intersymbol interference (ISI).

Figure 4.84(a) shows a block diagram of a Gaussian filtered MSK (GMSK) modulator [13–15]. The GMSK modulator is identical to the VCO-based version of the MSK modulator, except for the lowpass modulation filter. The bandwidth of the Gaussian filter is often given in terms its relation to the bit rate. Recall

$$T \equiv T_b = \frac{1}{f_b} \tag{4.82}$$

where T or T_b is the bit period of the filter and f_b is the bit rate. If we let B equal the −3-dB bandwidth of the filter, then we can specify the filter response in terms of its *relative bandwidth*, or BT

$$BT = (\text{Filter Bandwidth}) \cdot (\text{Bit Period}) = \frac{\text{Filter Bandwidth}}{\text{Bit Rate}} \tag{4.83}$$

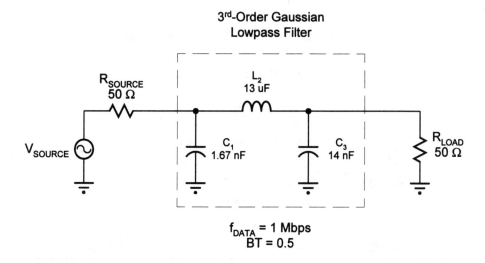

Figure 4.83 Actual third-order Gaussian lowpass filter (BT = 0.5, 50Ω system).

As we shall see shortly, GMSK has a very narrow transmitted spectrum, yet exhibits very little ISI and is tolerant to hardlimiting. Figure 4.84(b) shows a simple noncoherent demodulator, similar to that used in FSK, which can be used to demodulate GMSK. Applications for GMSK include the global system for mobile communication (GSM) with BT = 0.3, and the digital European cordless telephone (DECT) with BT = 0.5.

The time domain response of GMSK with BT = 0.5 (i.e., the filter bandwidth is equal to one-half the bit rate) is shown in Figure 4.85. The top trace of the figure shows the modulated signal. Just as in the case of MSK, GMSK is a constant envelope modulation. In contrast to MSK, the second trace shows that the phase now has a smooth slope. This is caused by the action of the Gaussian lowpass filter. Notice that the filter also smoothes out the input bit stream, removing the sharp transitions in the data.

The action of the filter has a radical effect on the transmitted spectrum. Figure 4.86 shows a plot of the power spectral density for various GMSK signals having different relative filter bandwidths (BT). Recall that the "M" in GMSK stands for minimum, so the modulation index, h, is equal to 0.5 in all cases. Should h not be equal to 0.5, then the modulations is referred to as GFSK. If no filter is used, then the BT = infinity. This is equivalent to MSK, and is shown in the plot for comparison purposes. The figure shows that a filter having a bandwidth equal to half the data rate (i.e., BT = 0.5) rolls off much more quickly in frequency than MSK (i.e., no filter). Using filters with smaller relative bandwidths causes even faster roll-off. Unfortunately, this faster spectral roll-off has a price.

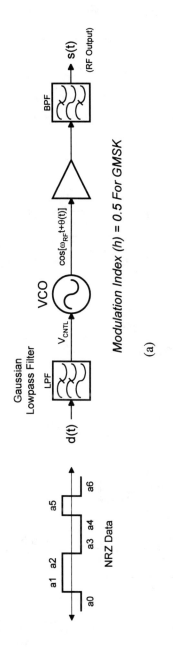

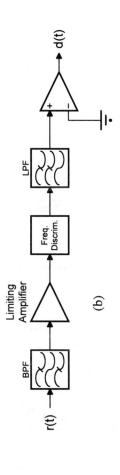

Figure 4.84 Gaussian lowpass filtered MSK (GMSK): (a) GMSK modulator (noncoherent), (b) GMSK demodulator (noncoherent).

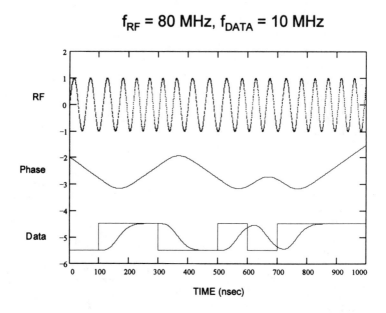

Figure 4.85 GMSK time domain waveforms.

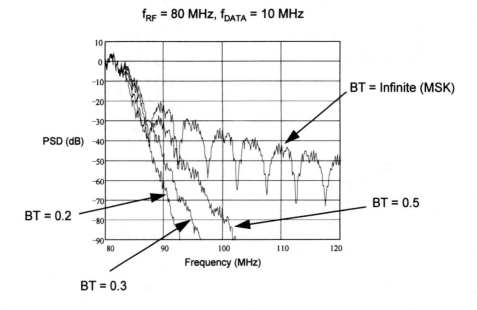

Figure 4.86 GMSK power spectral density for different relative filter bandwidths (BT).

As the relative filter bandwidth is lowered, more and more ISI is imparted to the waveform. One method by which we can get a qualitative indication of the amount of ISI on the waveform is by looking at an eye diagram. Figure 4.87 shows the test setup used to measure eye diagrams. The received, or recovered, data is fed to the vertical input of an oscilloscope. The scope is triggered on the rising edge of the recovered, or local, clock. This action causes multiple traces to overlap on the scope's cathode ray tube (CRT). Eye diagrams get their name by the way the signal resembles an "eye" on the CRT. As noise or ISI increases, the eye begins to close, increasing P_E.

Figure 4.88 shows a collection of waveforms with varying degrees of ISI. In Figure 4.88(a), very little filtering is used and all data transitions on the recovered waveform are rectangular in shape. Long sequences of 0s or 1s result in the horizontal lines at the bottom and top of the traces, respectively. In Figure 4.88(b), a special type of filter was used that rolls off the frequency response of the filter, causing the rectangular pulses to become sinusoidal in shape. However, at the sampling instant,

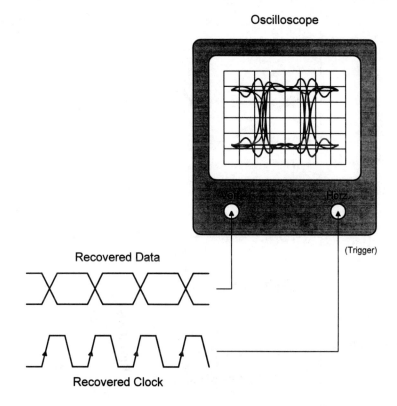

Figure 4.87 Measurement setup for observing eye diagrams.

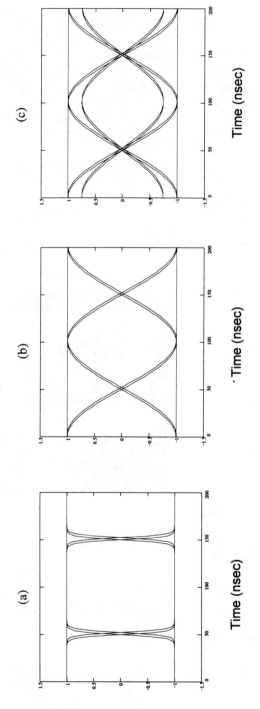

Figure 4.88 Intersymbol interference (ISI): (a) unfiltered eye pattern, (b) filtered eye pattern without ISI, (c) filtered eye pattern with ISI.

the signal has reached its full value, resulting in no ISI at the sampling instant. As we shall see later, Nyquist filters are one type of filter that possess this quality. On the other hand, the trace shown in Figure 4.88(c) *does* exhibit ISI at the sampling instant. ISI is created by data bits in periods preceding the present data period not fully settling out.

Figure 4.89(a) shows the time domain response of a GMSK modulation signal with a BT = 0.5 filter. Note that the amplitude of the modulation waveform is equivalent to the instantaneous frequency deviation of the carrier. Figure 4.89(b) shows an eye diagram of this same signal taken over many more bit periods. Notice that with a filter having a bandwidth of half the bit period (BT = 0.5), very little ISI is seen in the waveform. Figure 4.90(a) shows the time domain response for a BT = 0.25 filter (i.e., one-quarter the bit rate). Notice how alternating strings of 0s and 1s result in waveforms that are nowhere near the full value of the original, unfiltered signal. The eye diagram shows this effect much more clearly. At the sampling instant (near $t = 0$), the eye is almost closed. It is easy to see that very little noise will cause a bit error in this case.

When the GMSK signal is noncoherently demodulated, the output of the demodulator is equivalent to the instantaneous frequency deviation of the signal. We have just seen that with low values of BT, the eye is almost closed. Therefore, a relatively high SNR must be maintained if bit errors are to be kept low. Earlier, we discussed how MSK can be viewed as either phase or frequency modulation. We saw that MSK probability of bit error can equal that of BPSK and QPSK if the signal is processed over two bit periods when the signal is coherently demodulated. We might expect that we should be able to approach this level of performance if we coherently demodulate GMSK—given the fact that GMSK is a derivative of MSK.

Figure 4.91 shows a block diagram of a coherent GMSK demodulator. The demodulator is identical to that of MSK. In many cases, a predetection Gaussian bandpass filter is used at the input of the demodulator. The lowpass filters shown in the figure can be implemented as matched filters for better performance. Figure 4.92 shows a plot of the amount of degradation in performance for a GMSK signal for different values of BT. According to the plot, we need to increase E_b/N_0 by only 1 dB over that of coherently demodulated MSK in order to maintain the same P_E performance when BT = 0.215. This is surprising in view of how bad the eye diagrams look when BT is reduced.

The reason for this result is that when we coherently demodulate the GMSK signal, we are really looking at the *phase* of the signal over two or more bit periods. Figure 4.93 shows eye diagrams for both the I and Q portions of a coherently demodulated GMSK signal for BT = 0.5. As the figure shows, the eye diagrams are relatively open. If we reduce BT to 0.25, we get the eye diagrams shown in Figure 4.94. Although there is some closing of the eye, the waveforms are remarkably good—especially compared to the eye diagram shown for noncoherent demodulation. This

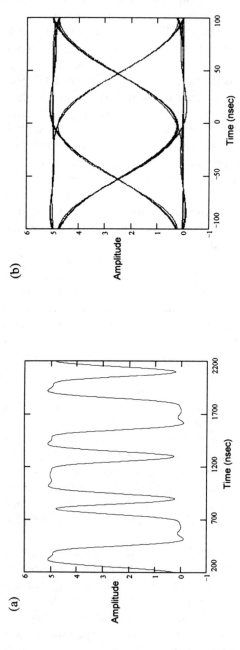

Figure 4.89 GMSK demodulated waveforms (noncoherent): (a) recovered bit stream (BT = 0.5), (b) eye diagram (BT = 0.5).

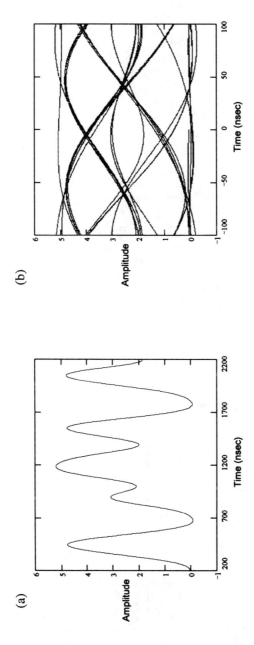

Figure 4.90 GMSK demodulated waveforms (noncoherent): (a) recovered bit stream (BT = 0.25), (b) eye diagram (BT = 0.25).

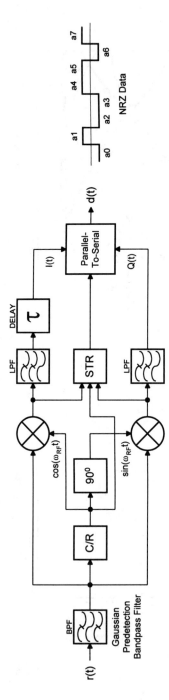

Figure 4.91 Coherent GMSK demodulator.

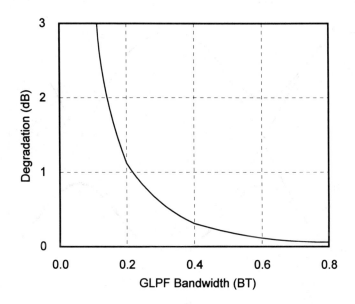

Figure 4.92 Theoretical E_b/N_0 degradation of GMSK as a function of BT compared to ideal coherent demodulation.

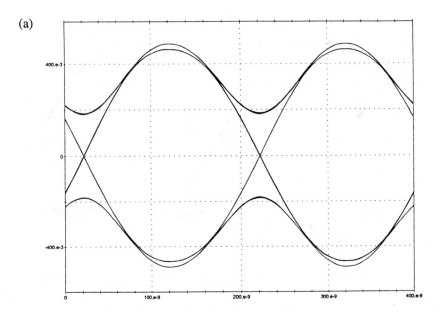

Figure 4.93 Eye diagrams for coherently demodulated GMSK (BT = 0.5): (a) *I*-channel, (b) *Q*-channel.

(b)

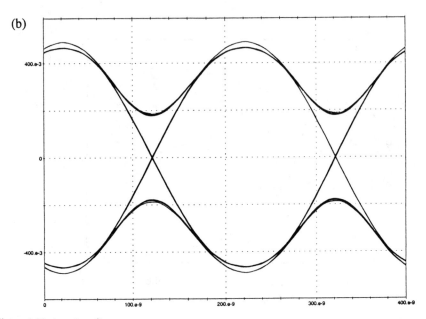

Figure 4.93 (continued)

(a)

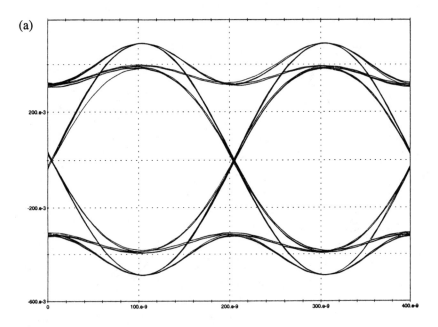

Figure 4.94 Eye diagrams for coherently demodulated GMSK (BT = 0.25): (a) *I*-channel, (b) *Q*-channel.

(b)

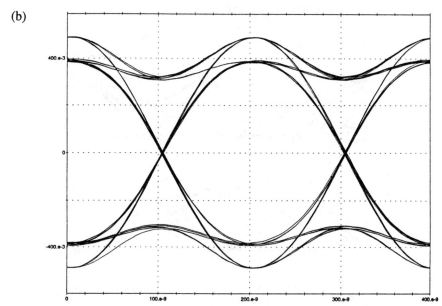

Figure 4.94 (continued)

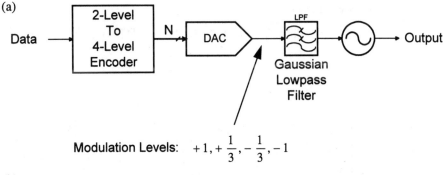

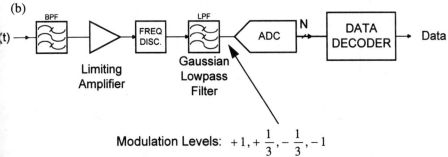

Figure 4.95 4-level GMSK: (a) modulator, (b) demodulator.

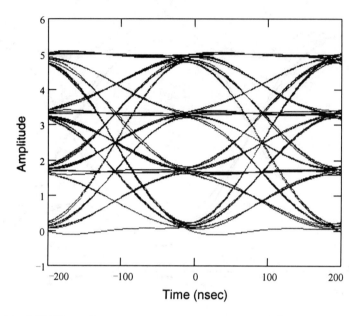

Figure 4.96 4-level GMSK eye diagram.

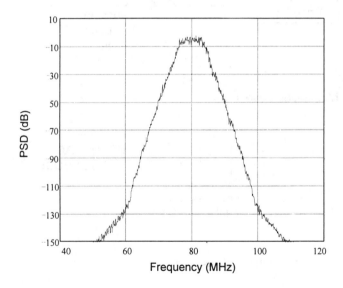

Figure 4.97 4-level GMSK power spectral density (h = 0.5, BT = 0.5).

is the reason the plot in Figure 4.92 shows little degradation over coherent MSK, even for low values of BT.

We can extend GMSK by using multiple frequencies to represent groups of bits in much the same way we did with FSK. Figure 4.95 shows a block diagram for a 4-level GMSK modulator and a 4-level GMSK demodulator. These circuits are identical to those used in FSK, with the exception of the lowpass filters following the DACs. An eye diagram for a 4-level GMSK signal is shown in Figure 4.96 with a BT = 0.5. Note that the relative filter bandwidth is in terms of the symbol rate and not the bit rate. Notice that even with four levels, there is a relatively small amount of ISI at the sampling instant $(t = 0)$.

The transmitted power spectral density for a 4-level GMSK signal with a carrier frequency of 80 MHz and a bit rate of 10 Mbps is shown in Figure 4.97. Again, the action of the lowpass filter caused the transmitted spectrum to fall off very rapidly.

4.4.2 $\pi/4$-Shifted Differentially Encoded PSK ($\pi/4$-DQPSK)

Variations of $\pi/4$-shifted differentially encoded PSK ($\pi/4$-DQPSK) have been around for quite some time. Baker [16] first applied this technique to 2,400 bps wired telephone modems. More recently, however, $\pi/4$-DQPSK has received attention as the U.S. (North American dual-mode cellular, NADC) and Japanese digital cellular (JDC) cellular telephone standards [17–21]. This modulation allows for a threefold increase in channel capacity over the existing analog FM cellular systems. Receivers can employ either coherent demodulators, differential demodulators, or discriminators (followed by integrate-and-dump filters). The ability to noncoherently demodulate the $\pi/4$-DQPSK signal results in cost and complexity reductions. Discriminator detection can also be used to demodulate the existing analog FM signals, allowing dual-mode operation.

$\pi/4$-DQPSK is a modulation that is a cross between QPSK and 8-PSK. Like QPSK, $\pi/4$-DQPSK separates the incoming bit stream into groups of two-bit symbols. In QPSK, each symbol would correspond to one of four different phase angles: +45 deg, −45 deg, +135 deg, or −135 deg. In $\pi/4$-DQPSK, however, every other symbol is part of a second QPSK constellation, which is shifted (or rotated) from the original QPSK constellation by $\pi/4$ (or 45 deg) with each symbol lying on either the I or Q axis. The inclusion of this second, $\pi/4$-shifted QPSK constellation means that symbols are represented by one of eight equally spaced phase angles, as in

8-PSK. However, unlike 8-PSK, only selected phase angles are available for the next symbol, depending on the phase angle of the present symbol.

Figure 4.98 illustrates the action of $\pi/4$-DQPSK modulation. Let the present symbol correspond to +45 deg (i.e., one of the white dots). In $\pi/4$-DQPSK, the only allowable symbols for the next symbol are *on* the I and Q axes (i.e., one of the black dots). This means that the phase of the next symbol can be either +90 deg, 180 deg, −90 deg, or 0 deg.

Continuing on, let's assume that we do in fact jump to +90 deg from our original symbol, which was at +45 deg. This means that the next symbol (the one after our *now* present value of +90 deg) must lie *off* the I and Q axes. In short, if the present symbol corresponds to one of the white dots in Figure 4.98 (i.e., *off* the I and Q axes), then the next symbol must be one of the black dots in the figure (i.e., *on* the I and Q axes). If the present symbol corresponds to one of the black dots, then the next symbol must correspond to one of the white dots. Even when long strings of 0s or 1s are encountered in the input data stream, a phase transition will always occur in the transmitted signal. This feature makes it easier to generate symbol timing in the receiver.

Figure 4.99 shows the phase-modulated carrier versus I and Q data inputs. No premodulation filters were used in order to show the relative phase changes of the carrier. However, as we will discuss later, special premodulation filters are generally used to bandlimit the transmitted signal.

The maximum phase transition in $\pi/4$-DQPSK is limited to 135 deg (in either direction). Recall that QPSK has maximum phase transitions of 180 deg, while the maximum phase transition of OQPSK is 90 deg (both, again, can be in either direction). In effect, $\pi/4$-DQPSK is a compromise between QPSK and OQPSK. By limiting phase transitions to 135 deg, amplitude variation is reduced compared to QPSK, yet is still larger than the 90-deg transitions of OQPSK. Limiting the maximum phase transition per symbol allows the use of amplifiers with lower output back-off (OBO) without experiencing spectral regrowth. In reality, even though $\pi/4$-DQPSK has less amplitude variation than QPSK, the amount of OBO over QPSK is negligible when reasonably linear amplifiers are used.

The "D" in $\pi/4$-DQPSK comes from the fact that the next symbol phase angle is derived by adding a *differential* phase angle to the present symbol's phase. These differential phase angles are transmitted, rather than absolute phase angles. Using differential encoding in the transmitter allows the use of simpler differential demodulators in the receiver. It has been shown [21] that differential demodulators are more robust in high-Doppler, Rayleigh-faded cellular environments. Table 4.1 shows a mapping of the differential phase angle, $\Delta\phi_k$, which is added to the present phase angle, as a function of A_k and B_k (the parallel symbol bits) for the next symbol. For example, given a phase angle of $\pi/4$ (45 deg) for the present symbol, suppose the next set of bits is $A_k = 1$ and $B_k = 1$. This means $-3\pi/4$ (−135 deg) will be added to the present symbol's phase, resulting in a new symbol having a phase angle of $-\pi/2$ (−90 deg). Note that the values of A_k and B_k result in gray-coded differential phase angles. Again, suppose $A_k = 1$ and $B_k = 1$ and our present symbol is $\pi/4$ (45 deg). This means the next symbol we transmit will have a phase angle of $-\pi/2$ (−90 deg).

As the signal passes through the noisy channel, it is possible that the recovered symbol is estimated to be either 0, $+\pi/2$ (+90 deg), or $+\pi$ (180 deg)—our only choices. The Euclidean distance on the I/Q phase plane between $-\pi/2$ and 0 (or π) is less than the distance between $-\pi/2$ and $+\pi/2$. With Gaussian white noise, should an error occur, it is much more likely that either 0 or π are chosen by the demodulator as the estimate of received phase angle. In other words, it is more likely the resultant vector lands in an adjacent quadrant than in the opposite quadrant when additive noise causes an error. If an error does occur, A_k, B_k will most likely be estimated to be either 1, 0 or 0, 1. Remember, if there were no error, the received values of A_k, B_k would be 1, 1. Therefore, the most likely symbol errors result in only one bit being in error.

Figure 4.100 shows a block diagram for a $\pi/4$-DQPSK modulator. As the figure shows, the modulator is very similar to that used for QPSK or 8-PSK, where analog in-phase and quadrature signals are used to drive a quadrature modulator. However, in the $\pi/4$-DQPSK modulator, the in-phase (I) and quadrature (Q) data streams are first encoded. The values of I_k and Q_k driving the inputs of the quadrature modulator are derived from the present symbol phase angle, ϕ_k, and the differential phase angle, $\Delta\phi_k$ (which is a function of A_k and B_k), by the following relations:

$$I_k = \cos[\phi_k] = \cos[\phi_{k-1} + \Delta\phi_k] \tag{4.84}$$

$$Q_k = \sin[\phi_k] = \sin[\phi_{k-1} + \Delta\phi_k] \tag{4.85}$$

Alternatively, the present values of I_k and Q_k can be written as functions of the previous values of I_k, Q_k (i.e., I_{k-1}, Q_{k-1}) and the differential phase angle, $\Delta\phi_k$, as

$$I_k = I_{k-1}\cos[\Delta\phi_k] - Q_{k-1}\sin[\Delta\phi_k] \tag{4.86}$$

$$I_k = I_{k-1}\sin[\Delta\phi_k] + Q_{k-1}\cos[\Delta\phi_k] \tag{4.87}$$

Notice that a pair of premodulation lowpass filters are used to bandlimit analog inputs I_k and Q_k, before being fed to the quadrature modulator. In addition to specifying the use of differential phase angles, the NADC and JDC standards specify the use of special premodulation filters in the transmitter and receiver. Recall that raised cosine (Nyquist) filters have very sharp roll-off in the frequency domain (approaching ideal, brick-wall response) and exhibit no ISI at sampling intervals, making them ideal for PSK premodulation filters.

A variation of the raised cosine filter is used in $\pi/4$-DQPSK. The raised cosine filter is separated into two elements whose aggregate response, when connected in series, results in the raised cosine filter response. One filter is used in the transmitter and the other filter is used in the receiver. The demodulated signal will then be free of ISI, after passing through both the transmit and receive filters since the series cascade of the two filters has a raised cosine response. If we let the filter response be $S(\omega)$, then we can write

$$|H_{RC}(\omega)| = |S(\omega)| \cdot |S(\omega)| = |S(\omega)|^2 \tag{4.88}$$

where $H_{RC}(\omega)$ is the response of the raised cosine filter. We can then write the frequency response of each of the filters as

$$|S(\omega)| = \sqrt{|H_{RC}(\omega)|} = |S_{SRRC}(\omega)| \tag{4.89}$$

This type of filter is called a square-root raised cosine (SRRC) filter, whose frequency response can be written as the square-root of the raised cosine filter, or [18]

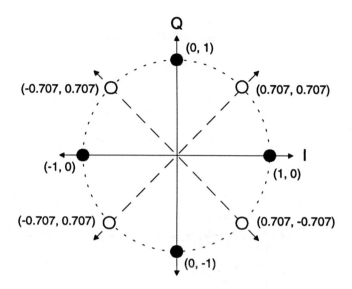

Figure 4.98 π/4-DQPSK phase domain.

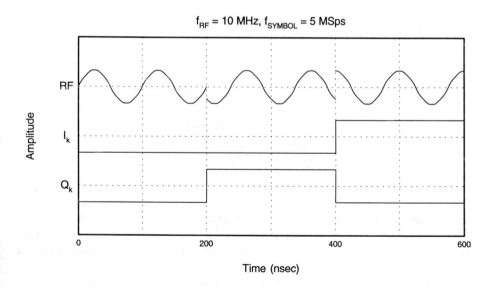

Figure 4.99 π/4-DQPSK time domain waveforms.

Table 4.1
Differential Phase Angle, $\Delta\phi_k$, Versus Symbol Bits A_k (In-Phase) and B_k (Quadrature)

A_k	B_k	$\Delta\phi_k$
0	0	$+\pi/4$ (+45 deg)
0	1	$3\pi/4$ (135 deg)
1	0	$-\pi/4$ (−45 deg)
1	1	$-3\pi/4$ (−135 deg)

$$
\begin{aligned}
|S_{\text{SRRC}}(\omega)| &= T; \quad 0 \le |f| \le \frac{(1-r)}{2T} \\[2mm]
&= T\sqrt{\frac{1}{2}\left\{\sin\left[\pi\frac{T}{r}\left(|f|-\frac{1}{2T}\right)\right]\right\}}; \quad \frac{(1-r)}{2T} \le |f| \le \frac{(1+r)}{2T} \qquad (4.90) \\[2mm]
&= 0; \quad |f| > \frac{(1+r)}{2T}
\end{aligned}
$$

where T is the symbol period and r is the roll-off factor. In terms of its impulse response, the SRRC filter can be expressed as

$$
\begin{aligned}
S_{\text{SRRC}}(t) &= 1 - r + 4\frac{r}{\pi}; \quad t = 0 \\[2mm]
&= \frac{r}{\sqrt{2}}\left[\left(1+\frac{2}{\pi}\right)\sin\left(\frac{\pi}{4r}\right) + \left(1-\frac{2}{\pi}\right)\cos\left(\frac{\pi}{4r}\right)\right]; \quad t = \pm\frac{T}{4r} \qquad (4.91) \\[2mm]
&= \frac{\sin\left[\pi(1-r)\dfrac{t}{T}\right] + 4r\dfrac{t}{T}\cos\left[\pi(1+r)\dfrac{t}{T}\right]}{\pi\dfrac{t}{T}\left[1-\left(4r\dfrac{t}{T}\right)^2\right]}; \quad \text{for all other } t
\end{aligned}
$$

An advantage to splitting the raised cosine filter into two elements is that the SRRC filter in the receiver now becomes a matched filter. A matched filter is a filter whose response is such that the desired signal passes without attenuation, while all other signals (such as noise and interference) are rejected. Matched-filter-based receivers generally have lower P_Es than other receivers, for the same level of channel noise.

One way to demodulate the $\pi/4$-DQPSK signal is to use a standard coherent I/Q quadrature demodulator as shown in Figure 4.101. However, the analog I and Q signals will alternate between two-level and three-level eye patterns, as shown in the figure. If the received symbol lies on either the I or Q axes, the values of I and

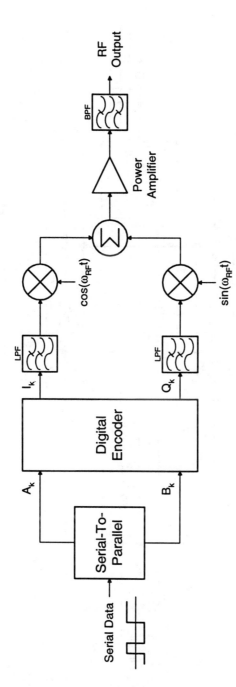

Figure 4.100 $\pi/4$-DQPSK modulator.

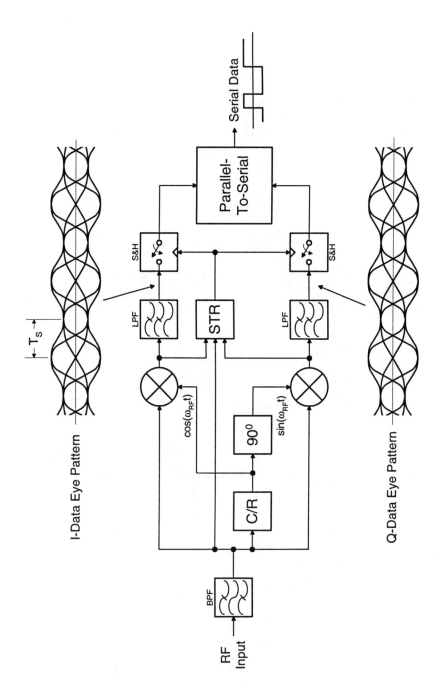

Figure 4.101 $\pi/4$-DQPSK coherent demodulator.

Q can be 0, +1 or −1 (three-level eye pattern). If the received symbol lies off the I and Q axes, the values of I and Q can be ±0.707 (two-level eye pattern). A disadvantage of direct coherent demodulation is that the resulting carrier recovery (C/R) and symbol timing recovery (STR) circuitry becomes rather complex.

Figure 4.102 shows a block diagram of a much simpler differentially coherent $\pi/4$-DQPSK demodulator. Notice that no carrier recovery circuit is needed in this implementation. An analog delay is used to produce a version of the received signal, delayed by one bit period. This delayed "reference" is used as the in-phase double-balanced mixer carrier input, while a $\pi/2$ shifted version is used as the quadrature mixer carrier input. The use of an analog delay greatly reduces the complexity of the demodulator. Lowpass filters are used to filter out high-frequency signals resulting from the mixing operation. An advantage of the differentially coherent demodulator is that the analog I and Q outputs are standard two-level eye patterns, as shown in the figure. Parallel bit streams can be easily obtained from the resulting two-level I and Q outputs via a pair of bit-slicing circuits (analog comparators whose decision level is adjusted to be in the center of the eye). A parallel-to-serial converter (multiplexer) is then used to interleave the I and Q data back into the original bit stream. Although the P_E of the differentially-coherent receiver is worse than the fully coherent demodulator, it is more robust in high-Doppler, fast-fading channels.

A major disadvantage of $\pi/4$-DQPSK is the need for linear amplifiers in the transmit chain. Most power amplifiers exhibit some degree of nonlinearity. In general, maximum efficiency is obtained from power amplifiers when the output signal is close to the maximum power the amplifier can produce (i.e., when the amplifier is at or near saturation). Unfortunately, the closer the output gets to the saturation point of the amplifier, the more distorted the signal becomes.

One way to overcome this problem is to use complex pre or postdistortion circuitry, which cancels out most of the distortion products of the amplifier, even when the power amplifier is run close to saturation. Figure 4.103 shows a block diagram of the predistortion circuitry used by Stapleton et al. [22], which is based on a slowly adapting feedback system. The "clean" $\pi/4$-DQPSK modulated input signal, $y_m(t)$, is applied to a quadrature modulator whose inputs are derived from $y_m(t)$ and a pair of correction signals. The complex modulator produces a distorted version of $y_m(t)$ and feeds this predistorted signal, $v_d(t)$, to the nonlinear power amplifier's input. The distorted transmitted output is applied to another mixer/bandpass filter combination, which extracts the out-of-band component. The action of the feedback loop produces a version of $v_d(t)$ that contains distortion products that minimize out-of-band spectral components out of the power amplifier.

Figure 4.104 shows a plot of the transmitted output spectrum with and without predistortion circuitry in the transmitter. As the figure shows, spectral regrowth can be kept below −60 dBc at the output when distortion canceling circuitry is used. Unfortunately, this solution results in higher complexity and, ultimately, higher cost.

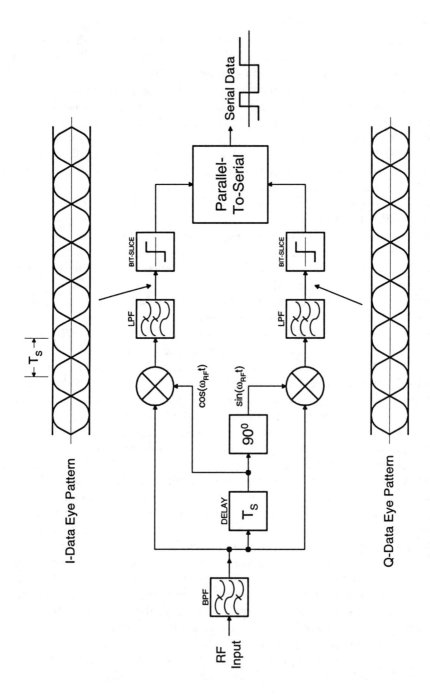

Figure 4.102 $\pi/4$-DQPSK differentially coherent demodulator using analog delay.

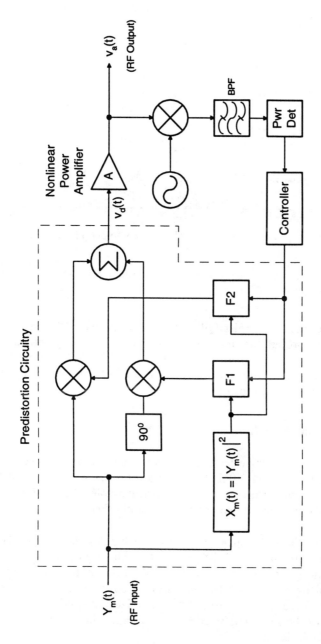

Figure 4.103 Predistortion circuitry to prevent spectral regrowth at power amplifier output for $\pi/4$-DQPSK [22].

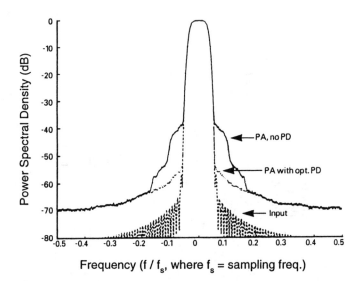

Figure 4.104 π/4-DQPSK transmit power spectral density with and without predistortion circuitry. (*Source:* Stapleton [22].)

As expected, the P_E of π/4-DQPSK is worse than that of QPSK. Figure 4.105 shows a plot of the probability of bit error versus E_b/N_0 for coherently demodulated π/4-DQPSK. For a probability of bit error of 10^{-4}, π/4-DQPSK is worse than QPSK by 0.7 dB, which is generally considered a small price to pay compared to the advantages π/4-DQPSK has to offer.

4.4.3 Feher's QPSK (FQPSK)

GMSK possesses the attractive features of a very narrow transmitted output spectrum, little ISI (at least for BT = 0.5), and a constant envelope. Other digital modulations possess these same features as well. One such example *is* Feher's QPSK (FQPSK). FQPSK was developed (and patented) by Dr. Kamilo Feher, now at the University of California, Davis.

Figure 4.106 shows a block diagram of an FQPSK modulator. The FQPSK modulator is very similar to an MSK modulator. The input data stream is demultiplexed into I and Q waveforms, each now at half the original data rate, with the Q signal delayed by a single bit period. Each signal is then sent to special pulse-shaping filters, which are the key to FQPSK modulation. After quadrature modulation, the output of the summing junction is fed to a hardlimiting circuit to produce a constant envelope signal.

The FQPSK pulse-shaping filters have the following impulse response

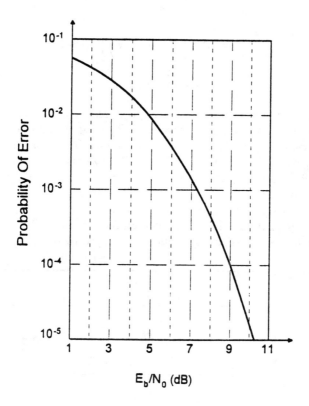

Figure 4.105 $\pi/4$-DQPSK probability of bit error.

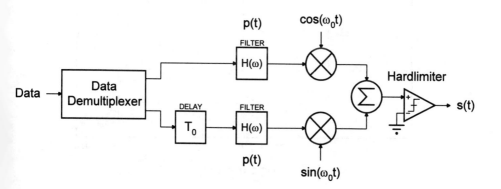

Figure 4.106 FQPSK modulator.

$$p(t) = 0.5 \cdot \left[1 + \cos\left(\frac{\pi t}{T_b}\right) \right]; \ |t| \le T_b \tag{4.92}$$

$$p(t) = 0; \ \text{elsewhere} \tag{4.93}$$

As (4.92) and (4.93) show, the filters produce a cosine shaped pulse response on each data bit. This response is shown in Figure 4.107. Notice that the pulse extends over two bit periods. Figure 4.108 shows a block diagram of the pulse-shaping filter, which is implemented as an FIR lowpass filter having tap weights given by (4.92) and (4.93) above. Since the input to each of the filters is only a single bit, the complexity of the filters is greatly reduced.

Figure 4.109 shows the time domain response for the FQPSK modulator. The upper two traces in the figure are the demultiplexed *I* and *Q* data streams after passing through the pulse-shaping filters (and delayed by one bit period for the *Q*

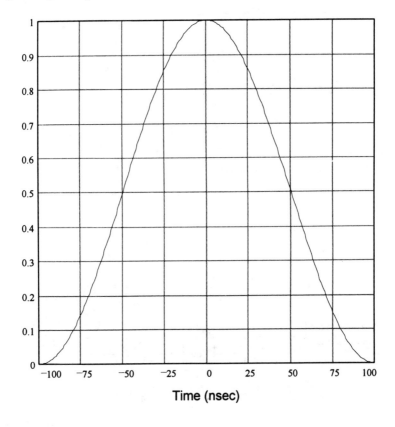

Figure 4.107 FQPSK premodulation filter impulse response (for 10-Mbps data).

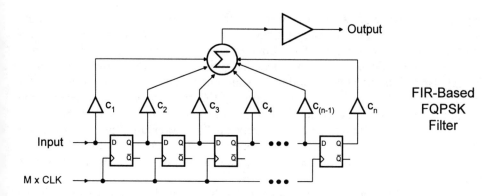

Figure 4.108 FIR-based FQPSK premodulation filter.

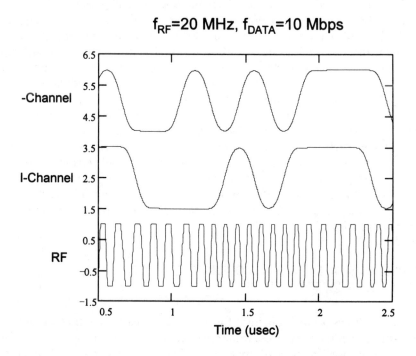

Figure 4.109 FQPSK time domain waveforms.

signal). The lower trace shows the output of the hardlimiter. It is interesting to note that the output of the summing junction is not constant envelope. Therefore, a hardlimiting circuit is required in order to produce a constant envelope signal. However, unlike QPSK, which experiences significant spectral regrowth when non-linearly amplified, the FQPSK modulator summing junction output does not exhibit spectral regrowth. This is due to the action of the pulse-shaping filters feeding the quadrature modulator. A virtue of having a constant envelope signal at the output of the modulator is that we need not rely on the power amplifier in saturation (compression) to produce a constant envelope signal. This avoids AM-to-PM modulation errors, which can be caused by the power amplifier's nonlinear phase versus input signal-level response.

Figure 4.110 shows eye diagrams for both the I and Q data streams. Notice that not only is there no ISI at the sampling instants, but the zero crossings are jitter-free. The absence of ISI at the sampling instants greatly improves the P_E performance of the received signal. The absence of jitter on the signal causes less uncertainty in the sampling instants and also enhances P_E performance.

Figure 4.111 shows a plot of the power spectral density of FQPSK versus that of GMSK, with BT = 0.5. Notice that FQPSK falls off very quickly in the first 20 to 30 dB of the response. In addition, the main lobe of the FQPSK signal is narrower than that of the GMSK signal. Below the −30 dB point, GMSK falls off more rapidly than FQPSK. FQPSK is a very attractive modulation to consider in applications such as satellite transponders. The rapid roll-off in the first 20 to 30 dB, or so, helps to reduce ACI when all channels are received at approximately the same power level.

Figure 4.112 shows a block diagram of the FQPSK demodulator, which is identical to that of the MSK demodulator. Figure 4.113 shows a plot of probability of bit error versus E_b/N_0 for coherently demodulated FQPSK. The performance of FQPSK is only slightly worse than that of BPSK or QPSK.

FQPSK is a true phase modulation. If a string of 0s or 1s are input to the modulator, the resulting output is a steady-state *phase-shifted* version of the carrier. MSK or GMSK, in contrast, produce steady-state *frequency* differences for long strings of 0s or 1s on the input of the modulator, and can be noncoherently demodulated with simple circuitry. Depending on the particular application, this may or may not be important.

4.5 EXAMPLES OF INTEGRATED CIRCUIT DIGITAL COMMUNICATIONS CIRCUITS

Figure 4.114 shows a block diagram for a typical wireless digital communications system—the GEC/Plessey wireless local area network (WLAN) transceiver [23–25]. This transceiver uses frequency-hopped spread spectrum techniques. It operates in the 2.4-GHz industrial, scientific, and medical (ISM) band at a data rate of 625

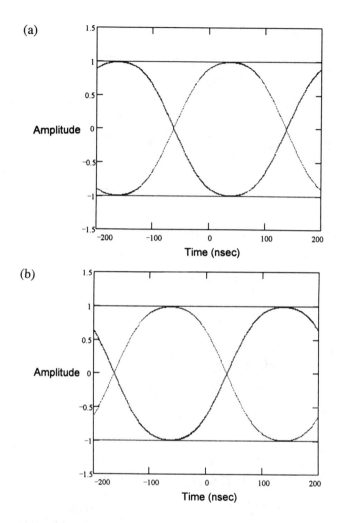

Figure 4.110 Coherently demodulated FQPSK waveforms: (a) *I*-channel, (b) *Q*-channel.

Kbps. The transceiver employs GFSK modulation. This transceiver was mentioned earlier in Chapter 3.

The circuit is divided into six main ICs. The first IC contains a low-noise amplifier (LNA), downconverter, power amplifier, upconverting mixer, RF switches, and a 2-GHz oscillator. This IC is fabricated in a Gallium Arsenide (GaAs) MESFET process and performs most of the RF front-end functions. Notice that this chip has a large number of RF I/O lines coming in and out of the chip, which can cause unwanted coupling and crosstalk problems if care is not taken in circuit partitioning.

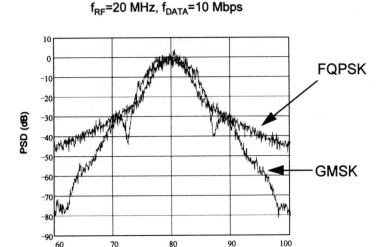

Figure 4.111 FQPSK versus GMSK power spectral density.

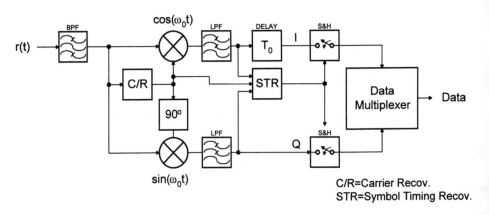

Figure 4.112 FQPSK demodulator.

GaAs is a good choice for this chip as it offers superior noise performance (i.e., low noise figure), as well as high-efficiency power amplifiers and low-loss switches.

After the first downconversion, the signal is passed to the IF and baseband processing circuits. Two additional ICs are fabricated using a high-speed silicon IC process with an f_T of 13 GHz. The first IC contains the second downconverting mixer, IF amplifiers, and a 312-MHz oscillator. This includes a received signal

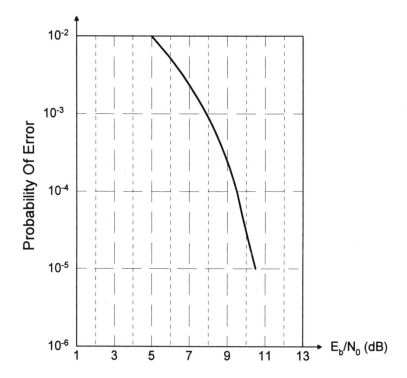

Figure 4.113 FQPSK probability of bit error.

strength indicator (RSSI) for measuring the level of the received signal as well as a quadrature detector to demodulate the GFSK signal. A second silicon bipolar IC contains another oscillator at 700 MHz and a number of high-speed digital dividers and prescalers that interface the synthesizer chips. Silicon bipolar is a good technology to use in the IF and baseband portion of the radio because of its high level of functionality and high yield. The use of a high-speed process greatly reduces the amount of power required for each function.

A number of PLLs are also shown in Figure 4.114. These circuits are fabricated in CMOS. The use of high-speed dividers in the bipolar chips allows the CMOS PLLs to run at much lower speeds, thus conserving power.

Prior to integration, the transceiver required 40 ICs, 200 passive elements and measured 7 inches by 5 inches. After integration (i.e., the system just described), the transceiver requires only 4 ICs and 50 passive elements and measures only 3 inches by 2 inches.

Another example of a digital wireless communication system is shown in Figure 4.115. This is the National Semiconductor DECT transceiver [26, 27]. The

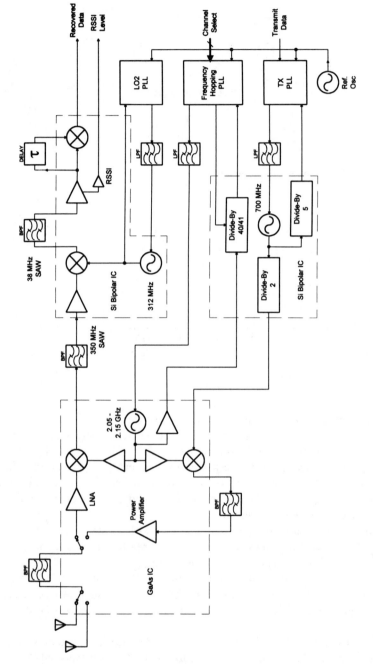

Figure 4.114 GEC Plessey WLAN transceiver.

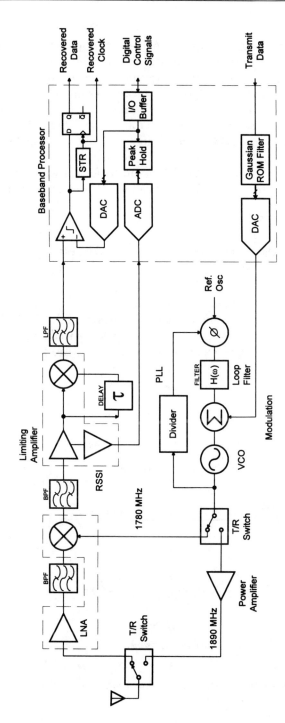

Figure 4.115 National Semiconductor DECT transceiver.

transceiver operates at a carrier frequency of 1.88 to 1.90 GHz with a data rate of 1.152 MHz. It is interesting to note that with the exception of the power amplifier and transmit/receive (T/R) switch, the entire transceiver is fabricated in National Semiconductor's BiCMOS silicon bipolar IC process. Both the power amplifier and T/R switch are fabricated in a GaAs MESFET process.

The downconverter has a relatively high total noise figure of 8.7 dB. However, this is sufficient to meet the DECT specifications. DECT uses GMSK modulation with a relative filter bandwidth, BT, of 0.5. The Gaussian data filter is done on-chip using a read-only memory (ROM) look-up table that drives a DAC, whose output is then filtered with a simple RC lowpass filter. The entire transceiver operates on only 3V.

4.5.1 Design of Modulators and Demodulators for Wireless Communications Systems

Let's turn our attention now to how some of the modulators and demodulators are actually designed in wireless communication systems. Recall that in 8-PSK we represent groups of three bits by eight different phase angles spaced uniformly about the unit circle on 45-deg increments. Figure 4.116 shows how we can use unit increments

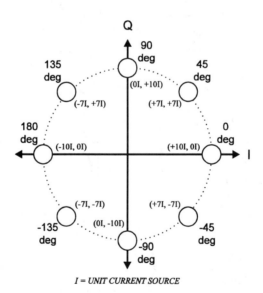

Figure 4.116 8-PSK phase domain showing symbols based on unit-weight currents (unit-weight currents for I, Q shown in parentheses).

of current or voltage to obtain our desired phase increments and yet still lie on the unit circle. Recall from our discussions on quadrature modulation

$$\theta = \arctan\left[\frac{Q}{I}\right] \tag{4.94}$$

$$A = \sqrt{I^2 + Q^2} \tag{4.95}$$

For example, let $I = 7$ and $Q = 7$. Then $q = 45$ deg and $A = 9.899$. If we let $I = 0$ and $Q = 10$, then $q = 90$ deg and $A = 10$. So we see that by using unit increments of current or voltage to derive the I and Q signals, we can get exactly the phase with less than 0.1-dB amplitude error! Furthermore, by using unit increments we can ensure this relation is maintained over process and temperature variations. Symbols that fall on the negative side to either the I or Q axis can be derived by simply inverting the polarity of the I and Q signals.

Figure 4.117 shows a block diagram of the 8-PSK modulator. The NRZ data and clock are applied to the data encoder portion of the chip, which produces the appropriate differential driver signals for the modulator. The modulator is driven

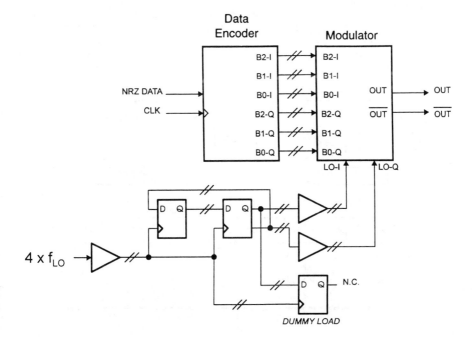

Figure 4.117 Data encoder and digital I/Q generator for 8-PSK IC modulator.

by a digital quadrature generator. Notice that all digital lines are differential in nature. This reduces the amount of digital noise radiated and received within the chip. The quadrature generator uses a $4 \times$ frequency clock. Notice the dummy gates within the digital quadrature generator. These dummy gates ensure that each of the critical nodes is equally loaded to ensure process and temperature invariant operation.

Figure 4.118 shows a schematic of the quadrature modulator itself. Notice the symmetry of the circuit. This helps to ensure tight balance between the I and Q signals. The I and Q signals are derived from identical, switched, unit current sources. A common bias line (VREF) ensures that all current sources are equally biased. Matching between each unit current source can be further improved by using a

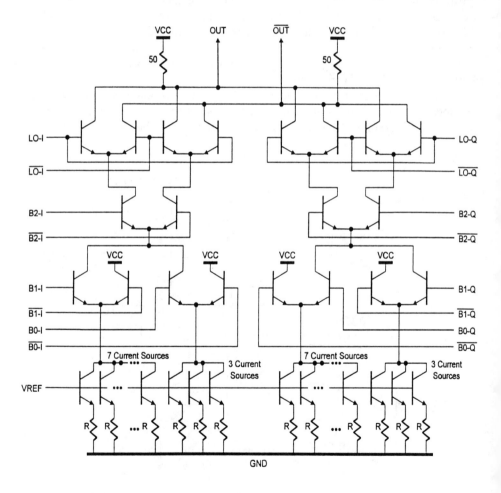

Figure 4.118 Schematic for monolithic 8-PSK modulator.

relatively large voltage drop across each emitter resistor. This helps to cancel out any V_{BE} offset voltage variations. In order to ensure high-speed operation, currents are simply steered from one portion of the circuit to another. All I/O lines are differential in order to improve noise margin and reduce crosstalk. Digital signal swings are kept to 250 mV to reduce slew limiting and unwanted coupling. Current sources are switched in units of 0, 7, or 10. Negative values of I and Q are produced by simply steering the current to the other side of the circuit. The LO input need only be large enough to fully switch the upper differential pairs. Usually, 100 to 200 mV is sufficient. The summing junction of the quadrature modulator is implemented by simple current summations at the collectors of the LO switching pairs.

Figure 4.119 shows a layout of the modulator. All critical differential pairs are quad-ed to reduce thermal and process gradients. All resistors are made from the same unit size and weight values to ensure matching. The overall layout is completely symmetric. The current sources which are tied together in groups of 7 or 3 (for 3 + 7 = 10) and are interspersed with one another for greater uniformity.

This concept can be taken one step further. The same unit current/voltage approach can be used to design a 16-PSK modulator, as shown in Figure 4.120 [28]. Groups of 4 bits are represented by one of 16 different symbols spaced uniformly about the unit circle at 22.5-deg increments. As mentioned earlier, 16-PSK reduces the symbol rate of BPSK by 4, making the resulting spectrum only one-quarter as wide. Figure 4.121 shows a schematic of the modulator used in Hewlett Packard's [28] version of their 16-PSK modulator, which contains the DACs, I and Q mixers, and summing junction in one circuit. As in the case of the 8-PSK circuit, the LO signals are derived digitally (though not shown here). This circuit is fabricated in a high-speed silicon bipolar IC process with an f_T of 13 GHz, which is ideal for obtaining the combination of speed, matching, and level of complexity required to meet the design goals. Notice the similarity of this circuit compared to the 8-PSK modulator. Here, currents are grouped in to units of 5, 4, 3, 1, and 0. The phase error of this circuit is only 0.12 deg and the amplitude variation is only 0.18 dB. The modulator is capable of operating at 110 MSymbols/sec with an IF output frequency of 400 MHz.

A more general modulator is shown in Figure 4.122 [29, 30]. This circuit is fabricated in a 0.5-μm GaAs MESFET process. It is a general purpose quadrature modulator that can accept modulation inputs from dc to 500 MHz and LO inputs from 700 MHz to 3 GHz. In this circuit, the 0-deg and 90-deg versions of the LO are derived from the outputs of two passive RC filters. Active FET resistors are used to trim the phase to its final value. Variable amplitude drivers are used to match the I and Q LO amplitudes before they are applied to the Gilbert-cell mixers. The currents at the mixer outputs are summed together and applied across resistive loads. A differential to single-ended driver is used to provide a back-terminated 50-ohm output.

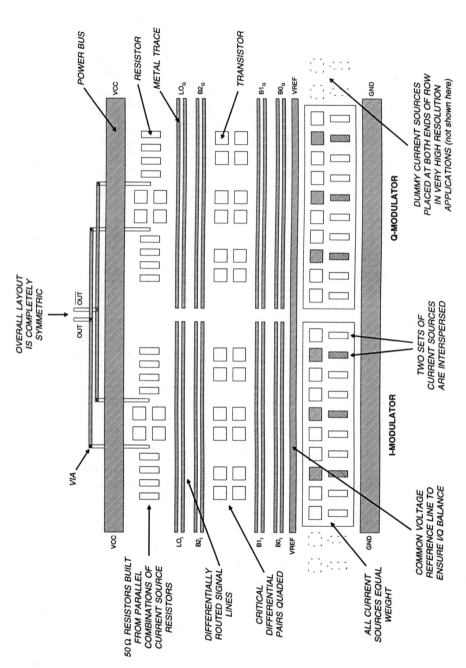

Figure 4.119 Layout for monolithic 8-PSK modulator.

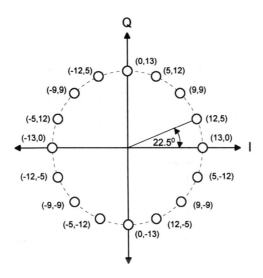

Figure 4.120 16-PSK phase domain showing symbols based on unit-weight currents (unit-weight currents for I, Q shown in parentheses).

4.6 SUMMARY

Modern digital wireless communication systems are requiring increasingly sophisticated modulation/demodulation circuits in order to meet customer and regulatory demands. Many different modulation techniques are available. The best choice is always determined by the constraints of the wireless system designer. We have examined both phase and frequency modulation schemes as well as a few of their derivatives. The use of integrated solutions is becoming more and more prevalent as the level of complexity of today's wireless systems increases and as the demand by the consumer for lower cost continues.

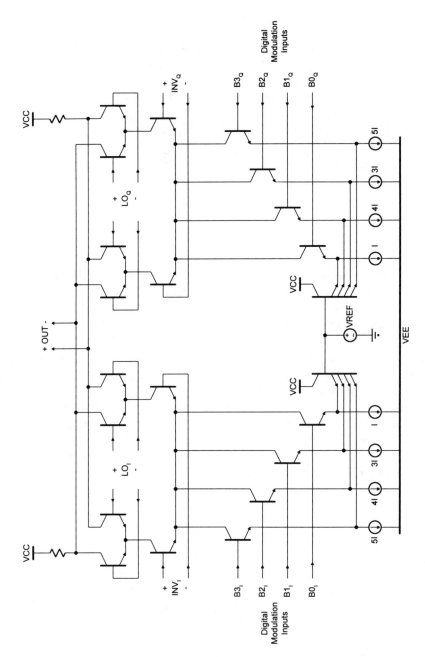

Figure 4.121 Schematic for monolithic 16-PSK modulator (Hewlett Packard).

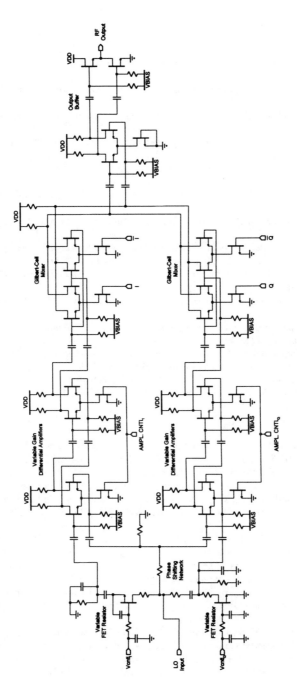

Figure 4.122 Schematic for GaAs I/Q direct IC modulator (Alcatel).

References

[1] Lathi, B. P., *Modern Digital and Analog Communication Systems*, Holt, Rinehart and Winston, New York, NY 1983.

[2] Feher, K., *Digital Communications; Satellite/Earth Station Engineering*, Prentice-Hall, Englewood Cliffs, NJ 1983.

[3] Feher, K., *Advanced Digital Communications; Systems and Signal Processing Techniques*, Prentice-Hall, Englewood Cliffs, NJ, 1987.

[4] Pasupathy, S., "Minimum Shift Keying: A Spectrally Efficient Modulation," *IEEE Communications Magazine*, July 1979, pp. 14–22.

[5] Samueli, H., "Digital Wireless Transceiver Architectures," ISSCC94 Short Course, San Francisco, CA, February 1994.

[6] Lindsey, W. C. and Simon, S. K., *Telecommunication Systems Engineering*, Prentice-Hall, Englewood Cliffs, NJ, 1988.

[7] Sklar, B., *Digital Communications: Fundamentals and Applications*, Prentice-Hall, Englewood Cliffs, NJ, 1988.

[8] Feher, K., *Wireless Digital Communications: Modulation and Spread Spectrum Applications*, Prentice-Hall, Englewood Cliffs, NJ, 1995.

[9] Wozencraft, J. M., and Jacobs, I. M., *Principles of Communication Engineering*, Wiley, New York, NY, 1965, Chapter 7.

[10] Nyquist, H., "Certain Topics in Telegraph Transmission Theory," *AIEE Trans.*, Vol. 47, April 1928, p. 817.

[11] Jeruchim, M. C., Balaban, P, and Shanmugan, K. S., *Simulation of Communication Systems*, Plenum Press, New York, NY 1992.

[12] Zverev, A. I., *Handbook of Filter Synthesis*, John Wiley & Sons, New York, NY, 1967.

[13] Murota, K., et. al., "GMSK Modulation for Digital Radio Telephony," *IEEE Transactions On Communications*, Vol. COM-29, No. 7, July 1981, pp. 1044–1050.

[14] Socci, J., "GMSK As A Modulation Scheme For A Frequency Hopped PHY," *IEEE 802.11 Wireless Access Method and Physical Layer Specifications*, July 1993.

[15] Rohde and Schwartz, Inc., "Signal Generators For Modern Communication Techniques," August 1990.

[16] Baker, P. A., "Phase-Modulation Data Sets for Serial Transmission at 2000 and 2400 Bits Per Second," Part I, *AIEE Trans. on Communications Electronics*, July 1962, pp. 166–171.

[17] Guo, Y., and Feher, K., "Modem/Radio IC Architectures For ISM Band Wireless Applications," *IEEE Transactions On Consumer Electronics*, Vol. 39, No. 2, May 1993, pp. 100–106.

[18] Chennakeshu, S. and Saulnier, G. J., "Differential Detection Of $\pi/4$-Shifted-DQPSK For Digital Cellular Radio," *IEEE Transactions on Vehicular Technology*, Vol. 42, No. 1, February 1993, pp. 46–53.

[19] Feher, K., and Mehdi, H., "Modulation/Microwave Integrated Digital Wireless Development," *IEEE Transactions on Microwave Theory and Techniques*, Vol. 43, No. 7, July 1995.

[20] McGrath, F., et. al., "A 1.9-GHz GaAs Chip Set for the Personal Handyphone System," *IEEE Transactions on Microwave Theory and Technique*, Vol. 43, No. 7, July 1995.

[21] Goode, S. H. Kazecki, and D. W. Dennis, "A Comparison of Limiter-Discriminator, Delay, and Coherent Detection for $\pi/4$-QPSK," *Proc. IEEE Veh. Technol. Conf.*, Orlando, FL, May 1990, pp. 687–694.

[22] Stapleton, S. P., Kandola, G. S., and Caver, K., "Simulation and Analysis of an Adaptive Predistorter Utilizing a Complex Spectral Convolution," *IEEE VT*, Vol. 41, Nov. 1992, pp. 387–395.

[23] Williams, D. A., "A Frequency Hopping Microwave Radio System For Local Area Network Communications," 1993 IEEE MTT-S Symposium, Atlanta, GA, June 14–15, 1993.

[24] Williams, D. A., "An Integrated Microwave Radio Transceiver For WLAN Applications," 1993 Wireless Symposium, San Jose, CA.

[25] Devlin, L. M., et. al., "A 2.4 GHz Single Chip Transceiver," IEEE 1993 Microwave and Millimeter-Wave Monolithic Circuits Symposium, Atlanta, GA, June 14–15, 1993.

[26] Madsen, B. and Fague, D. E., "Radios For The Future: Designing For DECT," *RF Design*, April 1993, pp. 48–53.

[27] Fague, D. E., "A Fully Integrated Modulator/Demodulator For DECT," RF Expo West, San Jose, CA, March 17–19, 1993.

[28] Wuppermann, B., et al., "A 16-PSK Modulator With Phase Error Corrections," *ISSCC93*, San Francisco, CA, February 1993.

[29] Boveda, A., et. al., "A 07–3 GHz GaAs QPSK/QAM Direct Modulator," ISSCC93, San Francisco, CA, February 1993.

[30] Boveda, A., et. al., "A 07–3 GHz GaAs QPSK/QAM Direct Modulator," *IEEE Journal Of Solid State Circuits*, Vol. 28, No. 12, December 1993.

CHAPTER 5
▼▼▼

MIXERS FOR WIRELESS APPLICATIONS

Stephen A. Maas
Nonlinear Technologies, Inc.
Long Beach, CA

5.1 INTRODUCTION

In the past twenty years, mixer technology has evolved from a major research topic into one of great maturity. The venerable diode mixer is now well understood and diode mixers exhibiting excellent performance over multioctave bandwidths are commonly available at low cost. Although their designs are somewhat less mature, active and passive FET mixers have special advantages in minimizing the complexity of a system or in reducing noise or intermodulation distortion.

Wireless applications create special needs for mixers. Because most wireless systems are commercial, they are manufactured in large quantities and must have low cost. Many wireless components are used in handheld and battery-powered systems, where small size and low dc power are often essential. Electrical performance is usually secondary to cost, although frequently one or two performance characteristics are critical. As with military and aerospace components, thermal stability and mechanical ruggedness are usually also important, but in wireless systems must be achieved at much lower expense. The role of integrated circuits (ICs) is not as clear in wireless as in other systems; certainly, silicon ICs are a proven, low-cost technology, but the cost effectiveness of more advanced technologies such as GaAs and heterojunction monolithics is somewhat less certain.

Most wireless applications are below 2.5 GHz, so the emphasis of this chapter will be largely in this frequency range. Wireless applications in the higher microwave frequencies are not uncommon, however, and we do not neglect them entirely. Low-frequency operation does not necessarily make life easier for the mixer designer; one of the most difficult problems posed by low-frequency operation is to make low-frequency components, especially mixer baluns, small enough for such applications as wireless telephones and handheld transceivers.

This chapter will cover all of the fundamental issues of mixer design, including preferred mixer topologies, choice of semiconductor devices, and passive component design for mixer implementation.

5.2 PROPERTIES OF MIXERS

Mixers have a variety of properties that will be discussed in this section.

5.2.1 Frequency Mixing

A mixer is fundamentally a multiplier. An ideal mixer multiplies a signal by a sinusoid, shifting it to both a higher and lower frequency, and selects one of the resulting sidebands. A modulated narrowband signal, usually called the *RF signal*,[1] represented by

$$S_{RF}(t) = a(t)\sin(\omega_s t) + b(t)\cos(\omega_s t) \qquad (5.1)$$

is multiplied by the function called the *local oscillator* (LO) signal

$$f_{LO}(t) = \cos(\omega_p t) \qquad (5.2)$$

to obtain the *IF signal*[2]

$$S_{IF}(t) = \frac{1}{2}a(t)(\sin((\omega_s + \omega_p)t) + \sin((\omega_s - \omega_p)t))$$

$$+ \frac{1}{2}b(t)(\cos((\omega_s + \omega_p)t) + \cos((\omega_s - \omega_p)t)) \qquad (5.3)$$

---

[1]RF was originally an abbreviation for radio frequency. Like many technical abbreviations and acronyms (e.g., radar, laser, FET), the term has evolved into a word in itself, and has lost most of its original meaning. Thus, we regularly use terms like RF frequency to distinguish from RF port, RF voltage, etc. Some of these terms may seem redundant or illogical to grammatical bluestockings who insist that RF have only its original meaning. For better or worse, language evolves; this is an example.

[2]IF originally meant intermediate frequency. The above footnote applies here as well.

In the ideal mixer, two sinusoidal IF components, called *mixing products*, result from each sinusoid in $s(t)$. In receivers, the difference-frequency component is usually desired, and the sum-frequency component is rejected by filters.

Unfortunately, real mixers do not follow this neat description. Even if the local oscillator voltage applied to the mixer's LO port is a clean sinusoid, the nonlinearities of the mixing device distort it, causing the LO function to have harmonics. Those nonlinearities can also distort the RF signal, resulting in RF harmonics. The IF is, in general, the combination of all possible mixing products of the RF and LO harmonics. Filters are usually used to select the appropriate response. The other so-called *spurious responses* must be eliminated, usually by filters.

Every mixer, even an ideal one, has a second RF frequency that can create a response at the IF. This is a type of spurious response, and is called the *image*; it occurs at the frequency $2f_{LO} - f_{RF}$. For example, if a mixer is designed to convert 10 GHz to 1 GHz with a 9-GHz LO, the mixer will also convert 8 GHz to 1 GHz at the same LO frequency. Although none of the types of mixers we shall examine inherently reject images, it is possible to create combinations of mixers and hybrids that do reject the image response.

It is important to note that the process of frequency shifting, which is the fundamental purpose of a mixer, is a linear phenomenon. Although nonlinear devices are invariably used for realizing mixers, there is nothing in the process of frequency shifting that requires a nonlinearity. Distortion and spurious responses other than the sum and difference frequency, which are often severe in mixers, are not fundamentally required by the frequency-shifting operation that a mixer performs.

5.2.2 Practical Characteristics of Mixers

Conversion Efficiency

Mixers using Schottky-barrier diodes are passive components and consequently exhibit conversion loss. This loss has a number of consequences: the greater the loss, the higher the noise of the system, and the more amplification that is needed. High loss contributes indirectly to distortion because of high signal levels that result from the additional preamplifier gain required to compensate for this loss. It also contributes to the cost of the system since the necessary low-noise amplifier stages are usually expensive.

Mixers using active devices often (but not always) exhibit conversion gain. High mixer gain is not necessarily desirable because it reduces stability margins and can increase distortion. Usually, a mixer gain of unity, or at most a few dB, is ideal.

Noise

In a passive mixer whose image response has been eliminated by filters, the noise figure is usually equal to, or only a few tenths of one decibel above, the conversion loss. In this sense, the mixer behaves as if it were an attenuator having a temperature equal to or slightly above the ambient.

In active mixers, the noise figure cannot be related easily to the conversion efficiency; in general, it cannot even be related qualitatively to the device's noise figure when used as an amplifier. The noise figure of an active FET mixer depends strongly on the characteristics of the design; these are discussed in Section 5.4.

Bandwidth

The bandwidth of a diode mixer is limited by the external circuit, especially by the hybrids or baluns used to couple the RF and LO signals to the diodes. In active mixers, bandwidth can be limited either by the device or by hybrids or matching circuits that constitute the external circuit; much the same factors are involved in establishing active mixers' bandwidths as amplifiers' bandwidths.

Distortion

It is a truism to say that everything is nonlinear to some degree and generates distortion. Unlike amplifiers or passive components, however, mixers often employ strongly nonlinear devices to provide mixing. Because of these strong nonlinearities, mixers generate relatively high levels of distortion. A mixer is usually the dominant distortion-generating component in a receiver.

Distortion in mixers, as with other components, is manifested as intermodulation distortion, (IM or IMD), which involves mixing between multiple RF tones and harmonics of those tones. If two RF excitations f_1 and f_2 are applied to a mixer, the nonlinearities in the mixer will generate a number of new frequencies, resulting in the IF spectrum shown in Figure 5.1. Figure 5.1 shows all intermodulation products up to third order; by nth order, we mean all n-fold combinations of the excitation

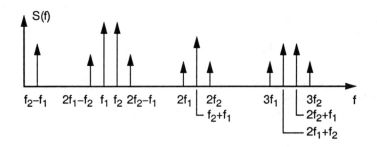

Figure 5.1 IF spectrum of intermodulation products up to third order. The frequencies f_1 and f_2 are the excitations.

tones (not including the LO frequency). In general, an nth order nonlinearity gives rise to nth (and lower) order distortion products.

An important property of intermodulation distortion is that the level of the nth-order IM product varies n decibels for every decibel change in the levels of the RF excitations. The extrapolated point at which the excitation and IM levels are equal is called the nth-order IM intercept point, abbreviated IP_n. This dependence is illustrated in Figure 5.2. In most components, the intercept point is defined as output power; in mixers it is traditionally input power.

Given the intercept point IP_n and input power level, the IM level P_{IM} can be found from

$$P_{IM} = nP_{lin} - (n - 1)IP_n \tag{5.4}$$

where P_{lin} is the level of each of the linear RF tones (which are assumed to be equal). By convention, P_{lin} and P_{IM} are the powers of a single frequency component, not the total power of all components. The intercept point can be defined in terms of either input or output power; usually, input power is used for mixers and output power for other types components.

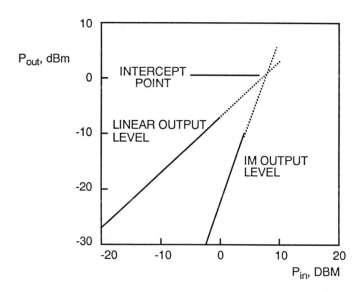

Figure 5.2 The output level of each nth-order IM product varies n dB for every dB change in input level. The intercept point is the extrapolated point at which the curves intersect; it can be specified in terms of either input or output power.

Spurious Responses

A mixer converts an RF signal to an IF signal. The most common transformation is

$$f_{IF} = f_{RF} - f_{LO} \tag{5.5}$$

although others are frequently used. Our discussion of frequency mixing indicated that harmonics of both the RF and LO could mix. The resulting set of frequencies is

$$f_{IF} = mf_{RF} + nf_{LO} \tag{5.6}$$

where *m* and *n* are integers. If an RF signal (which may or may not be in-band) creates an in-band IF response other than the desired one, it is called a *spurious response*. Usually the RF, IF, and LO frequency ranges are selected carefully to avoid spurious responses, and filters are used to reject out-of band RF signals that may cause in-band IF responses. IF filters are used to select only the desired response.

Many types of balanced mixers reject certain spurious responses where *m* or *n* is even. (It is possible to create a circuit that rejects responses where *m* or *n* are odd, but this usually rejects the desired response as well.) Most singly balanced mixers reject some, but not all, products where *m* or *n* is even; doubly balanced mixers reject all responses where *m* or *n* (or both) are even.

We designate a spurious response by its (*m, n*) indices, where *m* and *n* are by convention the RF and LO harmonics, respectively. If a mixer rejects a certain response having an even *m* or *n*, it will reject all other responses having the same *m* or *n* even. For example, if a mixer rejects the (2, 1) it will also theoretically reject the (4, 1), (4, 3), (6, 1), and so forth. For simplicity, in this chapter we will describe the (2, 1), (1, 2), or (2, 2) rejection of various types of mixers, and the reader should understand that all similar responses are theoretically rejected as well.

Other Rejection Phenomena

Balanced mixers reject AM noise from the LO, some even-order IM products, and provide port-to-port isolation. These phenomena depend on the cancellation of IF current or voltage components that are equal in magnitude and have a phase difference of 180 deg. Phase and magnitude errors are introduced by imbalance in the circuit, and these cause the cancellation to be imperfect.

In spurious response or IM cancellation, phase errors occurring at the RF or LO frequencies are multiplied by the harmonics involved in the spurious response. For example, if there is a 5-deg error in the phase balance of the RF hybrid, the

(6,1) spurious-response components will have a 30-deg error in the IF. For this reason, high-order spurious responses and IM products, although theoretically rejected, are not canceled by balanced mixers as effectively as low-order ones.

Spurious Signals

It is important to distinguish spurious *signals* from spurious *responses*. Spurious signals usually result from unwanted frequency components in the LO signal that are converted to the IF or simply leak through the mixer into the IF circuit. Mixers rarely create spurious signals; they often create spurious responses. Frequency synthesizers and phase-locked oscillators are notorious generators of spurious signals. Removing spurious signals is usually the job of the designer of the LO source.

dc Power

Mixers used in battery-powered applications must use minimal dc power. Of course, passive diode and FET mixers use no dc power, but their loss necessitates the use of more amplifier stages and balanced mixers require more LO power than single-device mixers; these indirectly affect dc power consumption. Some types of single-device mixers have performance characteristics that are similar to those of balanced mixers, and do not require high LO power.

Thermal and Mechanical Characteristics

Diode mixers and resistive FET mixers have very good thermal stability. The thermal stability of active FET mixers is similar to that of FET amplifiers. In many applications, mixers are subject to considerable mechanical and thermal stress as well as environmental dust and moisture. Modern low-cost fabrication techniques and device packages address these concerns adequately.

Cost

Most wireless applications require low-cost, large-quantity manufacturing. In these, design cost is usually a small portion of the cost of each component, and the cost of parts and assembly are dominant. These costs can be minimized by minimizing parts count and handling, and through automated assembly. Most types of mixers use few components and are amenable to automated assembly techniques.

5.3 SEMICONDUCTOR DEVICES FOR MIXERS

The primary devices used for mixers are Schottky-barrier diodes and FETs. Bipolar-junction transistors (BJTs) are also used occasionally, primarily in Gilbert-cell multiplier circuits (see Section 5.6.5), but because of their superior large-signal handling ability, higher frequency range, and lower noise, FET devices are usually preferred. Schottky-barrier diodes have the advantage of low cost and do not need dc bias. Unlike FETs and bipolar transistors, diodes are two-terminal devices and thus can be reversed; this allows them to be used in configurations that are impractical for three-terminal devices.

The Schottky-barrier diode is the dominant device used in mixers. Because Schottky diodes are inherently capable of fast switching and have very small reactive parasitics, they can be used in very broadband mixers. Schottky-diode mixers usually do not require matching circuits, so no tuning or adjustment are needed.

Although mixers using Schottky-barrier diodes always exhibit conversion loss, transistor mixers are capable of conversion gain. This helps simplify the architecture of a system, often allowing the use of fewer amplifier stages than necessary in diode-mixer receivers. The local oscillator (LO) and RF signals can be applied to separate gates of dual-gate MOSFETs, allowing good RF-to-LO isolation to be achieved in a single-device mixer. Dual-gate devices can be used to realize self-oscillating mixers, in which a single device provides both the LO and mixer functions.

5.3.1 Schottky-Barrier Diodes

A Schottky-barrier diode consists of a rectifying metal-to-semiconductor junction. The semiconductor consists of a thin epitaxial layer grown on a heavily doped substrate. The metal contact is the anode, and an ohmic cathode contact is made to the substrate either directly to the bottom of the chip or to the top side by etching away the epitaxial layer. Figure 5.3 shows the structure of a typical Schottky-barrier diode.

Equivalent Circuit

The equivalent circuit of the intrinsic diode is shown in Figure 5.4; real diodes may require additional elements to account for lead inductance, intermetallic capacitance, or package parasitics. The junction I/V characteristic, $I(V)$, is the dominant nonlinearity of the diode. The junction capacitance is more weakly nonlinear, and its nonlinearity rarely has a significant effect on the mixer's operation. The series resistance, R_s, is the resistance of the undepleted epitaxial layer under the junction. The resistance of the substrate and the ohmic contact may also be a minor contributor to R_s.

The I/V characteristic is given by the exponential expression

$$I(V) = I_0\left(\exp\left(\frac{qV}{\eta KT}\right) - 1\right) \tag{5.7}$$

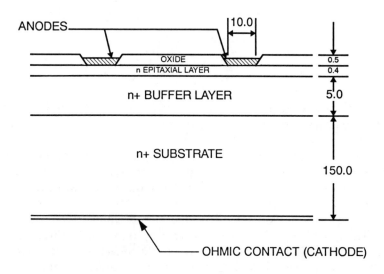

Figure 5.3 Cross section of a Schottky-barrier diode chip. Dimensions (in microns) and doping levels are typical for microwave diodes; low-frequency diodes may have larger junction areas and thicker epitaxial layers.

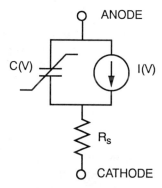

Figure 5.4 Equivalent circuit of a Schottky junction. The equivalent circuit of a practical diode often must include other parasitics, such as lead inductance and package capacitance.

where q is electron charge, K is Boltzmann's constant, and T is absolute temperature in kelvins. η is the ideality factor, usually in the range of 1.05 to 1.25, and accounts for the nonideality of the junction. I_0 is the current parameter, proportional to junction area. It is affected in a complex manner by temperature and the materials used in the junction.

The capacitance $C(V)$ is given by the expression

$$C(V) = \frac{C_{j0}}{\left(1 - \dfrac{V}{\phi}\right)^{\gamma}} \tag{5.8}$$

where C_{j0} is the zero-voltage junction capacitance and ϕ is the built-in potential of the junction. γ depends on the doping profile of the epitaxial layer; it is 0.5 if the epilayer is uniformly doped.

The series resistance R_s depends strongly on the geometry of the junction and the doping level of the epilayer: the longer or narrower the current path through the semiconductor, or the lower the doping level, the greater will be R_s. The value of R_s is lowest in diodes that use a cathode connection to the underside of the substrate; it is greatest in diodes that have the cathode connection on the top of the substrate. The value R_s is approximately inversely proportional to junction area (but C_{j0} is proportional to junction area). Because of the higher mobility in GaAs, GaAs diodes have lower R_s than silicon. R_s is usually on the order of a few ohms.

A figure of merit for a Schottky diode is its cutoff frequency, f_c:

$$f_c = \frac{1}{2\pi R_s C_{j0}} \tag{5.9}$$

where R_s is measured at dc and C_{j0} at any convenient low frequency. f_c is usually a few hundred to several thousand GHz, depending on the type of diode and semiconductor material. As a general rule, f_c should be at least five and preferably ten times the RF frequency of the mixer.

Diode Types

Schottky diodes are manufactured in many configurations. Chip diodes, similar to the structure in Figure 5.3, are used frequently. The diode chip is usually soldered or epoxy-bonded to a mounting surface, which serves as the cathode connection. The anode is connected to the circuit via a thermocompression-bonded gold wire or ribbon. Chip diodes can be mounted in a wide variety of packages, from miniature "pill" packages, consisting of a ceramic cylinder with metal end caps, to inexpensive epoxy-filled packages with ribbon leads.

One of the most useful structures is a beam-lead diode, in which the diode chip and its gold ribbon leads are an integral part of the structure. Figure 5.5 shows a beam-lead diode. The cathode lead is connected to the substrate through the top of the chip via an etched hole in the epilayer. The epilayer is removed entirely in the vicinity of the cathode connection, allowing the cathode to contact the substrate directly. The anode is insulated from the substrate by an oxide layer.

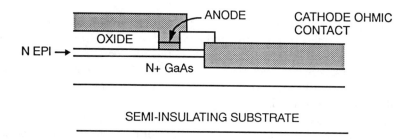

Figure 5.5 Cross section of a conventional beam-lead diode. The "overlay" capacitance between the anode lead and the epilayer is a significant parasitic; the cathode ohmic contact surrounds the anode to minimize series resistance.

The overlay capacitance between the anode lead and the substrate is a significant parasitic reactance. To minimize it, the anode lead is often located near the edge of the chip, resulting in a fragile structure. Other methods can be used to minimize the overlay capacitance while allowing a more rugged anode connection. These include air gaps, boron implants (which reduce the conductivity of the substrate), and glass-filled trenches under the anode lead.

A structure often used in monolithic circuits is an air-bridge mesa diode (Figure 5.6). The use of an air bridge to the anode minimizes overlay capacitance. The cathode is connected to the top of the structure; series resistance is minimized by wrapping the cathode ohmic contact around the anode.

5.3.2 Field-Effect Transistors

A variety of types of field-effect transistors (FETs) are used in mixers. Since the 1960s, silicon MOSFETs (often dual-gate devices) have dominated mixer applications in communication receivers up to approximately 1 GHz. At higher frequencies, GaAs MESFETs are often used.

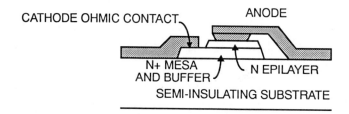

Figure 5.6 Cross section of a mesa diode used for monolithic circuits.

Although FETs can perform frequency mixing in several ways, the most common is to use the time-varying transconductance that results from applying the LO to the gate. Thus, in a FET used as a mixer, the transconductance is of primary importance. Most resistive and reactive parasitics are of secondary interest; the exception is the gate-to-channel capacitance, which limits the frequency range over which the device can be used.

Figure 5.7 shows the equivalent circuit of a FET. The topology is the same for all types of FETs; the differences are in the values of the elements, and especially in the *I/V* and *C/V* characteristics of the nonlinear elements. (The functional forms of these expressions will be presented in the sections describing the individual devices.) All FETs have resistances associated with the ohmic contacts to their drains and sources, R_d and R_s, respectively, and the resistance of the gate, R_g. The value R_{ds} is the drain-to-source resistance and R_i is the resistance of the undepleted epitaxial layer at the source end of the channel (R_i is usually negligible in MOSFETs and in all FETs used as resistive mixers; see Section 5.7.4). The value C_{ds} is the drain-to-source capacitance and C_{gd} is the drain-to-gate capacitance, caused primarily by the proximity of the gate and drain metalizations.

Silicon MOSFETs

Although silicon devices have distinctly lower transconductance than GaAs, they are useful up to at least the lower microwave frequencies. In spite of the inherent inferiority of silicon in comparison to GaAs, silicon MOSFETs do have some advan-

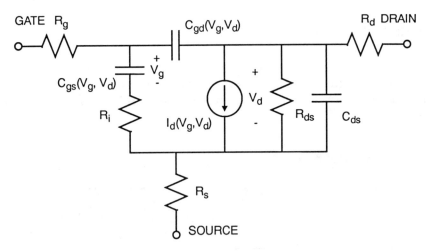

Figure 5.7 Equivalent circuit of a FET. Most types of FETs are well characterized by this equivalent circuit; only the values of the elements and their *I/V* or *C/V* characteristics are different.

tages. The primary advantage of silicon over GaAs is low cost, and the performance of silicon MOSFET mixers is not significantly worse than GaAs in the VHF and UHF ranges. The high drain-to-source resistances (R_{ds}) of silicon MOSFETs give them higher voltage gain than GaAs devices; in many applications this is a distinct advantage. Additionally, the positive threshold voltage (in an N-channel enhancement MOSFET), in comparison to the negative threshold voltage of a GaAs FET, is very helpful in realizing low-voltage circuits and circuits requiring only a single dc supply. Mixers using enhancement-mode silicon MOSFETs often do not require gate bias, and dual-gate MOSFETs offer convenient LO-to-RF isolation (when the LO and RF are applied to different gates).

A MOSFET is sometimes called an *insulated-gate FET*, because the gate is insulated from the channel by a very thin oxide layer. In an *enhancement-mode* device, a voltage applied to the gate generates carriers at the oxide-semiconductor interface, and the resulting channel becomes conductive. Increasing the gate voltage increases the surface charge and the channel current. A *depletion-mode* device has a lightly doped layer under the gate, so channel current is possible at zero gate voltage, and a reverse voltage is needed to deplete the layer and to turn off the channel. MOSFETs can be produced in either N- or P-channel form.

Figure 5.8 shows a cross section of a modern single-gate MOSFET [1]. This device is suitable for use up to at least 6 GHz but is intended primarily for 900-MHz applications. The device is built on a P+ substrate having a P– epilayer. A P– base is provided to adjust the threshold voltage, and P+ "sinkers" connect the source regions through the P+ substrate to the underside of the substrate for a ground

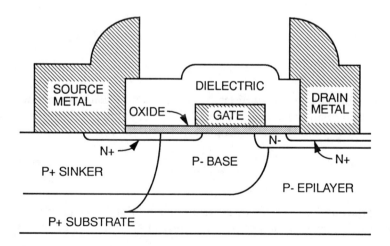

Figure 5.8 Cross section of a silicon MOSFET adequate for use in microwave mixers. (*After:* [1], used with permission).

connection. The gate length is 1.5 μm; gates of 0.6-μm length have been fabricated. A dual-gate version of this device exhibits 16-dB conversion gain and 1.5-dB noise figure at 900 MHz.

A simple and reasonably accurate expression for the channel current in a MOSFET is

$$I_d(V_g, V_d) = \beta\left((V_g - V_t)V_d - \frac{V_d^2}{2}\right) \tag{5.10}$$

where V_t is the threshold voltage, β is a constant having units of V^{-2}, and other voltages are as shown in Figure 5.7. Equation (5.10) is valid in the so-called *linear region* of the device, when $V_d < V_g - V_t$ and $V_g > V_t$. At higher drain voltages, the current is largely independent of V_d, and is given by (5.10) with $V_d = V_g - V_t$. This is called the *saturation region* of the device, because the current is "saturated" in the sense that it does not increase with gate voltage.[3] The current is therefore

$$I_d(V_g, V_d) = \frac{\beta}{2}(V_g - V_t)^2 \tag{5.11}$$

Equations (5.10) and (5.11) result from a relatively simple approximate analysis of the device. More accurate (and more complex) models have been developed, and are a subject of ongoing research. Equations (5.10) and (5.11) nevertheless describe one of the most important characteristics of a silicon FET: the square-law dependence of its drain current on gate voltage. This is responsible for the device's low third-order distortion.

Differentiating (5.11) gives the transconductance when $V_d > V_g - V_t$:

$$G_m(V_g) = \beta(V_g - V_t) \tag{5.12}$$

The transconductance is proportional to V_g and can be made to vary with time by the application of an LO signal to the gate. The magnitude of the transconductance variation is proportional to β; a simple expression for β is

$$\beta = \mu C_{ox}\frac{W}{L} \tag{5.13}$$

where μ is the mobility of holes or electrons (in cm^2/V s), C_{ox} is the oxide capacitance (the gate-to-channel capacitance per unit area, in F/cm^2), W is the gate width, and

---

[3]In GaAs devices, current saturation occurs primarily because electrons in the channel reach saturated drift velocity. Although velocity saturation also occurs in silicon devices, the effect on the *I/V* characteristic is much less, and current saturation is caused primarily by the pinch-off of the drain end of the channel.

L is the gate length, both in centimeters. Equation (5.13) shows that the peak transconductance is proportional to gate width and mobility and is inversely proportional to gate length. Because the peak transconductance is also proportional to C_{ox}, a thin oxide layer results in high transconductance.

GaAs MESFETs

A MESFET is a junction FET having a Schottky-barrier gate. Although silicon MESFETs have been made, they are now obsolete and all modern MESFETs are fabricated on GaAs. The gate's length is usually less than 0.5 μm, and may be as short as 0.1 μm; this short gate length, in conjunction with the high electron mobility and saturation velocity of GaAs, results in a high-frequency, low-noise device.

Figure 5.9 shows a cross section of a GaAs MESFET. The channel is a moderately doped epitaxial layer 1,000Å to 2,500Å thick, grown on an undoped substrate. Because N GaAs has higher mobility than P GaAs, all conventional GaAs MESFETs use N material for the channel. A thick, undoped buffer layer forms the lower boundary of the channel. This layer prevents substrate impurities from diffusing into the epitaxial layer during processing. Like the MOSFET, the drain and source metalizations form ohmic contacts at each end of the channel. The gate is often T-shaped in cross section to minimize its resistance while minimizing its length.

A dc voltage applied to the channel creates a longitudinal electric field. In normal operation, the field is strong enough to accelerate the electrons to their saturated drift velocity, creating an electron current from the source to drain. The Schottky barrier formed by the gate creates a depletion region that extends part way into the channel at zero gate bias. Varying the gate-to-source voltage modulates the depletion depth and hence the thickness of the conductive channel; the channel

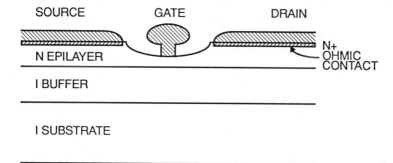

Figure 5.9 Cross section of a modern, recessed-gate MESFET. The recessed gate reduces the parasitic source resistance and minimizes the effective gate length. The mushroom-shaped gate cross section minimizes the resistance of the very short (< 0.5 μm) gate.

current varies accordingly. As in the Schottky diode, the gate voltage also varies the depletion capacitance. A negative gate-to-source voltage is necessary to fully deplete the channel, causing it to "pinch off," preventing channel current.

Because of the complex velocity-field characteristic of the GaAs FET and the short channels of modern MESFETs, it is not possible to develop a simple and accurate expression for the GaAs FET's I/V characteristic from the physics of the device. Instead, empirical formulae are usually employed. A popular expression is that of Curtice [2]:

$$I_d(V_g, V_d) = (A_0 + A_1V_1 + A_2V_1^2 + A_3V_1^3)\tanh(\alpha V_d) \tag{5.14}$$

where A_n and α are constant coefficients and V_1 is an intermediate variable,

$$V_1 = V_g(t - \tau)(1 + \beta(V_{d0} - V_d)) \tag{5.15}$$

β is an empirical constant, and τ is the time delay. V_{d0} is the value of V_d at which the A_n coefficients were determined.

Because GaAs MESFETs are considerably more expensive than silicon devices, they are usually used only at frequencies that are too high for silicon devices. GaAs devices can be used at lower frequencies if necessary; for example, as IF amplifier stages in monolithic microwave mixers.

High Electron-Mobility Transistors

A high electron-mobility transistor (HEMT) is a junction FET that uses a heterojunction (a junction between two dissimilar semiconductors), instead of a simple epitaxial layer, for the channel. The discontinuity of the band gaps of the materials used for the heterojunction creates a layer of charge at the surface of the junction; the charge density can be controlled by the gate voltage. Because the charge in this layer has a very high mobility, high-frequency operation and very low noise are possible. It is not unusual for HEMTS to operate successfully as low-noise amplifiers above 100 GHz. HEMTs require specialized fabrication techniques, such as molecular-beam epitaxy, and thus are very expensive to manufacture. HEMT heterojunctions are invariably realized with III-V semiconductors; AlGaAs and InGaAs are common.

HEMTs are used for mixers in the same way as conventional GaAs FETs. Because the gate I/V characteristic of a HEMT is generally more strongly nonlinear than that of a MESFET, HEMT mixers usually have greater intermodulation distortion (IM) than FETs. The noise figure (NF) of a HEMT mixer usually is not significantly lower than that of a GaAs FET, however.

Figure 5.10 shows the cross section of one type of HEMT. There are many degrees of freedom in the design of such devices; for example, the number of hetero-

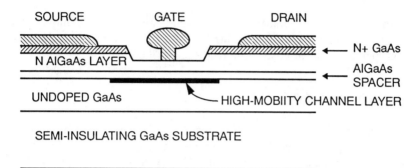

Figure 5.10 Cross section of a simple HEMT. The two-dimensional electron gas resides in the GaAs layer, adjacent to the AlGaAs spacer.

junctions, the thicknesses of layers, and the fraction of Al or In in AlGaAs or InGaAs devices are all variable, and can be used to optimize the device.

Like that of the GaAs FET, the *I/V* characteristic of a HEMT usually is described empirically. Because the *I/V* characteristics for HEMTs are very different from those of GaAs devices, *I/V* models appropriate for GaAs devices are usually inadequate for HEMTs. An expression that works well for many types of small-signal HEMTs is [3]

$$I_d(V_g, V_d) = I_p(1 + \tanh(\Psi))(1 + \lambda V_d)\tanh(\alpha V_d) \qquad (5.16)$$

where

$$\Psi = P_1(V_g - V_p) + P_2(V_g - V_p)^2 + \ldots \qquad (5.17)$$

The V_p is the gate voltage at peak transconductance, I_p is the current at the same point, and λ and α are constants.

Bipolar Transistors

Silicon bipolar-junction transistors (BJTs) find limited use in mixers. Because FETs generally provide better overall performance, silicon BJTs are rarely used in conventional single-device or balanced mixers. Bipolar devices are used as mixers when necessary for process compatibility, not for any inherent technical advantages. For this reason, we will be concerned with them only minimally.

Heterojunction bipolar transistors (HBTs), BJTs that use a heterojunction for the emitter-to-base junction, are also useful in wireless circuits. They are most useful as high-efficiency power amplifiers and low-distortion small-signal amplifiers.

Because they require III-V materials and special fabrication techniques, they are relatively expensive. Silicon-Germanium (Si-Ge) HBTs are a new technology that offers high performance at costs close to those of silicon BJTs.

5.4 FREQUENCY MIXING

The operation of all practical mixers is based on frequency conversion in a time-varying device parameter, usually conductance or transconductance. This time-varying conductance is created by applying an LO waveform to a nonlinear device. In a diode mixer, the junction conductance can be found from the derivative of (5.7)

$$g_j(t) = \frac{\partial}{\partial V} I(V) \bigg|_{V=V_{LO}(t)} = \frac{qI_0}{\eta KT} \exp\left(\frac{qV_{LO}(t)}{\eta KT}\right) \approx \frac{qI(V_{LO}(t))}{\eta KT} \qquad (5.18)$$

where $V_{LO}(t)$ is the junction voltage of the LO signal. The conductance waveform $g_j(t)$ can be expressed as a Fourier series

$$g_j(t) = \sum_{k=-\infty}^{\infty} G_k \exp(jk\omega_p t) \qquad (5.19)$$

where ω_p is the LO radian frequency. If a signal voltage $V_s \exp(j\omega_s t)$ is applied to this conductance, the resulting current is

$$i(t) = \sum_{k=-\infty}^{\infty} I_k \exp(j(k\omega_p + \omega_s)t) \qquad (5.20)$$

where the resulting frequencies $k\omega_p + \omega_s$ are the mixing products. In general, however, the voltage will not have only a single frequency, ω_s; the source impedance will generate junction-voltage components at the same mixing frequencies as the current components. Also, we customarily express the mixing products in (5.20) as $k\omega_p + \omega_0$, where ω_0 is the lowest mixing frequency; usually $\omega_0 = |\omega_s - \omega_p|$. With a change of the index k, this results in the same set of frequencies as in (5.20). Thus, we have

$$i(t) = \sum_{k=-\infty}^{\infty} I_k \exp(j(k\omega_p + \omega_0)t)$$

$$v(t) = \sum_{k=-\infty}^{\infty} V_k \exp(j(k\omega_p + \omega_0)t)$$

$$g_j(t) = \sum_{k=-\infty}^{\infty} G_k \exp(jk\omega_p t) \qquad (5.21)$$

where $v(t)$ is the small-signal junction voltage and $i(t)$ is the small-signal junction current. Substituting these into the relation

$$i(t) = g_j(t)\,v(t) \tag{5.22}$$

gives the result

$$\sum_{l=-\infty}^{\infty} I_l \exp(j(\omega_0 + l\omega_p)t) = \sum_{m=-\infty}^{\infty} \sum_{n=-\infty}^{\infty} G_m V_n \exp(j(\omega_0 + (n+m)\omega_p)t) \tag{5.23}$$

which can be expressed by the matrix equation

$$
\begin{bmatrix}
I_{-N}^* \\
I_{-N+1}^* \\
\cdots \\
I_{-1}^* \\
I_0 \\
I_1 \\
\cdots \\
I_{N-1} \\
I_N
\end{bmatrix}
=
\begin{bmatrix}
G_0 & G_{-1} & G_{-2} & \cdots & \cdots & G_{-N} & \cdots & G_{-2N} \\
G_1 & G_0 & G_{-1} & \cdots & \cdots & G_{-N+1} & \cdots & G_{-2N+1} \\
\cdots & \cdots & \cdots & \cdots & \cdots & \cdots & \cdots & \cdots \\
G_{N-1} & G_{N-2} & G_{N-3} & \cdots & \cdots & G_{-1} & \cdots & G_{-N-1} \\
G_N & G_{N-1} & G_{N-2} & \cdots & \cdots & G_0 & \cdots & G_{-N} \\
G_{N+1} & G_N & G_{N-1} & \cdots & \cdots & G_1 & \cdots & G_{-N+1} \\
\cdots & \cdots & \cdots & \cdots & \cdots & \cdots & \cdots & \cdots \\
G_{2N-1} & G_{2N-2} & G_{2N-3} & \cdots & \cdots & G_{2N-1} & \cdots & G_{-1} \\
G_{2N} & G_{2N-1} & G_{2N-2} & \cdots & \cdots & G_N & \cdots & G_0
\end{bmatrix}
\begin{bmatrix}
V_{-N}^* \\
V_{-N+1}^* \\
\cdots \\
V_{-1}^* \\
V_0 \\
V_1 \\
\cdots \\
V_{N-1} \\
V_N
\end{bmatrix}
\tag{5.24}
$$

where the infinite series in (5.24) is truncated, for obvious practical reasons, to $2N + 1$ terms. The subscript n refers to the frequency component

$$\omega_n = n\omega_p + \omega_0; \quad \omega_{-n} = -\omega_n \tag{5.25}$$

In (5.24), negative-frequency components have been represented as the conjugate of conventional positive-frequency phasors. Figure 5.11 shows the frequencies in (5.23) and (5.24).

The matrix in (5.24) relates the voltages to the currents at each mixing frequency. It can be treated in all ways like a multiport admittance matrix. Indeed, one can think of the junction of the mixing diode as a type multiport network in which the "ports" are voltages and currents at the various mixing frequencies instead of physically separate ports. The external circuit can then be treated as a set of circuits, one at each mixing frequency, connected to the diode. These circuits at the various mixing frequencies are interconnected via the diode.

A consequence of this representation is that the characteristics of the external circuit (often called the *embedding network*) at all mixing frequencies affect the

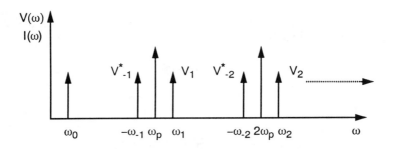

Figure 5.11 Mixing frequencies in a pumped nonlinear element.

properties of the mixer at the RF and IF frequencies. For this reason, mixer analysis must take into account the characteristics of the embedding network at all mixing frequencies. In theory, this is an infinite set of frequencies, so practical analysis requires truncating the set to some reasonable size. In most diode-mixer analyses, for example, a value of $N = 4$ (which requires 8 LO harmonics plus dc) is usually adequate.

Although we used diodes as an example, the same approach to analysis is applicable to active mixers. However, in active mixers the RF excitation voltage and IF output current are at separate ports that are only weakly coupled (e.g., the gate and drain of a FET), so it is often possible to simplify the analysis.

5.5 DIODE MIXERS

Semiconductor diode mixers are the most common of all mixer types. This section will discuss their design, implementation, and performance.

5.5.1 Diode Mixer Characteristics

Diode mixers are an example of a universal component in RF and microwave engineering: one that does nothing especially well, but does everything at least adequately. For this reason, diodes have been used as mixers from the earliest days of radio up to the present. It is safe to predict that they will be in common use far into the future.

In contrast to transistor mixers, which can provide conversion gain, diode mixers always exhibit conversion loss, usually 6 to 10 dB, depending on the frequency and type of design. Conversion loss and input/output impedance depend on LO level: conversion loss usually decreases monotonically with LO level and port impedance decreases, until the diode is very strongly pumped. At this point, further increases in LO level have progressively less effect on performance, until avalanche breakdown occurs.

Diode mixers can be very broadband. Because diodes can have very small junction capacitances (as little as 0.005 pF, if necessary) the junction capacitance rarely limits the bandwidth of a diode mixer; instead, bandwidth is usually established by the embedding circuit. The general approach to the design of a diode mixer is to avoid the use of a reactive matching circuit. Usually the diode is selected so that its junction capacitance is negligible, and the embedding circuit is designed to match the resistive diode.

Most practical diode mixers are balanced. Single-diode mixers are occasionally used for simple, low-cost applications where performance need not be high. Balanced mixers offer a number of advantages over single-diode mixers:

1. Balanced mixers usually have some degree of inherent port-to-port isolation.
2. AM noise on the LO signal is rejected by a balanced mixer.
3. Certain even-order spurious responses and intermodulation products are rejected by many types of balanced mixers.

Unfortunately, balanced mixers do not offer improved conversion loss or noise figure compared to single-device structures. These characteristics depend strongly on the balance of the hybrids or baluns used to realize the mixer; achieving good balance in a broadband hybrid is difficult. Finally, certain types of single-device mixers (for example, the series-switch FET mixer examined in Section 5.7.1) can offer performance that is comparable to that of balanced diode mixers.

5.5.2 Single-Diode Mixers

Although single-diode mixers have serious limitations, there are two good reasons for examining them. First, in spite of their limitations, single-diode mixers are occasionally used, and in some technologies (e.g., submillimeter applications) they are used exclusively. Second, all balanced diode mixers can be reduced to a single-diode equivalent circuit; thus, the single-diode mixer serves as a prototype for more complex mixers.

Figure 5.12 shows a simple circuit of a single-diode mixer. It illustrates a number of requirements for the design of practical mixers. First, the input and output circuits must be *noninteracting*; that is, they should present either a short or open circuit, depending on the configuration, to the other ports. In Figure 5.12, the parallel L-C resonators are ideal; they are open circuits at their resonant frequencies and short circuits at all other frequencies. Because of this property, at each frequency (RF, IF, and LO) the diode is connected to the respective port, and is isolated from all the others. Thus, at the RF frequency the LO and IF ports are disconnected; similarly, at the IF and LO frequencies the RF/LO and RF/IF ports, respectively, are disconnected.

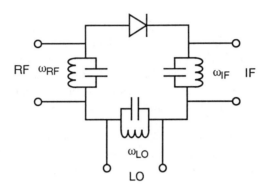

Figure 5.12 An ideal single-diode mixer. The parallel L-C resonators are ideal; each is a short circuit at all frequencies except the resonant frequency.

A second requirement is that the diode be matched at all important frequencies. As long as the terminations at all mixing frequencies are not reactive, the diode's impedance will be resistive. In this manner, the port's impedance can be selected to match the diode (transformers or other such structures can be used to obtain a standard port impedance (for example, 50Ω), if necessary). The diode's junction capacitance, if not negligible, can be absorbed into the capacitances of the resonators.

Third, the diode must be appropriately terminated at all unwanted mixing frequencies. In this case, the diode is short-circuited at all frequencies except the RF, IF, and LO. In general, the best overall performance is obtained when the diode is short-circuited; other terminations (for example, an open circuit) are theoretically possible, but are often difficult to implement or result in poor performance in some respects. In particular, an open-circuit at unwanted mixing frequencies results in greater intermodulation distortion than a short-circuit.

When the diode is short-circuited at all mixing frequencies except the RF and LO, (5.24) can be reduced to the form

$$\begin{bmatrix} I_{IF} \\ I_{RF} \end{bmatrix} = \begin{bmatrix} Y_{11} & Y_{12} \\ Y_{21} & Y_{22} \end{bmatrix} \begin{bmatrix} V_{IF} \\ V_{RF} \end{bmatrix} \tag{5.26}$$

This admittance matrix can be used in the same way as any two-port admittance matrix to find conversion loss and input or output impedance.

Fourth, conversion efficiency must be reasonably high. In most systems, the mixer's conversion efficiency need not be as high as possible; indeed, high conversion gain in an active mixer is often deleterious. Conversion loss of 6 to 8 dB is usually adequate for passive mixers, and approximately unity or a few decibels is usually optimum for active mixers. Only rarely should conversion efficiency be optimized at the expense of other performance characteristics.

The single-diode mixer in Figure 5.12 meets these requirements through the properties of the L-C resonators. This is usually the case; the performance of a single-device mixer depends strongly upon the characteristics of filters. Because practical filters have limited Qs and are often uncomfortably large, single-diode mixers often exhibit only modest performance. Indeed, even with perfect filters, single-diode mixers cannot be realized if the RF, IF, or LO bands overlap. For these reasons, balanced mixers, which employ the properties of hybrids or baluns to circumvent these difficulties, are more frequently used.

Figure 5.13 shows a practical microstrip implementation of a single-diode mixer. This mixer uses a ring resonator for LO injection. Such narrowband resonators are usually suitable only if the LO is fixed-frequency. The ring is $\lambda/2$ in circumference at the resonant frequency and is loosely coupled to the RF and LO microstrips. This coupling can be varied to adjust the Q of the resonator; for minimum LO loss, the bandwidth should be as wide as possible consistent with obtaining adequate RF-to-LO isolation. The ring generates a wave on the RF microstrip propagating toward the diode; if the diode is not well matched, LO power is reflected toward the RF port, and the LO-to-RF isolation is equal to the diode's LO reflection coefficient. The termination on the LO microstrip is necessary to ensure a low, broadband VSWR at the LO input; this is necessary for stable operation of the LO source. The termination is decoupled from the LO at the ring's resonant frequency, and therefore absorbs no power.

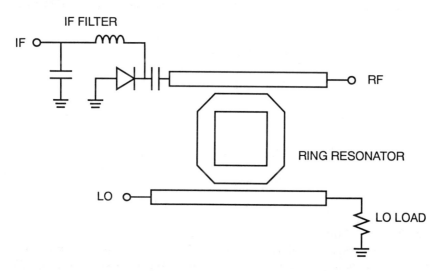

Figure 5.13 A practical microstrip realization of a single-diode mixer. A ring resonator is used to inject the LO; because this resonator is very narrow, only a single-frequency LO is practical. The LO load provides a good out-of-band match to the LO; it is decoupled at the ring's resonant frequency.

5.5.3 Baluns and Hybrids for Wireless Mixers

Before we discuss balanced mixers, we must examine one of the most important parts of the mixer: the balun or hybrid used in its realization.

Baluns and Hybrids

Although both baluns and hybrids are often used interchangeably in mixers, it is important to distinguish between the two. A balun is a structure that interconnects balanced and unbalanced transmission lines (*balun* is a contraction of the words *bal*anced and *un*balanced). One terminal of a balun's input port is grounded; both output terminals are ungrounded. Ideally, the outputs are "floating": the voltage between either and ground is undefined (just as the voltage between ground and the wires of a balanced, ungrounded transmission line are undefined). If an even-mode excitation is applied to the output, no power is transmitted to the input, and the even-mode output-port impedance is either an open or short circuit (most practical baluns exhibit an open circuit to even-mode excitation).

A hybrid is a special type of four-port junction. If a signal is applied to any port, it emerges from two of the other ports at half power; the fourth port is isolated. All four ports are matched. Only two fundamental types of hybrids are possible: the 180-deg and 90-deg forms. The ideal hybrids have the scattering matrices

$$S_{90} = \frac{1}{\sqrt{2}} \begin{bmatrix} 0 & 0 & -j & 1 \\ 0 & 0 & 1 & -j \\ -j & 1 & 0 & 0 \\ 1 & -j & 0 & 0 \end{bmatrix} \tag{5.27}$$

for the 90-deg hybrid and

$$S_{180} = \frac{1}{\sqrt{2}} \begin{bmatrix} 0 & 0 & 1 & 1 \\ 0 & 0 & 1 & -1 \\ 1 & 1 & 0 & 0 \\ 1 & -1 & 0 & 0 \end{bmatrix} \tag{5.28}$$

for the 180-deg hybrid.

If the input and output ports of a 180-deg hybrid are chosen so that the outputs are in phase, the input port is called the *sigma* or *sum* port, designated by the Greek letter Σ. The remaining isolated port is the *difference* or *delta* port, designated by Δ, because a signal applied to it will generate out-of-phase outputs. Because a 180-deg phase difference can be obtained at the output ports, a 180-deg hybrid often can be used in the same manner as a balun.

Indeed, certain components can be used as either baluns or hybrids. Figure 5.14 shows a transformer used as both. In Figure 5.14(a) the transformer is connected as a hybrid; in Figure 5.14(b), as a balun. This type of transformer is usually realized as a trifilar winding on a magnetic core; such a transformer can have a frequency range from a few megahertz to several hundred megahertz. The upper frequency response is limited by magnetic leakage flux and capacitance between the windings. Transmission-line transformers, which operate as transformers at low frequencies but as transmission lines at high frequencies, avoid this problem.

RF Transformers

RF transformers are used frequently in wireless components. Properly designed and configured, these can be used as either baluns or hybrids, and often have bandwidths of several decades. We use the term transformer somewhat loosely, as such compo-

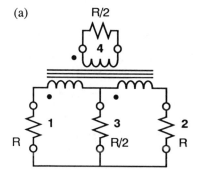

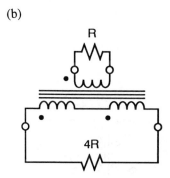

Figure 5.14 A transformer connected as (a) a hybrid and (b) a balun. The port numbers correspond to (5.28).

nents often operate as magnetic transformers at low frequencies, but as transmission-line structures at high frequencies.

A magnetic transformer consists of two inductors coupled by their magnetic fields. The inductors have self inductances L_1, and L_2, the inductance of one inductor when the other inductor's current is zero, and mutual inductance, M, loosely defined as the inductance common to both inductors. When the entire magnetic field generated by one inductor is contained by the other, the coupling is perfect and

$$M = \sqrt{L_1 L_2} \qquad (5.29)$$

Figure 5.15 shows a transformer connected to a load, treated as a two-port. When (5.29) is satisfied, the ratio of the port voltages is

$$\frac{V_2}{V_1} = \sqrt{\frac{L_2}{L_1}} \qquad (5.30)$$

which, for solenoidal inductors, is equal to the ratio of the turns of wire on the two coils. The current ratio is

$$\frac{I_1}{I_2} = -\frac{R + j\omega L_2}{j\omega M} \qquad (5.31)$$

which is the negative inverse of (5.30) when $R \ll \omega L_2$.

In mixers, transformers are often used for impedance transformation. The input impedance at L_1 is

$$Z_1 = R\frac{L_1}{L_2} \qquad (5.32)$$

when $R \ll \omega L_2$.

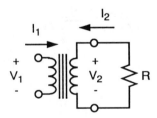

Figure 5.15 An RF transformer, configured as a two-port, with a terminating resistance R.

Equation (5.29) represents an idealization. In practice, there is always some leakage inductance, so the inductors are not perfectly coupled. In this case, the coupling coefficient k is defined as

$$k = \frac{M}{\sqrt{L_1 L_2}} \tag{5.33}$$

Clearly, when the inductors are perfectly coupled, $k = 1$; $k = 0$ implies $M = 0$, and there is no coupling. When $k < 1$ and $R \ll \omega L_2$, the input impedance at port 1 becomes

$$Z_1 = R\frac{L_1}{L_2} + j\omega L_1(1 - k^2) \tag{5.34}$$

The imperfect coupling causes the input impedance to be reactive. This reactance limits the bandwidth of the transformer. The bandwidth limitation of a magnetic transformer becomes apparent: on one hand, to satisfy the requirement that $\omega L_2 \gg R$, the inductance of the coils must be large; however, imperfect coupling causes part of that large inductance to appear in series with the transformer.

Another problem that limits the bandwidth of a transformer is the stray capacitance between the inductors. This capacitance degrades the performance of the transformer in a number of ways, affecting the phase characteristics and introducing spurious resonances. Because of these effects, simple magnetic transformers are limited in bandwidth to a few hundred megahertz.

Even so, simple magnetic transformers are adequate for many applications. Customarily, they consist of two or three bifilar or trifilar windings on a toroidal magnetic core. Leakage inductance can be minimized by the use of cores that completely enclose the windings.

Bandwidth can be increased through the use of a structure that is equivalent to a transformer at low frequencies, but is equivalent to a transmission-line transformer at high frequencies. In this way, at high frequencies the stray capacitance between the windings becomes part of the transmission-line capacitance and, because the transformer is not operating as a magnetic transformer, magnetic leakage is inconsequential. Such transformers may have bandwidths above 2 GHz, and are limited primarily by the practical difficulties in making low-parasitic connections at microwave frequencies. Many of the transmission-line transformers in the following section consist of twisted-pair transmission lines wound on a magnetic core to realize such transformers.

Transmission-Line Baluns and Hybrids

A wide variety of baluns and hybrids can be realized with transmission lines. Most of these require transmission-line sections one-quarter or one-half wavelength long.

This makes them uncomfortably large for use in wireless applications at frequencies below 2,400 MHz. Often, however, the structures can be modified to make them practical at low frequencies by either meandering the strips or loading them with dielectric material.

Figure 5.16 shows a simple transmission-line balun. It consists of a $\lambda/4$ section of transmission line whose characteristic impedance Z_0 is chosen to act as a transformer between the source and load impedances:

$$Z_0 = \sqrt{Z_s Z_L} \tag{5.35}$$

This balun is very sensitive to stray capacitance between the transmission line and ground. To account for this capacitance, we treat the conductors as coupled transmission-line sections, and their even-mode impedance accounts for the stray capacitance. Ideally, the capacitance is zero, causing the even-mode impedance to be infinite. The odd-mode impedance is half the transmission-line impedance; that is,

$$Z_{0o} = 0.5Z_0 = 0.5\sqrt{Z_s Z_L} \tag{5.36}$$

Analysis shows that the even-mode impedance should be at least ten times the odd-mode. This is often accomplished in microwave mixers by the use of a suspended substrate for the balun. At low frequencies, the even-mode impedance can be increased by using a twisted pair of conductors for the transmission line and inserting the conductors into a magnetic core to increase the even-mode inductance. This has two effects: first, it reduces the size of the balun; second, the balun operates as a 1:1 transformer at low frequencies, improving the low-frequency response. (Of course, to have a good VSWR in this low-frequency range, the source and load impedances must be equal.)

The transmission-line balun can be shortened by loading the input and output ends with capacitors. The optimum capacitance depends on the mode impedances and is best determined by numerical optimization. As an initial value, the line length should be $\lambda/8$ and the capacitors' reactances should be $0.707Z_0$.

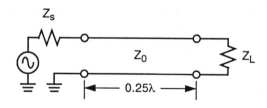

Figure 5.16 A simple transmission-line balun.

Other transmission-line transformer combinations can be used as baluns. Figure 5.17 shows a balun having a 1:4 impedance ratio with $Z_0 = Z_s$. This balun is not fundamentally different from that of Figure 5.16; it consists of two such baluns connected in parallel at the input and in series at the output. By selecting $Z_0 = 0.5Z_s$, one could obtain a 1:1 transformation at the high-frequency end of its bandwidth. If the balun were wrapped on a magnetic core to improve its low-frequency response, the low-frequency impedance transformation would still be 1:4.

Figure 5.18 shows another 1:4 balun. Both outputs of this balun are at ground potential for dc, and provide a convenient dc return for the mixer diodes.

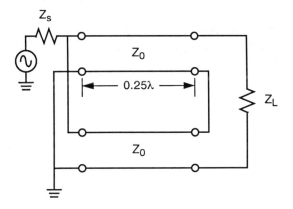

Figure 5.17 A 1:4 impedance-transforming transmission-line balun; $Z_0 = Z_s$, $Z_L = 4 Z_s$.

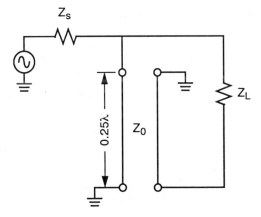

Figure 5.18 A 1:4 impedance-transforming balun; $Z_0 = Z_s$.

Lumped-Element Hybrids

It is possible to create a hybrid from lumped elements. Figure 5.19 shows such a 180-deg hybrid. The relation for the element values is very simple:

$$\omega L = \frac{1}{\omega_C} = 1.414Z \tag{5.37}$$

where Z is the port impedance. This hybrid has a bandwidth of approximately 20%.

Ninety-degree lumped hybrids are usually very narrowband. Figure 5.20 shows two circuits having bandwidths of approximately 1 to 2%, depending on the balance required. Figure 5.20(a) consists entirely of L-C elements and can be realized in an integrated circuit. Its design equations are as follows:

$$C_1 = \frac{1}{\omega Z}$$

$$L = \frac{R}{\omega\sqrt{1 + \omega C_1 Z}}$$

$$C_0 = \frac{1}{\omega^2 L} - C_1 \tag{5.38}$$

where Z again is the port impedance. Figure 5.20(b) shows a quadrature hybrid using a transformer [4]. The transformer is a 1:1 realization, consisting of a twisted pair wound on a magnetic core. It is assumed that coupling is perfect, so the self-inductance of the windings L is equal to the mutual inductance. Then,

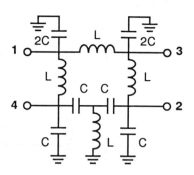

Figure 5.19 A lumped-element 180-deg hybrid.

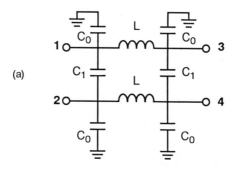

(a)

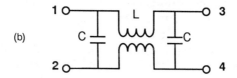

(b)

Figure 5.20 Two realizations of a narrowband quadrature hybrid: (a) using lumped L-C elements, (b) using a transformer and capacitors. The bandwidth is only 1 to 2%.

$$Z = \sqrt{\frac{L}{C}}$$

$$f_0 = \frac{Z}{2\pi L}$$

(5.39)

In theory, the hybrid is well matched at all frequencies, but the coupling is relatively narrowband and is 3 dB at the center frequency, f_0.

Multiple sections of the coupler shown in Figure 5.20(b) can be cascaded with either intervening transmission lines or lowpass filter sections to improve bandwidth [4]. However, the resulting coupler is relatively complex, requires high Q elements to achieve reasonably low loss, and is not much smaller than a coupled-line hybrid.

A set of uncoupled transmission lines loaded by capacitors can also create a quadrature hybrid that is only $\lambda/8$ long. The reactances of the capacitors are equal to the transmission-line impedances, which are also equal to the port impedances. The bandwidth of this hybrid is approximately 10%. The circuit is shown in Figure 5.21.

5.5.4 Singly Balanced Diode Mixers

A singly balanced diode mixer consists of two diodes interconnected by a balun or hybrid. Two fundamental realizations are possible: the 90-deg and 180-deg forms.

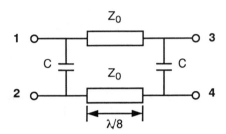

Figure 5.21 A quadrature hybrid realized from two uncoupled transmission lines and capacitors. The length of the line is $\lambda/8$, and the bandwidth is approximately 10%.

Figure 5.22 shows both types of mixers. Each consists of two diodes, a hybrid, and a lowpass IF filter. The diodes are connected to mutually isolated ports of the hybrids and the RF and LO are applied to the other pair of isolated ports. Because there is no inherent isolation between the RF or LO and IF ports, a filter is needed.

One of the most important reasons for using a balanced mixer is its inherent rejection of spurious responses (see Section 5.2.2). Most types of balanced mixers reject certain spurious responses where m or n (or both) are even. This phenomenon depends on the even or odd character of m and n; for example, if a mixer rejects the (2, 1) response, it will also reject the (4, 1) and (6, 1) responses.

For several reasons, the 180-deg balanced mixer is most practical for wireless applications:

1. As we have seen, 180-deg hybrids are usually more broadband than 90-deg.
2. The 180-deg-hybrid mixers have better even-order spurious response and IM rejection, and 180-deg mixers reject all (2, 2) responses. They reject the (2, 1)

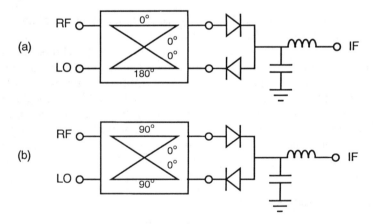

Figure 5.22 A (a) 180-deg and a (b) 90-deg singly balanced mixer.

spurious response if the Δ port is the LO, and the (1, 2) if the Σ port is the LO. The 90-deg hybrid mixer rejects only the (2, 2); it has no appreciable rejection of the (2, 1) or (1, 2).

3. The 90-deg hybrid mixers have a number of odd characteristics. Most important is the fact that the performance depends critically on both the RF and LO ports being well matched over both the RF and LO bands, and the diodes being well matched to the ports. If these requirements are not met, the balance suffers and most of the benefits of a balanced structure—AM noise rejection, spurious-response rejection, and port-to-port isolation—are lost.

Beyond these concerns, the design of the two types of mixers is virtually identical. Thus, we concern ourselves only with the 180-deg mixer.

Figure 5.23 shows a 180-deg mixer that uses a transformer as a hybrid. In this configuration, the RF uses the Δ port of the hybrid and the LO uses the Σ port. The diodes are connected to the remaining pair of mutually isolated ports. The IF port is connected to the diodes through a lowpass filter. This filter is necessary for rejecting the LO; it should also present a short circuit to the diodes at the LO frequency, so the right ends of the diodes are effectively connected to ground. Because the RF voltages at the diodes have a 180-deg phase difference, the diodes' connection point is a virtual ground for the RF. The inductor at the LO port is a current return for the IF; without this inductor, the IF is open-circuited. If the frequency separation between the IF and LO frequency is not large, a more complex filter may be necessary.

An important and often overlooked consideration in balanced mixers is the diodes' dc voltage and current. A dc bias is sometimes used in mixers to improve conversion loss and to reduce LO power requirements. A dc bias is most often used in single-diode millimeter wave mixers; it is often impractical to apply dc bias to a balanced mixer, so no external dc bias is used in the mixer of Figure 5.23. Even

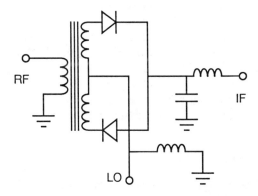

Figure 5.23 A singly balanced, 180-deg hybrid mixer using a transformer as a hybrid.

when possible, use of dc bias in balanced mixers is usually not advisable because inadvertent differences in bias at the two diodes can unbalance the mixer; in most wireless applications, maintaining the mixer's balance is usually more important than achieving a minor improvement in conversion loss.

In Figure 5.23 the diodes are in an antiparallel configuration at dc, so dc current circulates through them and the transformer unimpeded, and the dc voltage across the diodes must be zero.

The mixer of Figure 5.23 includes no RF or LO matching circuits. As with dc bias, matching circuits introduce a means whereby the mixer could become unbalanced, and therefore are usually avoided. Instead, the diodes are selected so that their junction capacitances are negligible, and the transformer, diode I/V characteristics, and LO level are selected to achieve a match. Silicon diodes having various barrier heights are available and can be selected to help achieve a match at a satisfactory LO level. The barrier heights of such diodes are adjusted by the selection of doping levels and anode metals, which result in various values of I_0 (5.7). Through these techniques, the knee voltage of the diode can be made as low as 0.3V, although these "low-barrier" diodes have relatively low breakdown voltages and high series resistances.

Diodes for both singly and doubly balanced mixers (which we shall consider in the next section) are often realized as two or four diodes on a single chip. Diodes fabricated in this manner have well-matched characteristics and identical parasitics. The use of such diodes helps to maintain the balance of the mixer.

The diodes' input impedance at the RF or IF port is defined as

$$Z_n = \frac{V_n}{I_n} \tag{5.40}$$

where n is the frequency component at mixing frequency ω_n in (5.25) (if $n < 0$, Z_n is the conjugate of the input impedance). Usually $n = 0$ for the IF and $n = 1$ for the RF. The RF input impedance of the mixer depends, as we have noted, on the IF termination as well as the diode characteristics and LO level. Even so, it is possible to define simultaneous conjugate match values for this impedance; the relation is the same as for conventional two-ports [5]. In general, the optimum IF output impedance is greater than the RF input impedance. Both impedances are high at low LO levels and decrease as LO level is increased.

Because the diodes are in parallel at the IF, each diode sees a 100Ω IF load impedance. Fortunately, as long as adequate LO power is available, this is usually near the optimum value.

5.5.5 Doubly Balanced Diode Mixers

Doubly balanced mixers use at least four diodes, most often in a ring configuration but occasionally in a star. Doubly balanced mixers have several advantages over singly balanced mixers:

1. Doubly balanced mixers reject all (m, n) spurious responses where m, n, or both are even.
2. The IF, RF, and LO ports are mutually isolated in all doubly balanced mixers. Therefore, with properly designed baluns, doubly balanced mixers can have overlapping RF, IF, and LO bands.

Because they use twice as many diodes, doubly balanced mixers require twice the LO power and have 3-dB greater IM intercept points (IP) than singly balanced mixers.

The Ring Mixer

Figure 5.24 shows the most common configuration for a doubly balanced diode mixer, a ring mixer. It is best understood as a switching mixer, in which the diodes are viewed as switches controlled by the LO. In Figure 5.24 points A and A' are virtual grounds for the LO; points B and B' are similarly virtual grounds for the RF. For one-half of the LO cycle, while point B' is positive and B is negative, diodes D1 and D2 are turned on and D3 and D4 are off. During the second half of the LO cycle, the voltage reverses and D3 and D4 are on, while D1 and D2 are off.

When D1 and D2 are switched on, the left half of the RF transformer's secondary is connected to the IF port through the secondary of the LO transformer. The current is in opposite directions in the two halves of the secondary, so no RF voltage is generated across the LO transformer's secondary; it is effectively a short circuit. When diodes D3 and D4 are turned on, the right half of the transformer is connected to the IF, and the polarity of the RF-to-IF connection is reversed. In this manner, the RF is connected to the IF but the polarity is switched every half-cycle.

This process can be described as

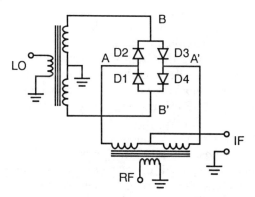

Figure 5.24 A doubly balanced diode ring mixer.

$$V_{IF} = s(t) V_{RF}(t) \qquad (5.41)$$

where $s(t)$ is a symmetrical square-wave switching function. As such, $s(t)$ has no dc component, so the IF contains no RF component. Furthermore, $s(t)$ has no even harmonics, so mixing with even harmonics of the LO is impossible; similar considerations can be used to show that mixing with even RF harmonics also cannot occur. Mixing occurs between the fundamental frequency of $s(t)$ and $V_{RF}(t)$; the resulting IF current excites both transformers in an even mode, so no IF voltage is developed across their secondaries and the transformers are "invisible" at the IF.

Ring mixers are often realized precisely as shown in Figure 5.24, with two transformers and four diodes. Such transformers can be realized in practice at frequencies up to about 500 MHz or, with care, perhaps 1 GHz.

For design purposes, the ring mixer can be reduced to a single-diode equivalent circuit [5]. When this is done, the source impedance seen by each diode is found to be $n^2 Z_{RF}$, where n is the turns ratio of the transformer (in a conventional trifilar transformer, $n = 2$). At the IF, the diode is terminated in $4 Z_{IF}$, where Z_{IF} is the IF load impedance. Finally, at odd LO harmonics the impedance seen by each diode is $n^2 Z_{LO}$, at even harmonics it is zero. This analysis assumes, of course, that the ideality of the transformer is maintained at all harmonics.

In an untuned, adequately pumped single-diode mixer, the optimum RF impedance is usually in the range 50 to 100Ω and the optimum IF load impedance is usually 100 to 200Ω. Thus, if a 50Ω source and load are used, and the diode's junction capacitance is negligible, the ring mixer with a 1:1 transformer usually has port VSWRs below 2.0. Conversion loss in this case is typically 6 to 8 dB.

Microwave Ring Mixer

Above 1 GHz, transformer hybrids are impractical, and distributed (transmission-line) baluns must be used. Microwave baluns differ from the ideal transformers in that they present an open circuit, not a short circuit, to IF currents. Therefore, microwave baluns have nothing that approximates the IF center tap of the transformer, and some other structure must be used to extract the IF. One common technique is simply to use IF blocking capacitors in the baluns and inductive connections from the diodes to the IF [5]; this may be adequate if wide IF bandwidth is not needed.

A better approach is to use a four-wire IF balun [6]. This allows the IF to overlap the RF or LO bands, and provides a very broad IF bandwidth [6]. Figure 5.25 shows a ring mixer using a pair of transmission-line baluns similar to those of Figure 5.16. No IF blocking capacitors are shown at the LO or RF ports because the balun should reject the IF inherently. If the IF bandwidth extends to very low frequencies, however, blocking capacitors may be necessary.

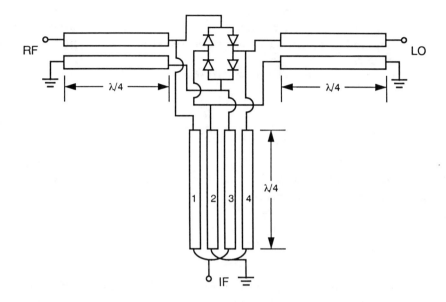

Figure 5.25 A realization of the doubly balanced ring mixer using transmission-line baluns and a four-wire IF balun. Balun lengths are λ/4 at the center of the RF/LO band. The four-wire balun allows the IF to overlap the RF or LO bands.

The IF balun consists of four coupled strips, λ/4 long at the center of the RF and LO frequency ranges. It operates as a transmission line for IF excitation, but as a set of stubs for RF and LO. Strips 1 and 3 in Figure 5.25 realize one side of the IF transmission line, and strips 2 and 4 realize the other. Thus, the odd-mode impedance, which is half the transmission-line impedance, should be equal to half the IF load impedance. Usually, the IF load is 50Ω, so the odd-mode impedance is 25Ω; as with other baluns, the even-mode impedance should be as high as possible. A Lange-coupler analysis can be used to determine the strip width and spacing.

Strips 1 and 3 are excited in an odd mode by the RF; strips 2 and 4 are similarly excited by the LO. Because these strips are shorted at the load end, they behave as high-impedance shorted stubs, their input impedance is high, and the IF balun has little effect on the RF or LO.

Effects of Imperfect Balance

Cancellation of AM LO noise and spurious responses, and the inherent port-to-port isolation of balanced mixers, occur because the currents generated by these phenomena ideally have equal amplitudes and phase differences of 180 deg. These currents cancel at the IF port. Because of imbalances in the hybrid, differences in the diodes' characteristics, or asymmetries in the layout or physical structure of the mixer, the amplitudes or phases may differ from the ideal. The currents then do not cancel completely.

The rejection R in dB under these circumstances is [5]

$$R = -10\log\left(\frac{1 - 2\sqrt{G}\cos(\theta) + G}{1 + 2\sqrt{G}\cos(\theta) + G}\right) \tag{5.42}$$

where G is the gain imbalance (a ratio) and θ is the phase imbalance. This expression is valid for all rejection phenomena.

5.5.6 Image-Rejection Mixers

The image-rejection mixer (Figure 5.26) is realized as the interconnection of a pair of balanced mixers. It is especially useful for applications where the image and RF bands overlap, or the image is too close to the RF to be rejected by a filter. The LO ports of the balanced mixers are driven in-phase, but the signals applied to the RF ports have a 90-deg phase difference. A 90-deg IF hybrid is used to separate the RF and the image bands. A full discussion of the operation of such mixers is a little complicated; such a discussion can be found in [5]. Equation (5.42) can be used to determine the image rejection.

The most difficult part of the design of an image-rejection mixer is the IF hybrid. If the IF frequency is fairly high, a conventional RF or microwave hybrid can be used. However, if the mixer requires a baseband IF, the designer is placed in the problematical position of trying to create a Hilbert-transforming filter, a theoretical impossibility. Fortunately, it is possible to approximate the operation of such a filter over a limited bandwidth; for specific design information see [7].

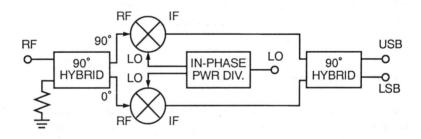

Figure 5.26 Image-rejection mixer.

5.6 ACTIVE FET MIXERS

Active FET mixers have several advantages, and some disadvantages, in comparison to diode mixers. Most significantly, an active mixer can achieve conversion gain, while diode and other passive mixers always exhibit loss. This allows a system using an active mixer to have one or two fewer stages of amplification; the resulting simplification is especially valuable in circuits where small size and low cost are vital. A precise comparison of distortion in diode and active FET mixers is difficult to make because the comparison depends on the details of the system. Generally, however, it is fair to say that distortion levels of well-designed active mixers are usually comparable to those of diode mixers.

Paradoxically, it is usually easy to achieve good conversion efficiency in active mixers even when the design is fundamentally poor. Thus, many research papers and trade journal articles have reported active mixers having good conversion gain but poor noise figures or high distortion. As a result, active FET mixers have gained a reputation for low performance. Nevertheless, achieving good overall performance in active FET mixers is not difficult; it involves only the observation of a few simple rules. We will discuss these in Section 5.6.2.

Because FETs cannot be reversed, as can diodes, balanced FET mixers invariably require an extra hybrid at the IF. This can be avoided only by using a *P*-channel device instead of an *N*-channel device, or vice versa; however, this is possible only in silicon circuits, and even then the characteristics of *P*- and *N*-channel devices are likely to be significantly different. We examine this further in Section 5.6.4.

5.6.1 Single-Device Active FET Mixers

As with diode mixers, the single-device FET mixer is minimally useful, but is a valid prototype for a balanced mixer. In a single-device FET mixer, filters must be used to separate the RF, IF, and LO, and it is rarely possible to have overlapping bands. As we shall see, LO-to-IF isolation is a special concern in FET mixers, and requires careful IF design.

Transconductance Mixers

Although a number of configurations are possible, the best mode of operation for a FET mixer is as a transconductance mixer. In this mode, the LO, applied to the gate of the device, modulates the FET's transconductance. In a conventional downconverter, the fundamental-frequency component of the transconductance is used for mixing. Maximizing this frequency component of the transconductance optimizes the gain, noise figure, and IM intercept points.

When the peak transconductance is limited, as it is in any FET, the fundamental LO-frequency transconductance component is maximized when the transconductance waveform is a rectangular pulse train having a peak value equal to the maximum transconductance and a 50% duty cycle. To realize this waveform, the FET must remain in current saturation throughout the LO cycle. This requires that (1) the drain voltage remain invariant at the dc value throughout the cycle, and (2) the FET's gate be biased at pinchoff. The first condition dictates that the drain termination for the LO be a short circuit; the second is easily satisfied through the use of external dc bias. When these conditions are met, the transconductance waveform will approximate a half-sinusoidal pulse train, which is as close to the optimum waveform as one can achieve in practice.

The RF signal is combined with the LO and applied to the gate. Mixing then occurs between the transconductance $g_m(t)$ and the RF signal. The small-signal drain current is

$$i_d(t) = g_m(t)v_{\mathrm{RF}}(t) \tag{5.43}$$

Usually the fundamental-frequency component of the transconductance is used.

Equation (5.43) is identical in form to (5.22), so one might expect a full conversion matrix to be necessary to describe an active FET mixer. In general, this is the case; however, in active FET mixers it is often possible to obtain a reasonably accurate analysis without so much analytical complexity. Unlike the diode, the FET inherently separates $v_{\mathrm{RF}}(t)$ at the gate and $i_d(t)$ at the drain; the FET's drain-to-gate coupling is usually very weak, especially when the input and output are tuned to different frequencies. Consequently, one can assume with reasonable accuracy that the small-signal gate voltage consists only of $v_{\mathrm{RF}}(t)$ and that the device is unilateral. The drain current $i_d(t)$ is then given directly by (5.43), and the IF component of the current can be found in a straightforward manner.

If feedback effects are ignored, the FET mixer's conversion gain is

$$G_c = \frac{G_{m,\mathrm{max}}R_L}{16\omega_{\mathrm{RF}}^2 C_{\mathrm{gs}}^2 (R_s + R_i + R_g)} \tag{5.44}$$

where $G_{m,\mathrm{max}}$ is the peak transconductance, R_L is the IF load resistance, and the other parameters are as shown in Figure 5.7. In the derivation of (5.44), we have assumed that the FET's gate is conjugate matched at the RF frequency. The minimum LO power required to pump the mixer when the gate is conjugate matched to the LO is

$$P_{\mathrm{LO}} = 0.5V_p^2\omega_p^2 C_{\mathrm{gs}}^2 (R_s + R_i + R_g) \tag{5.45}$$

We have assumed in deriving (5.45) that the gate is biased at the pinchoff voltage, V_p, and is pumped to a peak voltage of 0V. It is often difficult to match the gate successfully, especially at low frequencies where the gate's input impedance is very high. In this case, the required LO power may be greater than the value given by (5.45).

Equation (5.44) implies that the FET mixer's gain can be made arbitrarily large by selecting a high value of R_L. This conclusion is valid, within certain limits. The pumped FET's output impedance is remarkably high, on the order of a few thousand ohms, and for all practical purposes the FET's drain can be considered a current source. The value of R_L is limited by practical considerations, primarily realizability, bandwidth, and stability.

A practical difficulty in using FET mixers is that the FET is, after all, an amplifying device. As such, it amplifies the LO. To prevent high levels of LO leakage into the IF, the IF filter in a FET mixer must have considerably more rejection than would an LO filter in a diode mixer.

Another practical concern is amplifier-mode gain at the IF. Again, because the FET is inherently an amplifier, it can amplify an IF-frequency signal applied to or leaking into the input. If this signal is out-of-band noise from a low-noise amplifier, or even noise from a bias circuit or some other spurious noise source, the noise will be amplified and transmitted to the IF, making the mixer appear to have an extraordinarily high noise figure. Amplifier-mode gain is one of the major causes of high noise figures in FET mixers; it can be prevented by careful design of the input circuit.

Other Types of FET Mixers

Sometimes the LO is applied to the source of the FET instead of the gate. This reduces the level of the LO at the IF port slightly (although not as much as one might hope) and provides a modest degree of LO-to-RF isolation. It also provides a better LO match, especially at low frequencies, because the LO input impedance is approximately $1/<G_m>$, where $<G_m>$ is the time-average transconductance. This configuration requires careful bypassing of the source at the IF frequency or instability may result. See Section 5.6.2 for an example of a source-driven FET mixer.

The final option is, of course, to apply the LO to the drain. This is a distinctly different mode of operation, because varying the drain voltage of a FET in current saturation does not vary its transconductance. In order to achieve mixing, the LO voltage must be great enough to drive the drain voltage very low over part of the LO cycle, below the knee of the drain I/V characteristic, reducing both transconductance and drain-to-source resistance. This low average drain-to-source resistance reduces the FET's conversion gain significantly.

Because a large LO voltage is impressed on the drain, LO-to-IF isolation is likely to be poor. Furthermore, the IF filter must present an open circuit to the drain, not the optimum short circuit. This has a number of deleterious effects: it exacerbates intermodulation distortion, degrades stability, and creates amplifier-mode gain at the IF. This is not a good way to use a FET as a mixer.

5.6.2 Single-Device Mixer Circuits

Gate-Driven Transconductance Mixer

Figure 5.27 shows a single-device FET mixer. The RF/LO diplexer must combine the RF and LO and also provide matching between the FET's gate and both ports. The IF filter must provide an appropriate impedance to the drain of the FET at the IF frequency, and must short-circuit the drain at the RF and especially at the LO frequency and its harmonics.

Especially if the RF and LO are well separated in frequency, it may be virtually impossible to match the gate effectively at both the RF and LO frequencies. In this case, it is usually better to match the gate at the RF and accept a mismatch at the LO; the only cost will be increased LO power. Usually, it is simplest to combine the RF and LO with a separate filter diplexer. To prevent amplifier-mode gain, the RF matching circuit must reject the IF; an RF-frequency shorted stub at the FET's gate is adequate. The gate bias circuit often can provide the IF short.

As well as short-circuiting the drain at unwanted frequencies, the IF circuit must transform the IF load impedance (usually 50Ω) to an appropriate value at the FET's drain. At IFs below approximately 1 GHz, a 2:1 transformer provides an IF load of 200Ω; this provides good gain even with low-transconductance silicon MOSFETs, and, if stray capacitances are minimized, broad bandwidth. At higher frequencies, lumped or distributed matching circuits are necessary; with these, a high IF load impedance is more difficult to achieve. The RF/LO short circuit at the drain must be designed carefully. A single microstrip stub usually does not have adequate Q to provide an effective short circuit; a more complex filter circuit is necessary. Textbook filter designs are also inadequate; even if they provide high rejection out-of-band, the port impedance may be reactive, not necessarily a short circuit. A good

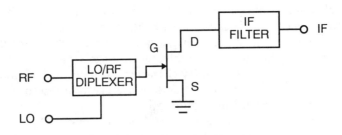

Figure 5.27 A single-device FET mixer.

approach is to use a textbook filter design as a starting point, and to optimize it numerically to achieve the desired terminations.

The only remaining dilemma in the design of a single-gate FET mixer is the determination of the input reflection coefficient. This can be found from nonlinear analysis or can be estimated. A good initial estimate of the input reflection coefficient can be found from S parameters of the device at approximately 50% I_{dss}, with a short-circuit load. The input reflection coefficient Γ_{in}, in terms of the FET's S parameters, is

$$\Gamma_{in} = S_{11} - \frac{S_{12}S_{21}}{1 + S_{22}} \tag{5.46}$$

Source-Driven Mixer

Figure 5.28 shows a single-device source-driven mixer that has been used successfully at frequencies below 2 GHz. The FET has a relatively large source resistor, R_s, that keeps the FET biased just above pinchoff when no LO power is applied. The value of this resistor is not critical; because the LO input impedance at the FET's source is much lower than R_s, the resistance has little effect on LO-port VSWR. The resistor R_g keeps the dc gate voltage at ground potential. The negative-going half-cycle of the LO waveform drives the FET's source negative, causing the FET to become active; the positive half-cycle keeps the FET turned off. In this way, the FET is switched in much the same manner as the gate-driven mixer.

In this circuit, the LO port is in series with the RF circuit; as a result, there is no significant RF-to-LO isolation. LO-to-RF isolation, however, is inherent in the

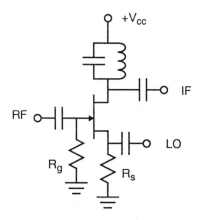

Figure 5.28 A single-device FET mixer having the LO applied to the source. This results in improved LO-to-RF isolation.

circuit. As with the gate-driven mixer, LO-to-IF isolation is inherently poor; the drain resonator must be designed carefully to reject LO leakage. Because the source is not bypassed for RF currents, the conversion gain of this mixer is lower than that of the gate-driven mixer.

5.6.3 Dual-Gate FET Mixer

Although several configurations and modes of operation of dual-gate mixers are possible [8], the configuration we shall examine, shown in Figure 5.29(a), usually provides the best performance in most receiver applications. In this circuit, the LO is connected to the gate closest to the drain (gate 2), while the RF is connected to the gate closest to the source (gate 1). An IF bypass filter is used at gate 2, and an

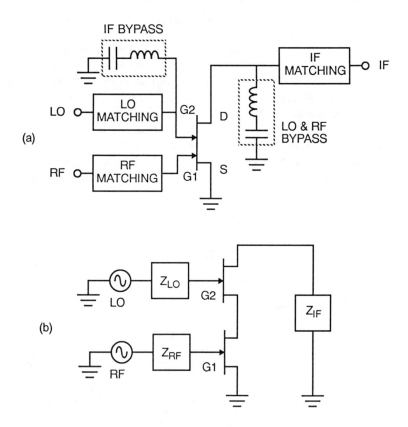

Figure 5.29 (a) A dual-gate FET mixer; (b) the FET mixer's equivalent circuit, in which the dual-gate FET is represented by two single-gate FETs in a cascode connection.

LO/RF filter, similar to those used in single-gate mixers, is used at the drain. Although not shown in the figure, dc bias must be provided to both gates and to the drain.

The operation of a dual-gate mixer is very different from that of a single-gate transconductance mixer; it is similar to that of a drain-pumped single-gate mixer. To understand the operation, we view the dual-gate FET as two single-gate FETs in a cascode connection (Figure 5.29(b)). In this view, it is evident that the upper FET acts as a source follower, applying LO voltage to the drain of the lower FET. The lower FET then provides mixing, operating as a drain-pumped mixer. We noted in Section 5.6.1 that mixing occurs in the drain-pumped mixer because the LO drives the drain voltage low enough, over part of the LO cycle, to push the FET into its linear region, and the FET provides mixing by a combination of time-varying transconductance and drain-to-source resistance. The average drain-to-source conductance is a parasitic resistance in parallel with the lower FET's drain, reducing the conversion efficiency. Because of this parasitic element, the dual-gate FET mixer is inferior in terms of conversion gain and noise figure to the active single-gate FET mixer. This inferiority is evident in published results: few reported dual-gate FET mixers have noise figures or gain as good as single-gate. The dual-gate mixer should be viewed as a tradeoff between a modest noise and gain performance and the low cost that results from the use of a single-device, unbalanced circuit, and fewer filters.

Even with lower gain and higher noise, properly designed dual-gate mixers have achieved third-order intercept points that modestly exceed those of single-gate mixers. It is not clear whether this is inherent in the dual-gate device or simply good fortune on the part of the designers. A reason often given is that the dual-gate mixer's I/V characteristic is more linear than the single-gate; however, dual-gate FET mixers usually do not operate in the most linear part of the characteristic. Whatever the reason, it is fair to state that dual-gate mixers have distortion at least as low as single-gate designs.

The IF current generated in the channel of the lower FET must pass through the channel of the upper FET; the latter acts as an inefficient IF amplifier stage. To operate as efficiently as possible, the gate of this FET must be grounded at the IF frequency; this is the purpose of the IF bypass filter in Figure 5.29(a). Furthermore, the drain voltage must remain high enough to keep the FET in its saturation region throughout the LO cycle; this requires the use of a filter to short-circuit the drain at the LO frequency. Of course, to provide good RF-to-IF as well as LO-to-IF isolation, this filter should short-circuit the RF as well.

5.6.4 Singly Balanced FET Mixers

In a singly balanced diode mixer, the LO and RF are applied to mutually isolated ports of a hybrid and the diodes are connected to the other pair of ports. No hybrid is needed in the IF, because the diode polarities are reversed. Unfortunately, no such reversal is possible in most types of FETs, so an IF hybrid is required. The only way to realize the analog of a reversed diode is to use a *P*-channel device for one of the FETs and an *N*-channel device for the other. Even in technologies where both *P*- and *N*-channel devices can be realized, this may not be a good thing to do. Because of the differences in *P* and *N* materials, *P*- and *N*-channel devices can never be identical, and in fact are often substantially different. Unfortunately, to achieve the benefits of a balanced mixer, the characteristics of the FETs must be closely matched.

Figure 5.30 shows one type of singly balanced FET mixer. Although this mixer uses a transformer hybrid for the LO and RF, any appropriate type of hybrid can be used. A matching circuit is needed at the gates of both FETs; to prevent unbalancing of the mixer, the circuits should not be tuned independently. Although the bias circuitry is not shown in the figure, dc bias must also be applied to the gates. The IF filters provide the requisite short circuits to the drains at the LO and RF frequencies, and additionally provide IF load impedance transformations.

The singly balanced mixer of Figure 5.30 is effectively two single-device mixers interconnected by hybrids. Thus, the design considerations for single-device mixers apply equally to this type of doubly balanced mixer.

Differential Mixer

Figure 5.31 shows a mixer that is similar in topology to a differential amplifier. The RF is applied to the lower FET, and the LO is applied through a balun or hybrid to the upper FETs.

This mixer operates as an alternating switch, connecting the drain of the lower FET alternately to the inputs of the IF balun. The node connecting the sources of

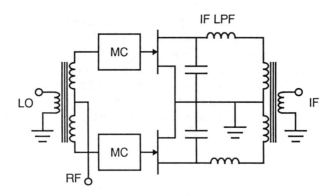

Figure 5.30 A singly balanced FET mixer. MC is the RF/LO matching circuit, and IF LPF is the IF output lowpass filter. In contrast with the singly balanced diode mixer, a transformer or hybrid is needed at the output.

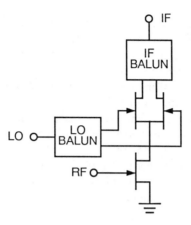

Figure 5.31 Differential mixer. This is a type of singly balanced mixer. Matching circuits and an IF filter may be necessary; these are not shown.

the upper FETs is a virtual ground for the LO, and therefore the LO simply switches the two upper FETs on and off on alternate half cycles. The upper FETs are biased slightly above pinchoff, so each is turned on over at least 50% of the LO cycle. In this way one of the upper FETs is always on, so the lower FET remains in its saturation region throughout the LO cycle. When on, the upper FETs remain in their linear regions. To accomplish this, the lower FETs must be biased between pinchoff and approximately $0.3\ I_{dss}$; higher bias may cause the lower FET to drop into its linear region over part of the LO cycle.

An LO matching circuit may be needed. Because the RF and LO circuits are separate, the gates of the upper FETs can be matched at the LO frequency and there is no tradeoff, as there often is in single-device mixers, between effective LO and RF matching. Similarly, the lower FET can be matched effectively at the RF. An IF filter, not shown in the figure, is necessary to reject LO current; in theory, the output balun rejects the RF, but in practice its bandwidth may not be great enough to do so. Because the upper FETs are not biased in current saturation, the LO short circuit at the FETs' drains, which is extremely important in other active mixers, is not quite as critical in this circuit.

The design of this mixer is straightforward. Because the source node of the upper FETs is a virtual ground, the input impedance is simply that of the input loop of the FET. The LO input impedance Z_{LO} is approximately

$$Z_{LO} \cong R_s + R_g + R_i + \frac{1}{\omega_{LO} C_{gs}} \qquad (5.47)$$

where ω_{LO} is the LO radian frequency and the other variables are defined in Figure 5.7. The lower FET is effectively an amplifier. Its load impedance is the impedance looking into the sources of the upper FETs, approximately the "on" resistance of the channels. Its input impedance can be found from S parameters. Finally, the IF circuit is designed in the usual manner: to present a moderately high IF impedance to the drains of the FETs, but a short circuit to LO and RF frequencies. The design of these matching and filtering structures is discussed in Section 5.6.2.

A version of this mixer that does not require an LO balun is shown in Figure 5.32. In this circuit, the source node of the upper FETs is no longer a virtual ground, so, to prevent imbalance, the lower FET must have a high drain-to-source resistance. This is usually not the case in GaAs FETs, but in MOSFETs and silicon junction FETs the resistance is usually adequate.

5.6.5 Doubly Balanced FET Mixers

Gilbert Multiplier

The venerable Gilbert analog multiplier [9] is frequently used as an RF or microwave mixer.[4] The classical Gilbert mixer depends on a cancellation between the exponential I/V and logarithmic V/I characteristics of bipolar transistors to produce an accurate

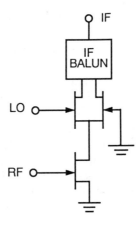

Figure 5.32 A version of the differential mixer that does not require an LO balun. Because the source node of the upper FETs is no longer a virtual ground, this works well only if the lower FET has a high drain-to-source resistance.

---

[4]Although Gilbert's original paper described a multiplier using bipolar transistors, FET implementations having the same circuit topology are in common use, and are named in honor of the original invention.

product characteristic. However, any circuit having the same topology, using either bipolar transistors or FETs, cancels the even-degree parts of the devices' *I/V* characteristics and preserves the odd-degree parts, producing multiplication. Any such circuit is therefore a balanced mixer.

Figure 5.33 shows a doubly balanced FET mixer based on the Gilbert multiplier. Like many doubly balanced mixers, this mixer consists of two of the singly balanced mixers shown in Figure 5.31. Each half of the mixer operates in the same manner as that of Figure 5.31 and is designed in exactly the same manner. The interconnection of the outputs, however, causes the drains of the upper four FETs to be virtual grounds for both the LO and RF, as well as for even-order spurious responses and IM products.

5.7 PASSIVE FET MIXERS

Passive FET mixers, often called *FET resistive mixers*, are used frequently in wireless applications. They have a number of advantages over diode and active FET mixers.

In a FET resistive mixer, the time-varying channel resistance is used for frequency conversion. The LO is applied to the gate, but no dc bias is applied to the channel (the gate may be dc biased, usually near pinchoff). The FET then operates entirely in its linear region, where the channel is effectively a gate-voltage-controlled

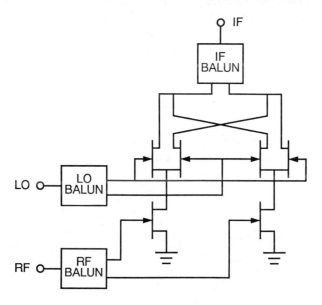

Figure 5.33 A doubly balanced mixer based on the Gilbert multiplier. It consists of two of the singly balanced mixers shown in Figure 5.31.

resistor. The RF signal is applied to the drain, and IF currents are filtered from the drain or source.

The primary advantage of the FET resistive mixer is its very low levels of intermodulation distortion. In diode or active FET mixers, the same nonlinearity that provides mixing causes intermodulation distortion. A strong nonlinearity is needed for efficient mixing, but that same nonlinearity causes distortion. In a FET resistive mixer, no such nonlinearity is employed; the channel resistance can be perfectly linear and mixing is still achieved. In fact, the channel resistance at low signal levels is very linear, so low distortion results; it is not unusual to achieve third-order input intercept points above 30 dBm in such mixers. Conversion loss and LO power are comparable to diode mixers.

The noise generated in the channel of a resistive FET is entirely thermal in origin, causing the mixer's single-sideband noise figure in decibels to be equal at room temperature to its conversion loss. This contrasts with diode mixers, which have substantial shot noise, and whose SSB noise figures usually exceed the conversion loss by at least a few tenths of a decibel.

Single-device resistive FET mixers are very practical components, especially at frequencies below approximately 2 GHz. At these low frequencies, the gate-to-channel capacitance may be negligible and LO-to-RF and LO-to-IF isolation is very good. Unfortunately, the gate-to-drain capacitance in an unbiased FET is as great as the gate-to-source capacitance, so at high frequencies the LO-to-RF isolation of a single-device mixer may not be adequate. Furthermore, the LO voltage on the drain increases the mixer distortion. Balanced FET resistive mixers also exist. In the best of these, both the drain and source terminals of each FET are virtual grounds, so the coupling problem is inherently solved.

5.7.1 Single-Device FET Resistive Mixers

Figure 5.34 shows the basic FET resistive mixer. As with other mixers, filters are used to separate the various mixing frequencies, to short-circuit unwanted mixing

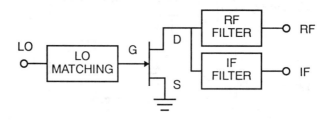

Figure 5.34 A FET resistive mixer. The drain is unbiased and mixing occurs because of the time-varying channel conductance.

products, and, if necessary, to provide matching. The design of a FET resistive mixer has a number of now-familiar requirements:

1. The drain of the FET must be short-circuited at the LO frequency. Short-circuiting it at LO harmonics is less important, because the capacitive nonlinearity generates only weak LO harmonics at the drain. The short circuit prevents LO voltages from appearing at the drain. This minimizes distortion by keeping the drain voltage at zero (except, of course, for the small-signal RF and IF voltages) where it is most linear.
2. The RF circuit must provide filtering to prevent IF leakage into the RF port, as well as matching. Matching is not particularly difficult; the real part of the RF input impedance at the FET's drain of a conventional 0.5 μm by 250 μm Ku-band FET, biased at pinchoff, is close to 50Ω.
3. Similarly, the IF must provide both filtering and matching. As with diode mixers, the IF output impedance is usually higher than the RF input impedance, approximately 100Ω.
4. In order to minimize LO power, the gate must be matched. If the LO short at the drain is well realized, the input impedance is given approximately by (5.47). Ideally, the gate should be short-circuited at the RF frequency, although at low frequencies RF-to-LO coupling may be low enough that this may not matter.
5. The optimum dc gate bias is usually at or below pinchoff. Distortion is lowest when the LO gate voltage variation is as great as possible and swings from the threshold of avalanche breakdown to the threshold of gate conduction. Such a large gate voltage may adversely affect reliability; fortunately, the voltage can be reduced a few decibels with minimal effect on distortion.
6. The drain must be short circuited at dc. A dc return path at the drain is usually required. This can be implemented as part of the RF or IF filter.

Figure 5.35 shows an example of a practical FET resistive mixer for use at frequencies below approximately 1 GHz. The RF is applied to the drain, and the IF is filtered from the source (although, in an unbiased FET there is little difference between the source and the drain). Because of the low frequency, the real parts of the input and output impedances are close to 50Ω and the reactive parts are negligible. In a monolithic circuit, the device width can be adjusted to optimize the port impedances. The parallel L-C tuned circuits ω_{RF} and ω_{IF} provide a short circuit to both the drain and source at the LO frequency; they also help to guarantee good LO-to-RF and LO-to-IF isolation.

The LO input impedance at the gate is virtually impossible to match at low frequencies. The value R_g, which is necessary for proper dc bias, can be selected to provide a good VSWR. If VSWR is not important, R_g can have a large value; in this

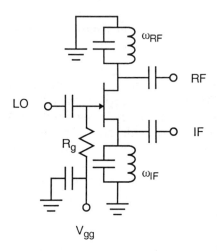

Figure 5.35 A series-switch FET resistive mixer.

case, the required gate voltage can be achieved with up to 6 dB less available LO power.

Even though this is a single-device mixer, its performance is in many ways similar to a balanced mixer. Because it uses a resistive FET, its distortion is very low; even-order distortion is often lower than that of a balanced diode mixer. AM LO noise is largely rejected by the weak gate-to-drain coupling. Port-to-port isolation, also attributed to the weak coupling, is inherently very good.

5.7.2 Balanced FET Resistive Mixers

Balanced FET resistive mixers have essentially the same advantages mixers as balanced diode or active-FET mixers: inherent port isolation, rejection of AM LO noise, and rejection of certain spurious responses and IM products. Figure 5.36 shows a singly balanced FET resistive mixer. At low frequencies, the LO balun can be realized by a transformer, as shown; at higher frequencies, a 180-deg transmission line or one of the structures in Figures 5.16 to 5.19 could be used. The LO pumps the two FETs 180-deg out of phase, but the RF is applied in phase at the drains. The IF currents in the FETs' channels therefore have a 180-deg phase difference, and an output hybrid or balun is necessary to subtract them. Of course, the two FETs generally require twice the LO power of a single FET, and have 3-dB greater IM intercept points. The mixer rejects all (2, 2) and (1, 2) spurious responses; (2, 1) responses are not rejected.

As with other types of gate-driven FET mixers, the LO input impedance at low frequencies is virtually an open circuit, and is thus almost impossible to match.

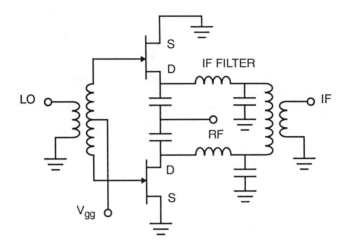

Figure 5.36 Singly balanced FET resistive mixer. The RF connection is a virtual ground for both the LO and the IF.

It is important to note in this case that the task of the LO circuit is not simply to provide a low LO VSWR, but also to maximize the LO gate voltage for a given available power. If matching is impossible, the next best option is to transform the source impedance to a high value, increasing the open-circuit LO voltage. The increase in source impedance is limited primarily by the gate capacitance of the FETs and stray capacitances of the circuit, and practical limitations in the realization of the transformer. Another advantage of an LO transformer is the short circuit it provides to the RF at the gate. This is the optimum termination.

The RF input impedance is half that of the single-device mixer. This may be uncomfortably low, on the order of 25Ω. To achieve an adequate VSWR, an RF matching circuit (not shown in Figure 5.36) may be necessary. The FETs are effectively in series at the IF, and the common 2:1 transformer (i.e., 4:1 impedance ratio) usually terminates the IF quite well.

An attractive feature of this mixer is that the drain-connection node is a virtual ground for the LO and IF; thus, no drain RF filter is needed. Similarly, the RF drives the IF transformer in an even mode, so RF leakage is inherently rejected from the IF port. An IF filter is still needed, however, to prevent LO leakage and to prevent the IF transformer from short-circuiting the RF signal at the FETs' drains.

5.7.3 FET Resistive Ring Mixer

The ring mixer in Figure 5.37 is a type of commutating mixer and operates in a manner analogous to the diode-ring mixer (Figures 5.24 and 5.25). As with the

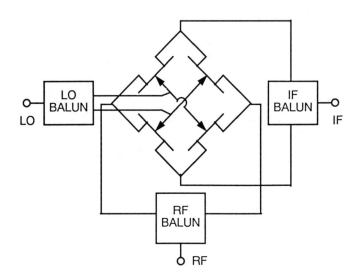

Figure 5.37 Commutating FET ring mixer. All ports are inherently isolated and no matching circuits are needed at low frequencies.

diode ring, the LO causes alternate pairs of devices to turn on, switching the polarity of the RF connection to the IF. This causes the RF to be multiplied by the switching waveform, and mixing between the RF signal and the fundamental frequency component of the switching waveform generates the IF output.

An important property of the FET ring mixer is that the nodes of the ring are all virtual grounds for the LO, the IF connection nodes are virtual grounds for the RF, and the RF connections for the IF. The gates are also virtual grounds for both the RF and IF. As with the diode mixer, the topology of the circuit isolates the ports, making filters unnecessary. The result is that the mixer's bandwidth is limited primarily by the RF, IF, and LO baluns.

The impedances of this mixer are not the same as those of the diode ring. The RF input and IF output impedances at the terminals of the ring are the same as those of a single device; thus, if transformers are used, a 1:1 is best for the RF and IF. (A slightly higher turns ratio might be better for the IF, but transformers having noninteger turns ratios usually have poor performance.)

Optimizing the distortion of a FET ring mixer is more or less automatic: the FETs effectively short-circuit each other at the unwanted mixing frequencies, inherently providing ideal terminations. No filtering or special efforts are required to achieve the optimum terminations. If GaAs devices are used, dc gate bias is necessary, and the gates should be biased near pinchoff. In enhancement MOSFETs having a threshold voltage around 1V, dc bias may be unnecessary.

5.7.4 Modeling FETs for Resistive Mixer Analysis

Conventional FET models, discussed in Section 5.3.2, are adequate for conversion-loss analysis of resistive FET mixers, but because the *tanh* function does not model the drain nonlinearities adequately, these models should not be used for IM analysis. The *I/V* function should be fit to measured data near zero drain-to-source voltage, because this is where the device is used; a poor fit to the *I/V* characteristic in the current-saturation region is tolerable. To achieve good LO power estimates, the equivalent-circuit element values used in (5.45) and the pinchoff voltage should be well characterized.

The equivalent circuit of a resistive FET need not be as complex as an active FET. Figure 5.38 shows the equivalent circuit of a FET used as a resistive mixer, including only the most important elements. A number of the elements in Figure 5.7 have been eliminated as unnecessary: R_{ds} is included implicit in I_d, and C_{ds}, which was part of the FET's biased channel, does not exist in the unbiased device (a small parasitic capacitance may remain; however, this is completely swamped by the low average value of R_{ds}). Similarly, R_i is usually unnecessary. The diodes used to model the gate-to-channel conduction in junction-gate devices should be included, as these present an important limit to achievable LO level. Note that the FET's gate-to-drain capacitance C_{gd}, when the drain is unbiased, is much greater than the biased value, and C_{gs} is somewhat less.

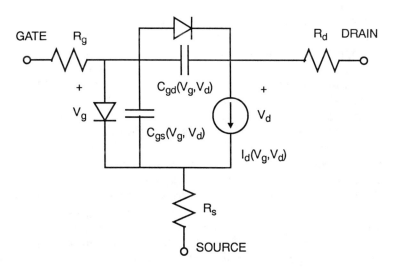

Figure 5.38 Equivalent circuit of a FET for resistive mixer analysis. A number of elements from Figure 5.7 have been eliminated; these are not necessary in resistive FET mixer analysis.

5.7.5 Mixer Comparison Table

Table 5.1 contains a brief comparison of mixer types and a qualitative description of their characteristics. Although such comparisons contain only a limited amount of information, they apparently are useful to readers. A similar table appears in [5].

5.8 HARMONIC-BALANCE SIMULATION OF MIXERS

A very important development for mixer design is the availability of general-purpose nonlinear-circuit simulators. These simulators, which use harmonic-balance analysis, are capable of analyzing and sometimes optimizing mixers having arbitrary circuit topology. As valuable as they are, their use requires some judgment, or invalid or inaccurate calculations may result.

Classical harmonic-balance analysis is applicable only to a circuit having a period excitation; the waveforms in a mixer arise from the combination of two noncommensurate frequencies, the RF and LO, and clearly the combination is not periodic. Fortunately, it is possible to modify the classical method to allow its application to mixers. Two approaches are possible. The first, called *multitone harmonic balance*, involves the use of aperiodic Fourier transforms to allow a limited number of noncommensurate excitations. The second, called *conversion-matrix analysis* (outlined in Section 5.4), is considerably more accurate and orders of magnitude faster for conversion efficiency and port-impedance calculations.[5] Most simulators offer both methods; because it is much faster and more accurate, only the conversion-matrix method should be used in mixer design.

Harmonic-balance analysis is an iterative method. It estimates a solution and iteratively improves the solution until the errors are acceptably low. As such, it is subject to convergence difficulties; the stronger the nonlinearity and the stronger the excitation, the greater the likelihood of convergence failure. Simulators often use a "source stepping" technique, wherein the excitation is increased stepwise, allowing convergence to be achieved more easily at each level. More sophisticated methods for improving convergence have been developed, but few have been implemented in commercial products.

Errors in models and circuit descriptions are a common cause of convergence failure. Chief among these are open-circuiting a diode at dc and using a value of threshold voltage in FET mixers that is not precisely the zero of $I_d(V_g)$. Any discontinuity in an I/V characteristic is likely to result in poor convergence.

The manner in which the design process is implemented does much to guarantee the success of a harmonic-balance simulation. In the first stage of the design, the circuit should be as simple as possible: ideal hybrids or transformers should be used,

---

[5]The author recognizes that this is a controversial statement, and is subject to strong objections, primarily from the marketing departments of software companies.

Table 5.1
Mixer Comparison Table

Mixer Type	VSWR			Port-to-Port Isolation			LO AM Noise Rej.	LO Spur. Sig. Rej.
	RF	LO	IF	RF/IF	LO/RF	LO/RF		
Single-diode	Depends on matching circuits	Depends on matching circuits	Depends on matching circuits	Depends on filters	Depends on filters	Depends on filters	None	None
Singly balanced diode (180 deg)		Same as single-diode		Depends on filters	Good; equal to hybrid isolation	Depends on filters	Good	Good
Singly balanced diode (90 deg)	Good	Good	Same as single-diode	Depends on filters	Poor	Depends on filters	Good	Good
Doubly balanced diode	Good	Good	Good	Good	Good	Good	Good	Good
Image rejection (diode)	Good	Good	Good	Good	Good	Good	Good	Good
Resistive FET ring	Good	Good	Good	Good	Good	Good	Good	Good

Mixer Type	Low-Order Spurious-Response Rejection	LO Power Requirements	Third-Order IM Intercept	1-dB Compression
Single-diode	None	Low	Low	Low
Singly balanced diode (180 deg)	(2, 2): good (2, 1): good if Δ port is the LO (1, 2): good if Σ port is the LO	Moderate	Moderate	Moderate
Singly balanced diode (90 deg)	(2,2): good (2, 1): none (1, 2): none	Moderate	Moderate	Moderate
Doubly balanced diode	(2,2): good (2, 1): good (1, 2): good	High	High	High
Image rejection (diode)	Same as doubly balanced mixer	High to very high (depends on type)	High	High
Resistive FET ring	Very good	High	Very good	Very high

Table 5.1 (continued)

Mixer Type	VSWR			Port-to-Port Isolation			LO AM Noise Rej.	LO Spur. Sig. Rej.
	RF	LO	IF	RF/IF	LO/RF	LO/RF		
Single-device active FET	Depends on matching circuits	Depends on matching circuits	Very high	Depends on filters	Depends on filters	Depends on filters; usually poor	None	None
Singly balanced active FET	Depends on matching circuits	Depends on matching circuits	Very high	Good (if RF port is Δ)	Good (if LO port is Δ)	Good (if LO port is Δ)	Good	Good
Single-device dual-gate FET	Depends on matching circuits	Depends on matching circuits; usually poor	Very high	Depends on filters	Good	Depends on filters; usually poor	None	None
Gilbert-cell FET	Depends on matching circuits	Depends on matching circuits	High	Good	Good	Good	Good	Good

Mixer Type	Low-Order Spurious-Response Rejection	LO Power Requirements	Third-Order IM Intercept	1-dB Compression
Single-device active FET	None	Low	Low	Low
Singly balanced active FET	(2, 2): good (2, 1): good if Δ port is the RF (1, 2): good if Σ port is RF	Moderate	Moderate	Moderate
Single-device dual-gate FET	None	Low	Moderate	Low
Gilbert-cell FET	(2., 2): good (2, 1): good (1, 2): good	High	High	High

bias circuits should be isolated by high-value RF chokes, and so forth. Real elements then can be substituted one by one for the ideal ones. Trying to begin with the analysis of a complete design usually is unsuccessful because the large number of circuit elements makes it difficult to debug a failed analysis.

References

[1] N. Camilleri, J. Costa, D. Lovelace, and D. Ngo, "Silicon MOSFETs, the Microwave Device Technology for the 90's," *1993 IEEE MTT-S International Microwave Symposium Digest*, June 1993, p. 545.

[2] W. R. Curtice, "A MESFET Model for Use in the Design of GaAs ICs" *IEEE Trans. Microwave Theory Tech.*, Vol. MTT-28, 1980, p. 448.

[3] I. Angelov, H. Zirath, and N. Rorsman, "Validation of a Nonlinear HEMT Model by Power Spectrum Characteristics," *1994 IEEE MTT-S International Microwave Symposium Digest*, June 1994, p. 1571.

[4] C. Y. Ho, "Design of Wideband Quadrature Couplers for UHF/VHF: Part I," *RF Design*, Nov. 1989, p. 58.

[5] S. A. Maas, *Microwave Mixers* (Second ed.), Norwood, MA: Artech House, 1993.

[6] D. Neuf, "M/W Mixer Using 4-Wire Transmission Line," *Microwave J.*, May, 1974, p. 48.

[7] M. G. Ellis, *Electronic Filter Analysis and Synthesis*, Norwood, MA: Artech House, 1994.

[8] C. Tsironis, R. Meierer, and R. Stahlman, "Dual-Gate MESFET Mixers," *IEEE Trans. Microwave Theory Tech.*, Vol. MTT-32, March 1984, p. 248.

[9] B. Gilbert, "A Precise Four-Quadrant Multiplier with Subnanosecond Response," *IEEE J. Solid-State Circuits*, Vol. SC-3, Dec. 1968, p. 365.

CHAPTER 6
▼▼▼

RF AND MICROWAVE FREQUENCY SYNTHESIZER DESIGN

Lawrence Larson and Steven Rosenbaum
Hughes Research Laboratories
Malibu, CA

6.1 INTRODUCTION

Frequency synthesizers are ubiquitous building blocks in wireless communications systems, since they provide the precise reference frequencies for modulation and demodulation of baseband signals up to the transmit and/or receive frequencies. Because of their crucial nature in the implementation of these systems, considerable effort has been expended over the years in an attempt to improve their performance and reduce their cost, to the point where they are now standard cells in completely integrated implementations.

A simple frequency synthesizer might consist of a transistor oscillator operating at a single frequency determined by a precise crystal circuit. Tunable transistor frequency sources rely on variations in the characteristics of a resonant circuit to set the frequency. These circuits can then be embedded in PLLs to broaden their range of operation and further enhance their performance. More recently, direct digital synthesis (DDS) has been developed as a technique for synthesizing a broad range of frequencies and waveforms using digital techniques. All of these different approaches will be covered in this chapter.

A representative view of a frequency synthesizer is illustrated in Figure 6.1, which shows a generic synthesizer producing a single tone of a given amplitude that has a delta-function-like characteristic in the frequency domain. However, the frequency characteristics of a realistic signal source will have a finite width in the frequency domain, as well as significant harmonic content at frequencies higher than the fundamental tone. The results of these imperfections in the frequency domain can result in degraded overall system performance, as the following simple example will illustrate.

A wireless communication system will usually consist of an information source, which is modulated up to RF/microwave frequencies and then transmitted. A receiver will take the modulated signal from the antenna, demodulate it, and send it to an information "sink," as illustrated in Figure 6.2. The rate at which information can be sent over the channel is determined by the available bandwidth (which is strictly regulated by national and international regulatory agencies), the modulation scheme, and by the integrity of the modulation/demodulation process.

This latter effect can be heavily influenced by the quality of the frequency synthesis, the noise floor of the synthesized signal, and the absence of distortion and

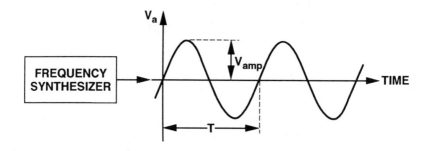

Figure 6.1 Block diagram of frequency synthesizer producing single-tone sinusoidal output.

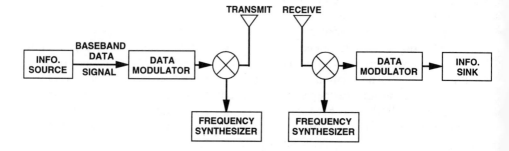

Figure 6.2 Block diagram of communications system, showing modulation and demodulation.

other spurious responses. For example, the bit error probability in a typical phase-shift keyed (PSK) system is approximately

$$P_e \approx \frac{1}{2}\operatorname{erfc}\left(\sqrt{S/N}\right) \tag{6.1}$$

where S is the signal power and N is the noise power in the frequency band of interest. The noise power is heavily influenced by integrity of the reference oscillator used in the transmitter and receiver.

The transmitted signal power (S) is usually fixed by the system. For example, a satellite communications system typically has a fixed transponder power limit. Thus, the most straightforward way to improve the bit error rate (BER) of the system is to reduce the noise power (N) in the band of interest. This may require improved noise performance of the frequency synthesizer portion of the receiver.

This chapter will cover the essential elements of frequency synthesizer design, utilizing both analog and digital techniques. The emphasis will be on practical design considerations, rather than theoretical depth. Where appropriate, we will refer the reader to other publications that cover the material more thoroughly.

6.2 FREQUENCY SYNTHESIZER FUNDAMENTALS

The multiplicity of frequency synthesizers makes straightforward classification difficult, but they can be generally categorized into direct and indirect architectures. Direct frequency synthesizer architectures utilize no feedback to generate the appropriate output frequencies. These direct synthesizers can be further subdivided into digital and analog approaches. Analog approaches utilize fixed precise frequency sources, combined with frequency multipliers, dividers, switches, filters, and mixers to generate the necessary output frequencies. A schematic of this generic approach is given in Figure 6.3. Direct analog synthesis is very straightforward conceptually, but has a number of drawbacks if multiple frequency and/or fine frequency control are required.

In particular, the physical size of the components required to implement the multipliers, dividers, mixers and filters can be daunting. Furthermore, these synthesizers can generate a menagerie of spurious responses, which must be carefully filtered if a clean output is required. One advantage of the direct synthesizer approach is that the fundamental waveform can be generated using an ultraclean source—for example, generated by a crystal source with a high Q—and the phase noise characteristics of the resulting higher frequency waveform can be excellent. This will be discussed in more detail in Section 6.3.

By contrast, direct *digital* frequency synthesizers (DDSs) rely on digital components to generate a variety of frequency and waveforms from a single system reference

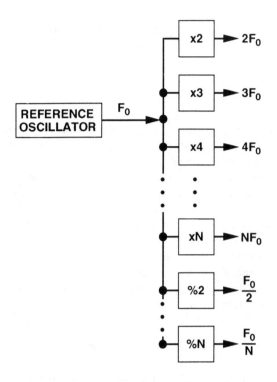

Figure 6.3 Direct analog frequency synthesizer utilizing a reference oscillator, along with frequency multipliers and dividers.

clock. The advantage of this approach is the precision available from an all-digital approach due to the lack of tuning required, as well as absence of drift and scalability with constantly improving digital IC technology. A block diagram of a typical direct digital frequency synthesizer is shown in Figure 6.4. There are four significant blocks

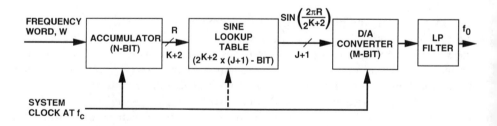

Figure 6.4 Direct digital synthesizer (DDS), utilizing an accumulator, sine lookup table, DAC, and smoothing filter.

for a typical DDS: a phase accumulator, sine lookup table (these first two together are often referred to as a numerically controlled oscillator (NCO)), digital-to-analog converter (DAC), and antialiasing filter.

By contrast, indirect frequency synthesizers rely on feedback, usually in the form of a PLL, to synthesize the frequency. A block diagram of a representative PLL frequency synthesizer is shown in Figure 6.5. Most PLLs contain three basic building blocks: a phase detector, an amplifier/loop filter, and a voltage-controlled oscillator (VCO). During operation, the loop will acquire (or lock) onto an input signal, track it, and exhibit a fixed phase relationship with respect to the input. The output frequency of the loop can be varied by altering the division ratio (N) within the loop, or by tuning the input frequency with an input frequency divider (Q). As such, the PLL can act as a broadband frequency synthesizer. Details of the design of the PLL will be developed in Section 6.4.

As with direct frequency synthesis techniques, indirect techniques can be implemented in both digital and analog forms. Purely digital PLLs are often implemented for clock generation on digital VLSI chips.

6.2.1 Frequency Synthesizer Figures of Merit

An ideal frequency synthesizer will produce a perfectly pure sinusoidal signal, which is tunable over some specified bandwidth. The amplitude, phase, and frequency of the source would not change under varying loading, bias, or temperature conditions. Of course, such an ideal circuit is impossible to realize in practice, and a variety of performance measures have been defined over the years to characterize the deviation from the ideal.

Noise

The output power of the synthesizer is not concentrated exclusively at the carrier frequency alone. Instead, it is distributed around it, and the spectral distribution on either side of the carrier is known as spectral sidebands. This is illustrated schematically in Figure 6.6. This noise can be represented as modulation of the carrier signal,

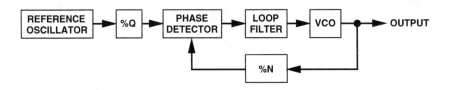

Figure 6.5 Indirect frequency synthesizer using a phase-locked loop.

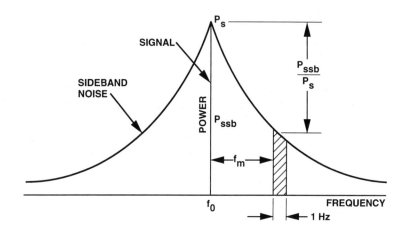

Figure 6.6 Phase noise specification of frequency source. The noise is contained in the sidebands around the signal frequency at f_0.

and resolved into amplitude modulation (AM) and frequency modulation (FM) components. The AM portion of the signal is typically smaller than the FM portion.

FM noise power is represented as a ratio of the power in some specified bandwidth (usually 1 Hz) in one sideband to the power in the carrier signal itself. These ratios are usually specified in the quantity of dBc/Hz at some frequency offset from the carrier. The entire noise power can be integrated over a specified bandwidth to realize a total angular *error* in the output of the oscillator, and oscillators are often specified this way. The calculation of the angular error from the FM noise performance is relatively straightforward, and a simple technique for accomplishing this was detailed in Chapter 2.

Frequency Pulling

Frequency pulling is the change in output frequency of an oscillator when the phase of the load VSWR is rotated by 360 deg. The allowed frequency range determines the amount of isolation required at the output of the oscillator. A load VSWR of 1.5:1 is typically used for test purposes.

A simplified expression for the total frequency deviation from the ideal can be given by [1]

$$\Delta \omega_T \cong \frac{\omega_0}{2Q_{ext}}(\text{VSWR} - \text{VSWR}^{-1}) \tag{6.2}$$

where Q_{ext} is the quality factor of the oscillator, and ω_0 is the radian frequency of oscillation. This expression demonstrates the importance of high Q in minimizing frequency pulling in oscillator design.

Power Output and Stability

The power output and stability of the oscillator are usually specified as a power output range in dBm at a given frequency and over a temperature range, say from 0°C to 60°C.

Tuning Range

The tuning range of an oscillator specifies the variation in output frequency with input voltage or current (usually voltage). The slope of this variation is usually expressed in MHz/V. In particular, the key requirements of oscillator or synthesizer tuning are that the slope of the frequency variation remain relatively consistent over the entire range of tuning and that the total frequency variation achieves some minimum specified value.

Frequency Stability

Frequency stability of an oscillator is typically specified in terms of parts per million per degree centigrade (PPM/°C) or ±PPM/°C. This parameter is related to the Q of the resonator and the frequency variation of the resonator with temperature. In a free-running system, this parameter is particularly important, whereas in a PLL it is less so since an oscillator that drifts may be locked to a more stable oscillator source.

Frequency Pushing

Frequency pushing is the effect on the output frequency due to changes in the dc power supply voltage, and is typically specified in terms of percent change in output frequency per volt. Values of less than 0.001 percent per volt are not uncommon for high Q oscillators.

Harmonics

Harmonics are outputs from the oscillator/synthesizer that occur at integral multiples of the fundamental frequencies. They are typically caused by nonlinearities on the transistor or other active device used to produce the signal. They can be minimized by proper biasing of the active device and design of the output matching network to filter out the harmonics. Harmonics are typically specified as a dBc below the carrier.

Spurious Outputs

Spurious outputs are outputs of the oscillator/synthesizer that are not necessarily harmonically related to the fundamental output signal. As with harmonics, they are typically specified as a dBc below the carrier.

Spurious responses are a major limitation in DDS, since the output of the DDS is a essentially a periodic repetition of an approximation to the ideal waveform. This effect, combined with aliasing and nonidealities in the DAC can create a significant spurious output. This will be covered in more detail in Section 6.5.

6.3 VOLTAGE-CONTROLLED OSCILLATORS

6.3.1 Introduction

Voltage-controlled oscillators are key components of all analog communications systems. They are essentially dc-to-RF converters; they produce an RF signal output with a only a dc signal input. An oscillator consists of two chief components: an active device that acts as an amplifier and a feedback network to provide positive feedback in the system. The feedback network is frequency-sensitive and includes some type of resonator to set the operating frequency. In addition, some type of variable reactance element must be present for control of the frequency. Normally, the variable reactance is controlled by a dc voltage, hence the term *voltage-controlled oscillator*. The typical design emphasis is on low noise, high efficiency, temperature stability, bandwidth, linear and wideband tunability, low cost, and reliability.

6.3.2 General Oscillator Design

A VCO acts as a sinusoidal frequency generator with a specified frequency versus voltage behavior. The types of oscillators under discussion here employ a positive feedback loop consisting of an amplifier and a frequency-selective feedback network. Such oscillators are known as *linear oscillators*. Although the oscillation amplitude is limited due to nonlinear effects in the amplifying device, the term linear oscillator is used to denote that the onset of oscillation is due to the embedded linear amplifier and to distinguish this from those oscillators that generate square, triangular, and other shaped waveforms (i.e., *nonlinear oscillators*).

A general block diagram of a linear oscillator is shown in Figure 6.7. Using standard notation for feedback amplifiers (of which this is a special case) the following characteristic equations describe the oscillator [2]:

$$x_o = Ax_i \qquad (6.3)$$

$$x_f = \beta x_o \qquad (6.4)$$

$$x_i = x_s + x_f \qquad (6.5)$$

$$A_f \equiv \frac{x_o}{x_s} \qquad (6.6)$$

Combining (6.3) through (6.5) with (6.6), we derive the well-known expression for the closed-loop gain

$$A_f = \frac{A}{1 - A\beta} \qquad (6.7)$$

The *loop gain* is defined as

$$L(s) \equiv A(s)\beta(s) \qquad (6.8)$$

where $s = j\omega$. If, at some frequency, ω_o, the loop gain is unity, then the closed-loop gain of the system will be infinite. In other words, there will be a finite output signal with zero input signal (implying oscillation). So, for a linear oscillator at some frequency ω_o, the magnitude of the loop gain is equal to one, the phase of the loop gain is equal to zero, and the system will oscillate at this frequency.

A generalized oscillator circuit is shown in Figure 6.8, where Q is a three-terminal amplifying device (transistor) in some configuration (common-base, -collec-

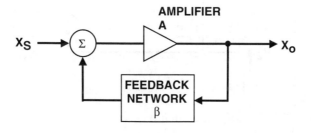

Figure 6.7 General block diagram of a linear oscillator.

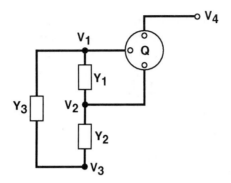

Figure 6.8 Generalized oscillator circuit.

tor, or -emitter, or common-gate, -drain, or -source). The transistor can be replaced with a voltage-controlled current source as in Figure 6.9. This configuration represents a generalization of the most commonly used circuit types: common-base, common-gate, common-emitter, and common-source. A slight variation in the configuration of the transistor model will enable analysis of the common-collector and common-drain circuit types. The circuit is shown without feedback; there must be some interconnection of nodes in order for it to oscillate. The exact feedback necessary to achieve oscillation will be derived next.

The phase of the feedback network necessary to cause oscillation is dependent on the phase shift through the amplifying device. For example, in the case of a common-emitter (or common-source) configuration, the phase of the feedback network must be 180 deg since the phase shift through the transistor is 180 deg, which yields a total loop phase shift of 0 deg. However, in the case of a common-base (or common-gate) configuration, the phase of the feedback network must be 0 deg since there is no phase shift through the transistor in that configuration.

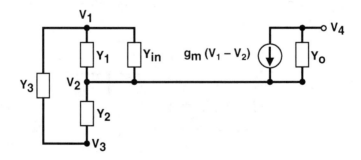

Figure 6.9 Generalized oscillator circuit with Q replaced by voltage-controlled current source.

Writing the nodal equations of the circuit of Figure 6.9 results in the following set of homogeneous equations. If a solution exists other than the trivial solution (i.e., all currents equal zero), then currents arise in the circuit with no forcing voltage, which implies the existence of oscillation. Therefore, as a condition of oscillation:

$$
\begin{bmatrix}
Y_1 + Y_3 + Y_{in} & -(Y_1 + Y_{in}) & -Y_3 & 0 \\
-(Y_1 + Y_{in} + g_m) & Y_1 + Y_2 + Y_{in} + Y_o + g_m & -Y_2 & -Y_o \\
-Y_3 & -Y_2 & Y_2 + Y_3 & 0 \\
g_m & -(Y_o + g_m) & 0 & Y_o
\end{bmatrix}
\begin{bmatrix}
V_1 \\
V_2 \\
V_3 \\
V_4
\end{bmatrix}
= 0
\tag{6.9}
$$

Equation (6.9) is the most general formulation of the oscillator circuit. For specific implementations, the oscillator matrix can be simplified. The rules for simplifying (6.9) are

1. If a node is connected to ground, then the corresponding row and column of that node are eliminated. For example, if node V_2 was connected to ground, row 2 and column 2 from the above matrix would be eliminated. We would also eliminate V_2 from the voltage matrix.
2. If two nodes are connected together, then the corresponding rows and columns are added together. For example, if node V_3 was connected to V_4, then rows 3 and 4 and columns 3 and 4 would be added together. V_3 and V_4 would be combined into one V in the voltage matrix.

For any system of homogeneous equations, the determinant of the system must be equal to zero to assure the existence of nontrivial solutions. Once (6.9) has been simplified based on the implementation desired, one need merely set the determinant equal to zero to find the conditions for oscillation.

6.3.3 Colpitts and Hartley Oscillators

A number of different configurations may be used to realize oscillator circuits; two of the most common are the *Colpitts* and *Hartley oscillators*. These oscillators may use a bipolar transistor (BJT) or a field-effect transistor (FET) as the active device. The common node may be any of the three transistor terminals or another circuit node. We will consider two example configurations below.

Example: Bipolar Transistor Common-Emitter Oscillator

An example of a bipolar transistor oscillator in the common-emitter configuration proceeds next. Since we are considering a common-emitter configuration, ground is placed at V_2 of Figure 6.9.

For a bipolar transistor, it is useful to make the following simplifying assumptions for the ideal case:

1. The output admittance Y_o is equal to 0.
2. The input admittance is purely real ($Y_{in} = G_{in}$).

For the common-emitter case, the feedback must be between the collector and the base, which will be achieved by connecting V_4 to V_3. As a result, the phase shift through the feedback network (collector-to-base) must be 180 deg, as required by the common-emitter configuration. By using the rules outlined in the previous section and setting the determinant equal to zero we have

$$\begin{vmatrix} Y_1 + Y_3 + G_{in} & -Y_3 \\ -Y_3 + g_m & Y_2 + Y_3 \end{vmatrix} = 0 \qquad (6.10)$$

Y_1, Y_2, and Y_3 are typically purely imaginary (inductors or capacitors), so that

$$Y_n = jB_n \qquad (6.11)$$

where $n = 1, 2, 3$. Also, the transconductance g_m is typically purely real. Since both the real and imaginary parts of the determinant must individually be equal to zero, the real part reduces to

$$B_1B_2 + B_2B_3 + B_1B_3 = 0 \qquad (6.12)$$

by dividing both sides of the equation by $B_1B_2B_3$ and using the relationship $X_n = 1/B_n$,

$$X_1 + X_2 + X_3 = 0 \qquad (6.13)$$

The imaginary part of the determinant reduces to

$$B_2 + \left(1 + \frac{g_m}{G_{in}}\right)B_3 = 0 \qquad (6.14)$$

which becomes

$$B_2 + (1 + \beta)B_3 = 0 \qquad (6.15)$$

for a bipolar transistor (since $\beta = g_m/G_{in}$). By dividing both sides of (6.15) by B_2B_3, combining with (6.13), and again using the relationship $X_n = 1/B_n$, (6.15) further simplifies to

$$X_1/X_2 = \beta \qquad (6.16)$$

Since β is real and positive, X_1 and X_2 must be of the same sign. Therefore, (6.13) shows that X_3 must be of opposite sign. In the case where X_1 and X_2 are negative (capacitors) and X_3 is positive (inductor), the configuration is called a *Colpitts oscillator*. In the case where X_1 and X_2 are positive (inductors) and X_3 is negative (capacitor), the configuration is called a *Hartley oscillator*. In fact, Colpitts and Hartley designations describe any possible configuration regardless of where ground in the circuit has been placed.

For a Colpitts oscillator, we can substitute the actual impedances at some frequency ω_o due to the capacitances C_1 and C_2 as well as that due to the inductance L into (6.13) to find

$$\omega_o = 1/\sqrt{L\left(\frac{C_1 C_2}{C_1 + C_2}\right)} \qquad (6.17)$$

and by substitution into (6.16)

$$\beta = C_2/C_1 \qquad (6.18)$$

If a Hartley configuration was employed for this oscillator, (6.13) would have yielded

$$\omega_o = 1/\sqrt{C(L_1 + L_2)} \qquad (6.19)$$

and (6.16) would have reduced to

$$\beta = L_1/L_2 \qquad (6.20)$$

Equations (6.18) and (6.20) seem to indicate that $C_2 \gg C_1$ and $L_1 \gg L_2$ since β is large for transistors. However, at high frequencies, β rolls off and it is common to find that $C_1 = C_2$ and $L_1 = L_2$. The final common-emitter bipolar Hartley oscillator is shown in Figure 6.10 with a bias network included [2].

Example: FET Common-Gate Oscillator

FET oscillators are often operated in the common-gate configuration, so ground is placed at V_1 of Figure 6.9. For a FET, it is useful to make the following simplifying assumptions for the ideal case:

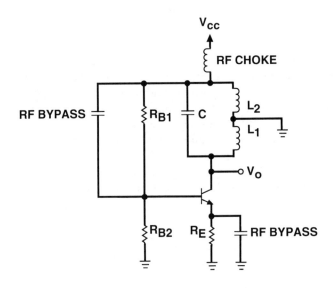

Figure 6.10 Common-emitter Hartley oscillator with bias network included.

1. The input admittance G_{in} is equal to 0.
2. The output admittance is purely real ($Y_o = G_o$).

Also, for a common-gate design, the feedback must be between the drain and the source, which will be achieved by connecting V_4 to V_3. One may also readily see that the phase shift through the feedback network (drain-to-source) is 0 deg, as required by the common-gate configuration. By using the rules outlined in the previous section, the determinant of (6.9) reduces to

$$\begin{vmatrix} Y_1 + Y_2 + G_o + g_m & -(Y_2 + G_o) \\ -(Y_2 + G_o + g_m) & Y_2 + Y_3 + G_o \end{vmatrix} = 0 \qquad (6.21)$$

In solving (6.21), Y_1, Y_2, and Y_2, are once again restricted to be purely imaginary. By setting the real part to zero and using the reciprocal relationship between B_n and X_n, the real part of (6.21) yields

$$X_1 + X_2 + X_3 = 0 \qquad (6.22)$$

which leads to (6.17) and (6.19) regarding the value of the frequency of oscillation ω_0. However, the imaginary part of the solution results in a slightly different equation from (6.18) and (6.20), due to the different characteristics of the FET, i.e.,

$$X_2/X_1 = g_m/G_o \qquad (6.23)$$

or, in the case of a Colpitts oscillator

$$C_1/C_2 = g_m/G_o \qquad (6.24)$$

and for the Hartley oscillator

$$L_2/L_1 = g_m/G_o \qquad (6.25)$$

As in the bipolar case, (6.24) and (6.25) indicate that $C_1 \gg C_2$ and $L_2 \gg L_1$ since g_m/G_0 is large for transistors. However, at high frequencies, g_m/G_0 rolls off and it is common to find that $C_1 = C_2$ and $L_1 = L_2$. The final common-gate FET Colpitts oscillator is shown in Figure 6.11 with a bias network included.

6.3.4 Clapp Oscillator

An additional configuration, called the *Clapp oscillator* [3] (shown in Figure 6.12), is very similar to the Colpitts in design and operation, except for the extra capacitor in series with the inductor. At the intended operating frequency, the series capacitor-inductor pair will present the same inductive reactance as the Colpitts, but if the frequency were to drift, that equivalent inductive reactance would change rapidly (much faster than a single inductor.) Figure 6.13 shows the resonant tank circuit for a Colpitts and a Clapp oscillator. At ω_o, the total series reactances are equal in the two tanks. At a frequency slightly offset from ω_o, however, the Clapp oscillator

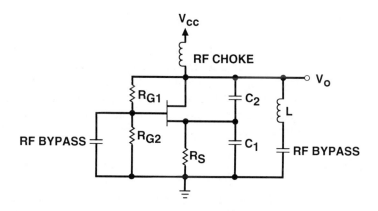

Figure 6.11 Common-gate Colpitts oscillator with bias network included.

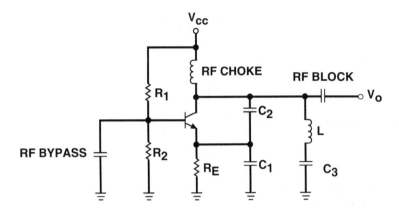

Figure 6.12 Common-base Clapp oscillator with bias network included.

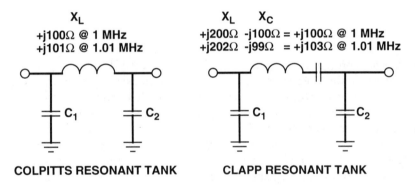

Figure 6.13 Resonant tank circuit for a Colpitts and a Clapp oscillator [3].

tank series reactance changes more than the Colpitts series reactance. As a result, the Clapp oscillator is more stable with frequency. It should be noted, however, that the operation is dependent upon a stable inductor. If the inductor drifts with temperature, the Clapp oscillator will drift at a *faster* rate than the Colpitts. So, the Clapp should only be used where it is possible to construct high-quality inductors and keep them at a constant temperature.

6.3.5 Crystal Oscillators

Additional frequency stability may be obtained by the use of a quartz crystal in the oscillator circuit. The quartz crystal is a high-Q resonator with unloaded Qs up to 100,000 and temperature drift of less than 0.001%/°C. The crystal has an equivalent circuit as shown in Figure 6.14. It should be apparent that the crystal has both a series and parallel resonant mode, which are near each other in frequency. The impedance of the crystal between resonances is inductive, with a rapid rate of change of impedance with frequency, which stabilizes the oscillator frequency because any significant change in operating frequency will cause a large change in the feedback loop phase shift to prevent off-frequency oscillations [4].

Crystals may be used as either parallel or series resonant circuits in oscillators [5]. The series resonant frequency of the crystal occurs when L_1 and C_1 are in resonance, which corresponds to the frequency

$$\omega_s = 1/\sqrt{L_1 C_1} \tag{6.26}$$

ω_s is referred to as the *series resonant frequency*. The crystal impedance will be infinite at the frequency

$$\omega_a = 1/\sqrt{L_1\left(\frac{C_o C_1}{C_o + C_1}\right)} \tag{6.27}$$

ω_a is referred to as the *antiresonant frequency* of the crystal.

Oscillators using crystal resonators are frequently designed in a parallel-mode configuration, where the crystal is used in the antiresonant mode. In this mode, the crystal actually serves as an inductor. A commonly used crystal oscillator circuit, called a *Pierce oscillator* (based on the common-emitter Colpitts oscillator), is shown in Figure 6.15. The crystal is used in place of the inductor in the regular Colpitts oscillator.

The main drawback to the crystal oscillator is its relatively small tuning range. One can introduce tunability into the crystal oscillator circuit by adding a capacitor in parallel with the crystal. In this mode, the antiresonant frequency can be decreased until it is equal to the series resonant frequency. The resultant so-called *pulling range* of the crystal oscillator is expressed by the difference between the antiresonance and series resonance of the crystal and is equal to

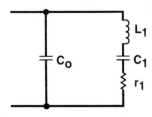

Figure 6.14 Equivalent circuit representation of crystal.

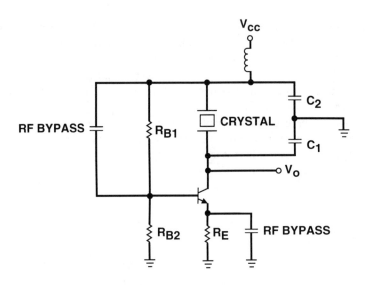

Figure 6.15 Pierce oscillator with bias network included.

$$f_{pull} = f_a - f_s = \frac{C_1}{2C_o} f_s \qquad (6.28)$$

6.3.6 Oscillator Configurations for Microwave Frequencies

A more general set of oscillator circuit configurations are often used for high-microwave frequencies, and they are shown in Figure 6.16 [6]. The first three are resonant circuits with series feedback. The second three are resonant circuits with shunt feedback. Typically, the ground node is at the load resistor. The element values in each circuit can be found in terms of the Z- or Y-parameters of the transistor by use of the equations below each circuit in the figure.

A is defined as the complex voltage ratio between the output voltage and the input voltage of the active device in the oscillator. The optimal value of A for maximum power transfer out of the transistor has been determined to be [7]

$$A = -\frac{y_{21} - y_{12}^*}{2Re(y_{22})} \qquad (6.29)$$

by setting the magnitude of the voltage at the input of the transistor to be a constant and optimizing for the highest power flowing out of the device. The value of A has been further optimized by determining the value that would give the maximum two-

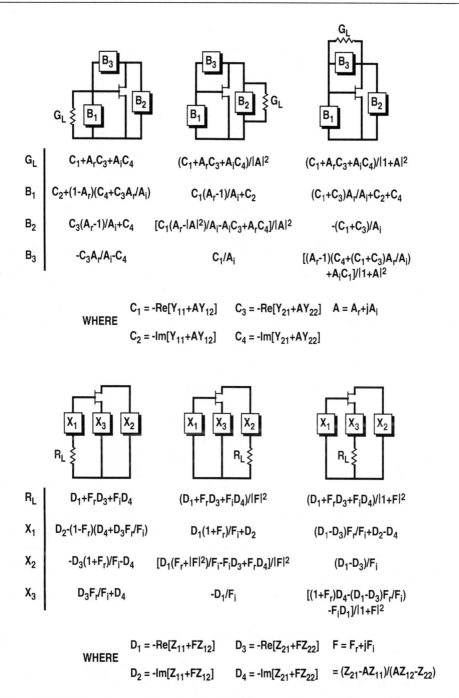

Figure 6.16 Oscillator circuit configurations for high-microwave frequencies [6].

port added power for the oscillating system. That value, written here as A_{opt}, is given by [8]

$$A_{opt} = \frac{-2g_{11}y_{21}}{\sqrt{N} + 2g_{11}g_{22} + j(g_{12}b_{21} + g_{21}b_{12})} \tag{6.30}$$

where

$$N = 4g_{11}g_{22}(g_{11}g_{22} + b_{12}b_{21} - g_{12}g_{21}) - (g_{12}b_{21} + g_{21}b_{12})^2 \tag{6.31}$$

A_{opt} differs from A in that its determination is independent of the input voltage to the transistor, whereas the determination of A assumes a constant input voltage magnitude, which is unlikely.

In general, it can be shown that a larger transistor must be used to achieve higher power outputs, due to its higher current handling capabilities. This must be balanced, however, by the resultant increase in transistor capacitances and conductances, which may begin to make the realization of the correct matching networks difficult at higher frequencies. A tradeoff may be made, however, by designing a lower power oscillator and following it with a higher power amplifier to act as a power boosting stage.

6.3.7 Dielectric Resonator Oscillators (DROs)

Dielectric resonators (DRs) are made of low-loss, temperature-stable, high-permittivity, and high-Q ceramic materials. They resonate in various modes determined by their material composition and their dimensions and shielding and are used in oscillators that employ transmission lines rather than lumped element components. In practice, the DR is placed near a microstrip transmission line so as to enable magnetic coupling between the two structures, as is shown in Figure 6.17 [9]. The lateral distance between the resonator and the microstrip conductor is the parameter that determines the coupling between the DR and the line. The coupling is accomplished by orienting the magnetic moment of the resonator perpendicular to the microstrip plane so that the magnetic lines of the resonator link with those of the microstrip line, as shown in the figure. The DR acts as a cavity, which reflects the RF energy at the resonant frequency. The resonant frequency of the DR is given by

$$f_r(\text{GHz}) = \frac{34}{a\epsilon_r}\left(\frac{a}{H} + 3.45\right) \tag{6.32}$$

where a (in millimeters) is the radius of the resonator and H is its height. Equation (6.32) is valid to about 2% accuracy in the range

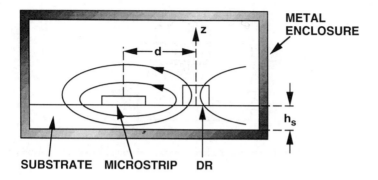

Figure 6.17 Coupling between a microstrip line and a dielectric resonator [9].

$$0.5 < \frac{a}{H} < 2 \tag{6.33}$$

$$30 < \epsilon_r < 50 \tag{6.34}$$

The equivalent circuit of the DR coupled to the microstrip line is shown in Figure 6.18. Typically, the manufacturer supplies the equivalent-circuit values of the DR. The resonant frequency and the unloaded Q of the resonator in terms of the equivalent circuit parameters are given by

$$\omega_o = 1/\sqrt{L_r C_r} \tag{6.35}$$

$$Q_u = \frac{\omega_o L_r}{R_r} \tag{6.36}$$

The circuit may be further simplified to the one shown in Figure 6.19, where

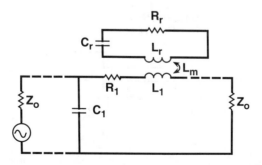

Figure 6.18 Equivalent circuit of a dielectric resonator coupled to a microstrip line [9].

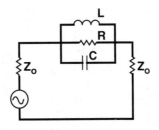

Figure 6.19 Simplified equivalent circuit of dielectric resonator coupled to a microstrip line [9].

$$L = \frac{L_m^2}{L_r} \tag{6.37}$$

$$C = \frac{L_r}{\omega_o^2 L_m^2} \tag{6.38}$$

$$R = \omega_o Q_u \frac{L_m^2}{L_r} \tag{6.39}$$

The coupling coefficient γ at the resonant frequency is defined by

$$\gamma = \frac{R}{R_{ext}} \tag{6.40}$$

where R_{ext} is the total resistance seen on both sides of the DR.

A FET common-source DRO is shown in Figure 6.20 [10]. The value X_3 of the series feedback element may be determined by the equations in Figure 6.16. The transmission line is terminated in its characteristic impedance. This transmission

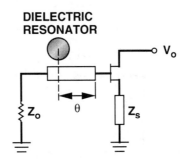

Figure 6.20 FET common-source DRO [10].

structure has a bandstop characteristic so the gate will see a reflection coefficient equal to unity at the resonant frequency and the characteristic impedance (reflection coefficient equal to zero) at all other frequencies. The transmission structure must present the correct impedance X_1 to the gate at resonance, which may be achieved by adjusting the phase of reflection coefficient at resonance by precisely placing the DR at some point θ (in electrical length at the resonant frequency) from the gate. The parameters for the placement of the DR are determined from the following equations:

$$\Gamma_t = \frac{\gamma}{\gamma + 1 + 2jQ_u\delta}\exp(-2j\theta) \tag{6.41}$$

where Γ_t is the desired reflection coefficient seen by the transistor and δ is the fractional bandwidth given by

$$\delta = (\omega - \omega_0)/\omega_0 \tag{6.42}$$

γ is the coupling coefficient between the DR and the transmission line and Q_u is the unloaded Q of the DR.

Since the general series feedback oscillator circuit diagrams are shown in terms of X rather than Γ, the following transformation is included for completeness:

$$X_3 = Z_3 = Z_0\frac{1 + \Gamma_t}{1 - \Gamma_t} \tag{6.43}$$

By using the above equations, we may determine the correct values for γ and θ to achieve the desired value of X_3.

Figure 6.21 shows another possible topology for the DRO that is similar to the previous example except the resonator is placed in the feedback path so that

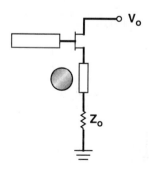

Figure 6.21 Alternate FET common-source DRO [10].

feedback is allowed only at the resonant frequency. This configuration ensures stability at all frequencies other than resonance. The circuit in Figure 6.22 uses the DR to complete a shunt feedback path between the drain and gate. In each of the above cases, as well as multiple other possible DRO configurations, the DR is used to satisfy one set of equations in Figure 6.16 to achieve the proper conditions for oscillation.

6.3.8 Practical Considerations for Oscillator Design

The previous sections have analyzed ideal oscillator cases only. In reality, there are a number of practical issues that must be considered. For example, in the preceding analyses, the elements Y_1, Y_2, and Y_3 were assumed to be purely reactive, while there will always be some finite Q associated with any capacitor or inductor, causing the admittance (or impedance) to be complex. Furthermore, the assumptions made about the bipolar transistor—that Y_{in} is purely real and Y_o is equal to zero—are only approximations. Likewise, the assumptions made about the FET are simplifications. In addition, the oscillator dependence on the output load was not considered in the above analyses, which could prove to be significant for high conductance loads. Finally, the bias networks and surrounding circuitry may impact the performance of the oscillator in complicated ways that will cause nonideal responses.

The previous assumptions need only be used to determine a starting point for the design, which may then be optimized considering the issues raised in this section. In particular, computer aided design (CAD) software may be employed to perform the computation required to include all nonideal effects in the full circuit. Most manufacturers of microwave components include full characterization data for each component that will be used in the circuit. This information, in the form of equivalent circuit models, S-parameters, or some large-signal model, can easily be incorporated into most time or frequency domain simulators for final simulation of the oscillator circuit.

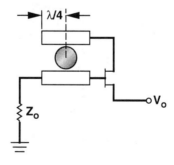

Figure 6.22 FET DRO using drain-to-gate feedback [10].

6.3.9 Voltage Control of Oscillators

In most configurations shown here, the frequency of oscillation can be varied with the addition or subtraction of capacitance in the resonant circuit. Diodes whose capacitance can be varied with application of dc voltage are called *varactor diodes* (or *varactors*) and serve to function as the controlling capacitance in oscillators. One commonly used VCO circuit technique is to shunt the inductor of a Colpitts oscillator with a varactor. The voltage applied changes the amount of capacitance in parallel with the inductor, which changes the frequency of oscillation. If the capacitor is in series with the inductor, the circuit is identical to the Clapp configuration, with the fixed capacitor replaced by the varactor. The advantage of this configuration is that a smaller capacitance can be used to change the frequency of oscillation. The transistor capacitance may also be used as the voltage-varying frequency-controlling reactance [11, 12].

A key measure of the quality of a VCO is the linearity of the voltage-frequency relationship, as will be seen in the next section on phase-locked loops. As such, it is imperative that the designer consider the voltage-capacitance relationship in the diode carefully when designing the VCO. For a resonator, ω is expressed by the relationship

$$\omega = \frac{1}{\sqrt{LC}} \tag{6.44}$$

If C is a function of voltage, as in the case of a varactor, then linear tuning with voltage is achieved when

$$C = \frac{K}{v^2} \tag{6.45}$$

where K is a constant and v is the tuning voltage. The capacitance versus voltage for varactors may be expressed by

$$C(v) = \frac{C_j(0)}{v^\gamma} \tag{6.46}$$

where

$C_j(0)$ = zero volt junction capacitance
γ = capacitance exponent

Hyperabrupt junction varactors have a capacitance exponent that is close to 2 over a wide range of tuning voltages, which make them ideal for linear voltage controlled oscillators.

6.3.10 Noise in Oscillators

Any noise sources within the active device in the oscillator will modulate the output signal such that noise sidebands are produced around the fundamental signal. These sidebands can be divided into three classes: AM noise, due to amplitude modulation by the noise sources; FM noise, due to frequency modulation by the noise sources; and PM noise due to phase modulation by the noise sources. FM and PM noise produce the same effects in the spectrum. These three noise sources produce spectra of the form shown in Figure 6.23.

The two primary types of noise in transistors are low-frequency $1/f$ noise due to trapping in the semiconductor material and thermal (white) noise. These noise sources act to modulate the fundamental oscillator signal (carrier) and produce AM and FM noise sidebands about the center frequency.

The following noise analysis is proposed by Sweet [13]. Consider the noise circuit shown in Figure 6.24. The thermal noise source is represented in the circuit by a current source at the input of the transistor, which acts as a randomly varying injected microwave signal. This signal produces both amplitude and frequency modulation. The low-frequency $1/f$ noise sources are not shown explicitly in the circuit,

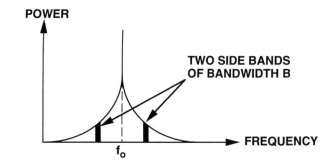

Figure 6.23 Oscillator output spectrum including effects of AM and PM noise [13].

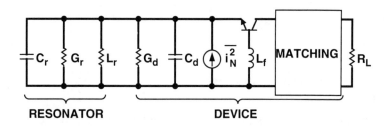

Figure 6.24 Noise equivalent circuit of transistor oscillator [13].

but they act to cause random variations in the transistor's equivalent-circuit elements, which randomly modulate the amplitude and frequency of the oscillator.

FM Noise

In analogy to diode negative resistance oscillators [14], the oscillator's mean square FM frequency deviation noise due to thermal noise of a transistor oscillator is

$$\overline{\Delta\omega^2} = \frac{\omega_o^2 k T_n B}{4 Q_r^2 P_{out}} \tag{6.47}$$

where

ω_o is the oscillator's center frequency;
k is Boltzmann's constant (1.38×10^{-23} J/K);
T_n is the transistor equivalent noise temperature in K;
B is the measurement bandwidth;
Q_r is the unloaded Q of the resonator;
P_{out} is the oscillator output power.

This noise is dominant for frequencies far from the carrier frequency.

The upconverted $1/f$ noise is dominant at frequencies near the carrier and the resulting mean square FM frequency deviation is given by

$$\overline{\Delta\omega^2} = \left[\frac{\omega_o^2 \left(\dfrac{\partial C_d}{\partial V_o}\right)^2}{4 Q_r G_r} \right] S_{\Delta V_o}(\omega_m) B \tag{6.48}$$

where

$\left(\dfrac{\partial C_d}{\partial V_o}\right)$ is the active device's capacitance change with dc voltage;

$S_{\Delta V_o}$ is the spectrum of the low-frequency noise fluctuation in V^2/Hz;

G_r is the resonator loss in terms of conductance;

ω_m is the low frequency from where the noise fluctuations are originating.

The spectrum of low-frequency noise fluctuations obey a pattern that may be expressed by

$$S_{\Delta V_o}(f) B = \frac{N}{f^\alpha} \tag{6.49}$$

where

N is an empirically derived strength parameter;
α is a spectrum shape coefficient (usually close to unity).

Since α is usually close to unity, the spectrum has a shape proportional to $1/f$.

The FM noise mean square frequency deviation may be converted to a noise-to-carrier ratio by using small-signal FM theory:

$$\frac{N}{C} = \frac{\overline{\Delta\omega^2}}{\omega_m^2} \tag{6.50}$$

AM Noise

The AM noise-to-carrier ratio in a bandwidth B at a given frequency ω, is

$$\frac{N}{C} = \frac{kT_n BG_r^3/P_{out}\left(\dfrac{\partial G_d}{\partial V_1}\right)}{1 + (\omega/\omega')^2} \tag{6.51}$$

where

$$(\omega')^2 = \frac{\omega_o^2 P_{out}\left(\dfrac{\partial G_d}{\partial V_1}\right)}{8Q_r^2 G_r^3} \tag{6.52}$$

ω' is referred to as the *AM corner frequency*

$\left(\dfrac{\partial G_d}{\partial V_1}\right)$ is the device's input terminal conductance sensitivity to RF signal level, also called the *oscillator nonlinearity factor*.

The AM upconverted $1/f$ noise is given by the expression

$$\frac{N}{C} = \left(\frac{1}{4}P_{out}^2\right)\left(\frac{\partial P_{out}}{\partial V_o}\right)^2 S_{\Delta V_o}(\omega)\frac{B}{1 + (\omega/\omega')^2} \tag{6.53}$$

where $\left(\dfrac{\partial P_{out}}{\partial V_o}\right)$ is the oscillator's power sensitivity to changes in dc voltage.

Additional analyses have produced models of oscillator noise based on the small-signal equivalent circuit of the transistor [15] and methods of designing for lowest phase noise [16, 17]. Various techniques have been used to lower oscillator noise, including a method to suppress $1/f$ noise that would be upconverted to near the carrier [18].

Figure 6.25 shows the typical shape of the AM and FM components of oscillator noise. Figure 6.26 shows a comparison of the $1/f$ noise spectra between Si bipolar transistors and GaAs FETs. Since the bipolar $1/f$ noise is so much lower than the FET's, bipolar oscillators have a distinct advantage where upconverted close to the carrier noise is important. When noise far from the carrier is more important, the GaAs FET is preferred due to its lower thermal noise.

Oscillator noise is typically specified in terms of decibels below carrier per Hertz (dBc/Hz) at some specified offset frequency.

6.3.11 Ring Oscillator VCO Designs

One drawback to the designs presented in the previous sections is the difficulty in realizing a monolithic resonator capable of sufficiently high Q to realize an acceptable phase noise. Some GaAs monolithic circuits are capable of realizing reasonably high-

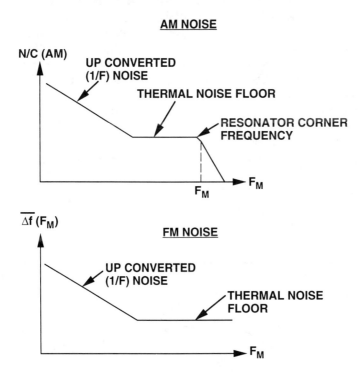

Figure 6.25 AM and FM oscillator noise spectra [13].

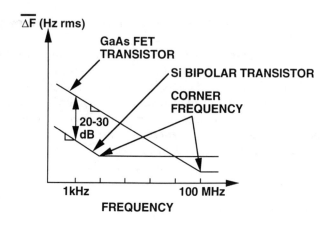

Figure 6.26 A comparison of $(1/f)$ FM noise spectra in Si bipolar transistors and GaAs FETs [13].

quality on-chip resonators, but most high-performance designs typically employ an off-chip resonator of some kind. Even in designs that do employ on-chip resonator circuits, the physical size required by the resonant circuit is often a significant portion of the overall chip area.

In order to avoid this limitation, a number of groups have explored the use of ring-oscillator-based VCOs as alternative approaches for the realization of voltage-variable oscillators [19, 20]. These circuits have the advantage that they do not require any kind of resonator, but their frequency and phase characteristics are somewhat poorer than those of resonator-based approaches. An illustrative example of a ring-oscillator VCO is shown in Figure 6.27.

The oscillation frequency of the circuit is determined by the gate delay in the inverting stages, which can be controlled by the current through the gate, and the number of gates in the oscillator, and is given by

$$f_{osc} = \frac{1}{2n\tau_d} \tag{6.54}$$

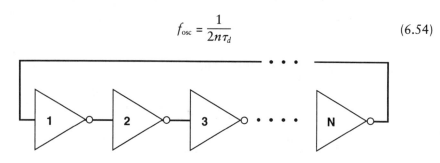

Figure 6.27 Ring oscillator voltage-controlled oscillator.

where n is the number of inverters in the ring oscillator, and is necessarily odd, and τ_d is the gate delay per stage.

One limitation of this approach is that maximum frequency of oscillation is typically limited by the minimum number of gates (three, typically) and the minimum gate delay, which is process dependent. A number of techniques have been developed to circumvent this limitation. One approach (shown in Figure 6.28) relies on the fact that at any given time in the ring oscillator, the output of $(n + 1)/2$ stages are high and $(n - 1)/2$ stages are low, or the output of $(n + 1)/2$ stages are low and $(n - 1)/2$ stages are high [21]. The oscillator toggles between these two conditions at a frequency $1/(2\tau_d)$. By sensing these transitions, we can generate a frequency of $1/(2\tau_d)$. The design of Figure 6.28 employs transconductance stages to convert the inverter output voltage to a current, and then the resulting currents are summed at a single output node. This approach has been employed to realize an 8-GHz oscillator in a 20-GHz BiCMOS process.

Another approach utilizes a four stage ring-oscillator, where one of the stages of the ring is noninverting. The outputs of the circuit are mixed with each other,

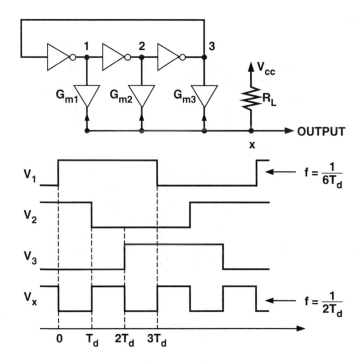

Figure 6.28 Ring oscillator VCO operating at $1/(2\tau_d)$ by sensing the transitions in the output of the inverters [21].

and twice the output frequency results. See Figure 6.29 for an example [22]. The advantage of this approach is that in-phase (I) and quadrature phase (Q) output are available simultaneously from the structure, which is a major consideration for quadrature modulation and demodulation in communication systems design (see Chapter 4). Furthermore, an output frequency of four times the ring is available if the I and Q signals are themselves multiplied. As a result, the final available output frequency is $1/2\tau_d$.

Noise in Ring-Oscillator VCOs

The noise in ring-oscillator VCOs is difficult to calculate exactly, because of the highly nonlinear nature of the circuit. However, the noise power spectral density can be estimated using techniques similar to those described in [23] for the analysis of a multivibrator, and are extended here for the case of a ring oscillator.[1] When this derivation is carried out, the resulting phase noise, in dBc/Hz, for the ring-oscillator VCO is given by

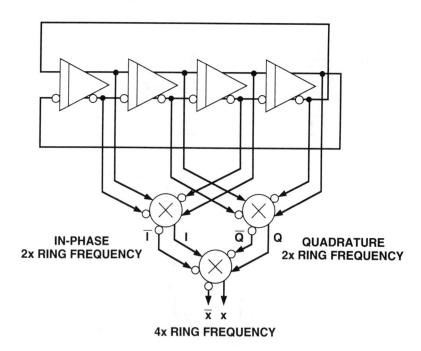

Figure 6.29 Frequency doubling and quadrupling ring-oscillator VCO [22].

[1]This analysis was provided to the author by Dr. Cynthia Baringer of the Hughes Research Laboratories.

$$S_\phi(f_m) \cong \frac{f_o}{f_m^2} \left(\frac{\Delta \tau_{\text{rms}}}{T_0} \right)^2 \tag{6.55}$$

where f_0 is the fundamental frequency of oscillation, T_0 is the fundamental oscillation period ($1/f_0$), and f_m is the offset frequency from the carrier.

The value $\Delta \tau_{\text{rms}}$ is the rms variation in the output waveform of the VCO. The jitter in the output waveform is defined as the ratio of the mean μ and rms variation of the output; that is,

$$\text{jitter} = \Delta \tau_{\text{rms}} / \mu \tag{6.56}$$

and is usually expressed in ppm.

A useful estimate for this rms variation can be given by

$$\Delta \tau_{\text{rms}} = \left(N/2 \left(C_P / R_L I_{\text{EE}}^2 \right) \nu_{\text{in}}^2 \right)^{1/2} \tag{6.57}$$

where N is the number of stages in the ring oscillator, I_{EE} is the bias current in the oscillator, R_L is the load resistance in the inverter, and C_P is the parasitic resistance at the output of the circuit. The input referred noise voltage, ν_{in}, is in volts/Hz$^{1/2}$, which can be calculated from a simple linear analysis of the circuit using a SPICE simulator or its equivalent.

These last two expressions demonstrate that the noise of the oscillator can be reduced by increasing the bias current and decreasing the parasitic capacitances, which has the overall effect of maximizing the slew rate of the circuit and minimizing the time period where the circuit is in the transition region, where it is amplifying the noise of the differential pair.

The noise of a ring-oscillator VCO can be very acceptable for some applications. The circuit described in [21] exhibited a phase noise of −75 dBc/Hz at a 1-kHz offset from the carrier at 6 GHz. The circuit described in [22] exhibited a phase noise of −100 dBc/Hz at a 100-kHz offset from the carrier at 6.8 GHz.

6.3.12 Multivibrator VCO Designs

Another approach for the design of resonator-less VCOs is the traditional multivibrator, an example of which is shown in Figure 6.30. This design has been used extensively for a number of lower frequency (<500 MHz) monolithic VCOs [23], and is a specific example of a class of oscillators known as *relaxation oscillators*.

The circuit operates with transistors Q1 and Q2 alternately turning on and off, with capacitor C being discharged through current sources *I*. Diodes D1 and

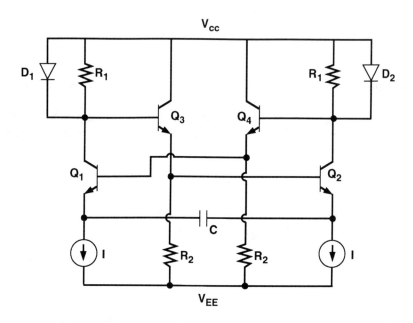

Figure 6.30 Multivibrator VCO. The circuit functions by charging and discharging capacitor C through current sources I.

D2 set the voltage swings at the collectors of Q1 and Q2, and transistors Q3 and Q4 are level-shifting devices. A square-wave output appears at the collectors of Q1 and Q2, and a triangle wave appears across the capacitor, whose frequency is given by

$$f_o = \frac{I}{4CV_{be(on)}} \tag{6.58}$$

A variety of techniques have been developed to improve the temperature stability and drift of the circuit, but they are all functionally equivalent to that of Figure 6.30. The oscillation frequency is controlled by varying the current I through the current sources, and a wide frequency tuning range can be achieved.

Another version of this oscillator type is shown in Figure 6.31. In this case, the clamping diodes D1 and D2 have been eliminated, and the circuit operates as a fixed oscillator of frequency

$$f_o = \frac{1}{8RC} \tag{6.59}$$

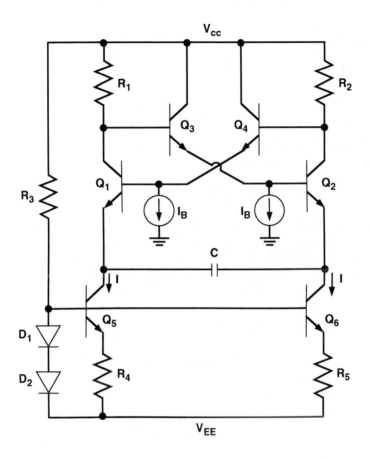

Figure 6.31 Multivibrator oscillator that oscillates at a fixed frequency $f = 1/8R_1C$.

Noise in Multivibrator VCOs

The noise process in relaxation oscillators was discussed in [23], and the results are summarized here. Like ring oscillators, the noise in a multivibrator is intrinsically higher than that of an oscillator that employs a high-Q resonator.

For a general relaxation oscillator, the jitter in the output waveform is given by

$$\text{jitter} = \frac{K \nu_{\text{nrms}}}{S} \tag{6.60}$$

where K is a constant, S is the slope of the waveform at a triggering point, and v_{nrms} is the equivalent input noise voltage to the circuit. For this particular multivibrator design, the jitter can be expressed as

$$\sigma(T) = \alpha \frac{RC}{I_o - I_R} \sigma(I_n) \qquad (6.61)$$

where I_R is $V_T/2R$ ($V_T = kT/q$), α is a constant approximately equal to 1, and $\sigma(I_n)$ is the standard deviation of the equivalent noise current generator at the output of the multivibrator (the collector of transistor Q2 of Figure 6.30). This last quantity can be calculated from SPICE modeling of the circuit over the desired current and temperature range of operation.

Equation (6.61) can be used to calculate the jitter of the resulting waveform, or it can be used in conjunction with (6.55) to calculate the theoretical phase noise of the output waveform. These results demonstrate that the output waveform jitter can be reduced by increasing the voltage swing in the circuit (increasing the slew rate S), and by reducing the noise sources within the circuit itself.

6.4 PHASE LOCK-LOOP DESIGN

The phase-locked loop concept has a long history [24] and has been used extensively for a wide variety of communications system designs, particularly in the satellite communications area [25]. One of the advantages of the PLL architecture is its ability to realize excellent phase noise performance over a wide tuning range, while simultaneously achieving exacting standards of frequency accuracy. This, combined with its ease of implementation in monolithic form make it nearly ideal for design in wireless communications systems.

This section will discuss the basic elements of PLL design, with an emphasis on analog techniques. There is an extensive body of literature on the design and implementation of PLLs, and the reader is encouraged to consult these for a more thorough treatment of the subject [26].

6.4.1 PLL Design Fundamentals

A basic block diagram of a linearized PLL architecture is shown in Figure 6.32. During operation, the loop acquires—or locks—to an input signal, and its output displays a fixed phase relationship to the input signal. A basic loop consists of a phase detector, amplifier/loop filter, and voltage-controlled oscillator. Phase detectors produce an output voltage whose dc level is proportional to the phase difference between the two input signals. The linearized small-signal gain of the phase detector is K_d, whose units are typically volts/radian.

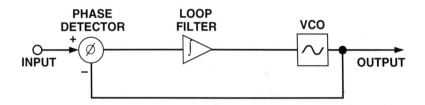

Figure 6.32 Fundamental PLL architecture.

The VCO produces an output frequency in response to an applied input voltage, and it is designed so that its output frequency range is at least as large as that required by the overall system. The ratio of change in output frequency to change in input frequency is known as the gain constant of the VCO, K_0.

The loop filter is typically a lowpass filter whose dc gain is K_f, and whose overall transfer function can be given by

$$F(s) = \frac{K_f(1 + s\tau_z)}{(1 + s\tau_p)} \qquad (6.62)$$

The loop filter is typically realized as an active filter, using operational amplifiers, resistors, and capacitors to set the transfer function of the circuit.

So, the overall *phase* transfer function of the loop is given by (recall that frequency is the time derivative of phase):

$$H_\theta(s) = \frac{\theta_{\text{out}}(s)}{\theta_{\text{in}}(s)} = \frac{K_d K_0 F(s)/s}{1 + K_d K_0 F(s)/s} \qquad (6.63)$$

which can be rewritten as [22]

$$H_\theta(s) = \frac{1 + \dfrac{s}{\omega_n}\left(2\xi - \dfrac{\omega_n}{\Omega_k}\right)}{1 + \dfrac{s}{\omega_n}2\xi + \left(\dfrac{s}{\omega_n}\right)^2} \qquad (6.64)$$

where $\Omega_k = K_d K_f K_0$, $\omega_n = \dfrac{\Omega_k}{\tau_p}$, and $\xi = \dfrac{1}{2}\left[\dfrac{\omega_n}{\Omega_k} + \omega_n\tau_z\right]$.

The value ω_n is known as the loop bandwidth, or natural frequency, of the PLL. The damping factor ξ determines the transient response of the system. Both factors are crucial in determining the overall frequency and time domain response of the system. Figure 6.33 plots the normalized step response as a function of ξ and ω [27].

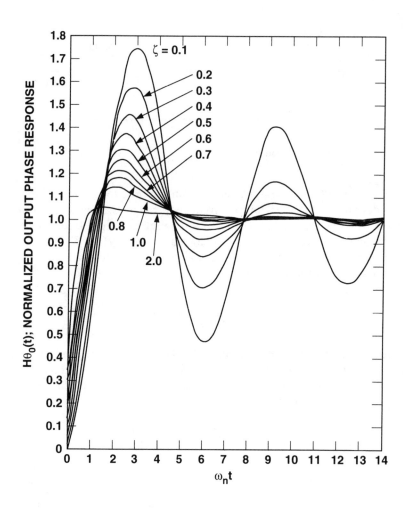

Figure 6.33 Normalized transient response of PLL.

The frequency response of the loop is identical for frequency-modulated signals (as opposed to phase-modulated signals) except for a constant term; that is,

$$H_f(s) = \frac{K_\phi}{K_0} H_\theta(s) \tag{6.65}$$

Equation (6.65) is a lowpass transfer function, which indicates that the output phase (and frequency) track the input phase and frequency up to the loop bandwidth.

Conversely, the phase error transfer function (the output phase error as a function of input phase) is given by

$$H_\theta(s) = \frac{\dfrac{s}{\omega_n}\left(\dfrac{s}{\omega_n} + \dfrac{\omega_n}{\Omega_k}\right)}{1 + \dfrac{s}{\omega_n}2\xi + \left(\dfrac{s}{\omega_n}\right)^2} \tag{6.66}$$

which is a highpass transfer function whose bandwidth is ω_n.

When configured as shown in Figure 6.32, the output frequency of the PLL simply tracks the input frequency. This feature of the loop can be used to great advantage in clock recovery circuits in digital communications receivers [28, 29]. A discussion of clock recovery in digital communications circuits appears in Chapter 4.

The PLL can also be used as a technique for locking an imprecise and noisy microwave oscillator to a less noisy, but lower frequency and more precise, signal source, as shown in Figure 6.34.

In this case, a frequency divider N has been inserted between the output of the VCO and the input to the phase detector. This circuit divides its input frequency down by a factor of N. The transfer function of the loop is the same as described in (6.65), except that K_0 has been replaced by K_0/N. The output frequency transfer function is then

$$H_f(s) = N\frac{K_\phi}{K_0}H_\theta(s) \tag{6.67}$$

The output frequency is now an integer multiple N of the input frequency. These frequency dividers can be fixed or programmable. If they are programmable, then the output frequency can be varied in discrete steps, and the PLL can function as a true broadband frequency synthesizer. This technique will be discussed in Section 6.4.2.

There are a variety of nonideal effects that can degrade the performance of the PLL. These effects can limit both the noise and dynamic range of the overall circuit.

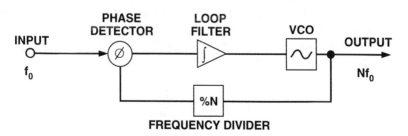

Figure 6.34 Phase-locked loop with frequency divider inserted within the loop.

Dynamic Range Limitations

The dynamic range of the PLL is limited by the tuning range of the VCO as well as the output voltage range of the loop amplifier. Since a significant amount of overshoot is possible because of the closed-loop operation of the system, the design of the loop amplifier must be able to account for the extra voltage required to accommodate the entire control voltage range. For example, if a 3V swing is required to tune the PLL over its steady-state frequency range, then the operational amplifier should be capable of producing at least a 6V swing to accommodate the transient overshoot conditions. At the same time, the VCO must also be able to accommodate a 6V swing at its input without experiencing a degradation in its control gain constant K_0.

The loss of loop gain due to transient overshoot can be a serious problem because it can decrease the damping factor ξ, which can result in even more overshoot. Major loop instability can result from this condition. Furthermore, in systems employing digital phase detectors, the transition pulses at the output of the phase detector can feed through the loop filter, causing an even larger voltage excursion at the input to the VCO. This last problem can be handled by additional filtering within the loop at a much higher frequency than the natural loop frequency [27].

Loop Lock Range and Acquisition Time

The PLL will begin its operation in an "unlocked" condition, where the input and output phases are not locked to each other. A simplified expression for loop lock-in range—the change in frequency that can be "pulled in" by the self acquisition of the loop—is given by Gardner [26]:

$$\Delta\omega_p \cong \Omega_k\sqrt{2\left(\tau_z/\tau_p\right)} \tag{6.68}$$

This expression assumes linear operation of all of the elements in the loop, and the actual available lock-in range can be substantially degraded due to nonlinearities within the loop itself.

Using a similar analysis, the time required to achieve lock assuming an initial frequency error $\Delta\omega_p$ was determined to be [22]

$$T_{\text{lock}}(\Delta\omega_L) \cong 2\tau_p\left(\frac{\Delta\omega_L}{\Delta\omega_p}\right)^2 \tag{6.69}$$

which demonstrates that the loop acquisition time is dependent on the square of the initial frequency error and the pole time constant of the loop filter.

Noise and Spurious Outputs of PLLs

There are a variety of factors that influence the noise and spurious outputs of a PLL, most of which are related to noise sources that modulate the VCO input in some way.

The first is the phase noise within the VCO itself. This can be modeled as an additive noise source at the input of the VCO, as shown in Figure 6.35. In addition to this noise source, there are noise sources associated with prescalers, both within and outside the loops, as well as the phase detector. Based on the linearized small-signal model of the PLL, the output noise power spectral density within the loop bandwidth is given by [30, 31]

$$S_o(f) = S_r(f)[N/Q]^2 + [S_{DQ}(f) + S_{DN}(f) + \{S_{PD}(f) + S_{LF}(f)\}/K_d^2(f)$$
$$+ S_{VCO}(f)/(K_d^2(f)K_v^2(f)K_f^2(f))]N^2 \qquad (6.70)$$

where

$S_o(f)$ is the phase noise density (in dBc/Hz) at the output of the PLL;
$S_r(f)$ is the phase noise density at the input to the PLL;
N is the division ratio within the loop;
Q is the division ratio preceding the loop;
$S_{DQ}(f)$ is the phase noise density at the output of the 1/Q divider, assuming an ideal input;
$S_{DN}(f)$ is the phase noise density at the output of the 1/N divider, assuming an ideal input;
$S_{PD}(f)$ is the phase noise density at the output of the phase detector;
$S_{LF}(f)$ is the phase noise density at the input of the loop filter;
$S_{VCO}(f)$ is the phase noise density at the output of the VCO;

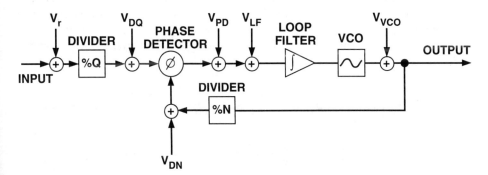

Figure 6.35 Linearized small-signal model of PLL showing noise sources.

and

K_d is the phase detector gain (V/rad);
K_f is the loop filter gain (V/V);
K_V is the VCO gain (rad/sec/V).

This expression demonstrates the importance of considering all of the noise sources when predicting loop performance, and the excellent noise-shaping properties of the loop at frequencies less than the loop bandwidth.

There are a variety of additional spurious outputs that can affect the performance of the PLL. These are principally introduced by noise sources on the VCO control line, and may arise from the power supply, package, or the pulse components of the phase detector. This latter effect can be especially troublesome because the phase detector transitions often occur near the extremes of the VCO input during a rapid excursion in the output frequency, causing even more noise to be coupled through during a transient response.

The relative level of spurious sideband components can be estimated using the following expression:

$$\frac{\text{sidebands}}{\text{carrier}} \cong \frac{V_{\text{spurious}} K_o}{2\omega_{\text{out}}} \tag{6.71}$$

where V_{spurious} is the level of spurious voltages at the input to the VCO.

6.4.2 PLL Applications

There are many different applications of the PLL presented in the previous sections. One example of their use is in "locking" a relatively noisy high-frequency oscillator (such as a typical VCO) to a relatively quiet lower frequency oscillator, such as a quartz oscillator. The resulting output has the phase noise characteristics of the quiet input reference, but the output frequency of the high-frequency VCO. This technique can be more generally employed to synthesize multiple output frequencies, as shown in Figure 6.36.

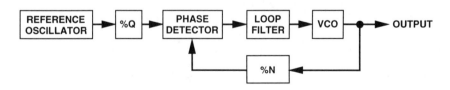

Figure 6.36 Multiple frequency synthesizer using PLL architecture.

In this case, the input frequency is divided by a factor Q, and the output frequency of the VCO is divided by N. Hence the output of the PLL is given by

$$f_{\text{out}} = \frac{Q}{N} f_{\text{in}} \tag{6.72}$$

If Q and N are programmable, then a wide variety of output frequencies is available at the output of the system.

This concept has been extended and elaborated on in several novel ways to realize an even greater range of output frequencies and programmability. Most of these approaches involve multiple loop frequency synthesizers. An excellent discussion of these circuits appears in [26].

Another example of the use of PLLs is in the area of demodulation of digital communications signals. Most digital modulation techniques require that the carrier frequency and phase be known precisely prior to demodulation. However, transmission of the carrier frequency along with the data is an inefficient use of bandwidth, and so a variety of techniques have been devised to recover the carrier and clock signals from the input. These circuits are known as *signal synchronizers*, and are discussed in more detail in Chapter 4.

For example, in binary phase-shift keyed (BPSK) data, the phase of the carrier waveform is shifted by +90 deg for a data bit = 1, and −90 deg for a data bit = 0. The resulting waveform can be represented by

$$V(t) = m(t)\sin(\omega_c t + \theta_i) \tag{6.73}$$

where $m(t)$ has an average value equal to zero if an equal number of zeros and ones is transmitted.

Figure 6.37 shows one technique for recovering the carrier from the resulting waveform. After squaring the input waveform, the resulting signal is

$$S(t) = V^2(t) = m^2(t)[1 - \cos 2(\omega_c t + \theta_i)] \tag{6.74}$$

This waveform contains a frequency component at twice the original input frequency, and a phase component of twice the original input phase. Now, the two input phases at ±90 deg are translated to ±180 deg, and they lie precisely on top of

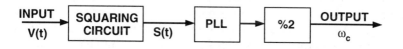

Figure 6.37 BPSK Demodulator using PLL for frequency tracking.

each other. So, the squaring of the input signal eliminates the digital information in the original input signal, leaving only twice the carrier frequency remaining.

The PLL functions as a tracking filter, acquiring the input waveform and the divide-by-two functions to recover the original carrier signal. One drawback of this approach is that the phase of the carrier is ambiguous (it is impossible to distinguish between a 1 and a 0). Special encoding techniques can be used to recover the original phase of the carrier, and a different demodulation scheme is required (see [32] for some examples).

Fractional-N PLL Synthesizers

The conventional approach to frequency synthesis using PLLs employs a frequency divider where the division ratio is programmable between 1 and an integer N, where the step size is equal to the reference frequency. This approach requires a very small reference frequency if fine frequency resolution is desired, and this will inevitably result in an increase in the loop settling time.

An alternative approach is to have the division ratio N take on fractional values so that the output frequency could be stepped in fractions of the reference frequency. The class of PLLs that accomplish this is known as fractional-N PLLs, and we will describe their operation very briefly here. The reader is directed to [33] and [34] for a more thorough treatment of the subject.

The most common method of accomplishing the fractional division is to have the divider divide the frequency down by $N + 1$ every $M \cdot K$ cycles (where the desired fraction is $1/M$) and divide by N the remaining $(1/M) \cdot K$ cycles. The effective division ratio is then $(N + 1/M)$. For example, if a PLL was desired with an output frequency of 835 MHz and a reference frequency 100 MHz, the integral part of the division would be N = 8/9, 1/M would be 0.35, with $M = 2.86$. So, the VCO output could be divided by 9 every 286 cycles (for $K = 1,000$) and divided by 8 every 714 cycles.

This approach is typically implemented using a *phase accumulation register*, as shown in Figure 6.38. In the above case, the quantity 1/M (0.35) would be added to an accumulator, which would overflow when its contents exceeded 1. Each time the accumulator overflowed, the divider would divide by 9, rather than 8. The accumulator itself holds the phase error that accumulates over time as the loop operates.

This accumulated phase error is itself useful in improving loop operation. The phase detector produces a sawtooth output, whose value increases each period. This voltage variation will modulate the VCO, and create an undesirable frequency modulation in the output. The contents of the phase accumulation register contain the negative of this error and, when summed with the phase detector output, produce a "corrected" error voltage to the loop filter and VCO free of frequency modulation.

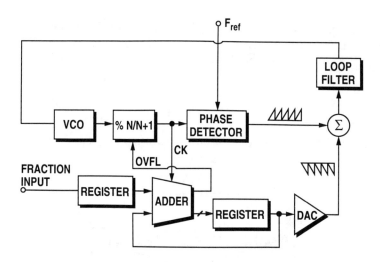

Figure 6.38 Block diagram of fractional-N frequency synthesizer.

The sawtooth generated by the D/A converter must be scaled to provide the correct voltage cancellation required by the VCO and loop filter.

6.4.3 Implementation of PLLs

Phase-locked loops are implemented using four basic components: VCOs, phase detectors, frequency dividers, and loop filters. The design of VCOs was discussed in detail in Section 6.3. This section will discuss the design of the other three components, beginning with phase detectors.

Phase Detectors

An ideal phase detector produces an output voltage proportional to the difference in phases of two input signals; that is,

$$V_{out} = K_d(\theta_1 - \theta_2) \tag{6.75}$$

Note that this output voltage is independent of the amplitude of the input voltage, so the inputs to the phase detector are usually digital signals.

This particular ideal transfer function is difficult to implement in practice, and a variety of circuit architectures have been developed to approximate it. They can be generally classified as exclusive-OR designs (sometimes referred to as "analog" designs) or digital edge-triggered designs.

Exclusive-OR designs typically utilize a balanced Gilbert multiplier, shown in Figure 6.39. The output waveforms of the circuit are also shown in the diagram. Through simple integration, the dc component of this waveform can be shown to be [35]

$$V_{dc} = I_{EE} R_C (2\phi/\pi) \qquad (6.76)$$

where ϕ is the phase difference between the two input signals.

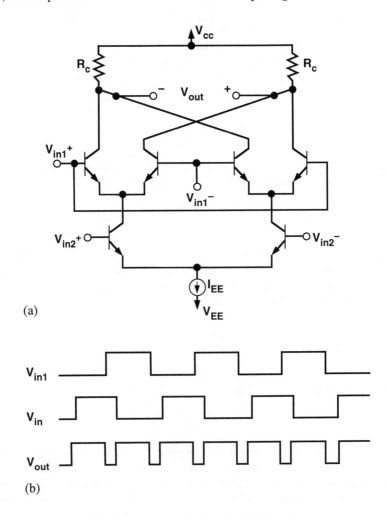

(a)

(b)

Figure 6.39 Gilbert-cell-based phase detector: (a) circuit diagram, (b) waveforms.

There are several drawbacks to this implementation of the phase detector. The major difficulty is in the region of 0 (2π) rad. In this case, the output voltage is symmetric about the origin, and the error voltage produced by the phase detector is the same whether the phase difference is 0.1 rad or −0.1 rad. In order to avoid this problem, PLLs using exclusive-OR type phase detectors are often operated near an input phase difference of $\pi/2$ rad. This can be accomplished by applying a low-noise dc offset voltage, equal to minus half-peak phase detector output, to the input of the integrator following the phase detector.

Another solution to this drawback is the use of so-called edge-triggered phase detector designs. A commercial example of this approach is the Analog Devices AD9901, whose block diagram is shown in Figure 6.40 along with a representative timing diagram. The phase detector is locked when the input and reference waveforms are 180-deg out of phase. When the two inputs are near lock, then only the input flip-flops and the exclusive-OR gate are active. The input flip-flops divide the input frequency and the reference frequency by a factor of two. This moves the region where the output voltage shifts abruptly from its most negative to most positive point from 180 deg (at lock) to 0 deg. The resulting transfer curve is illustrated in Figure 6.41. When the input and reference voltages are 180-deg out of phase, the output has a 50% duty cycle, and any shift in the phase between the signals will cause a change in the output duty cycle.

When the input and reference frequencies differ, then the circuit transitions to the frequency discriminator mode. Whenever the reference or oscillator input transitions twice before the other input (indicating that its frequency is higher), then the exclusive-OR gate is overridden and the set and reset inputs to the flip-flop push the output toward the correct frequency. When the frequency returns to the appropriate range, then the phase detector portion of the circuit takes over again. In this way, locking to a harmonic of the reference signal is avoided.

Frequency Dividers

Frequency dividers within PLLs are key components for frequency translation from high frequencies to lower frequencies. Most implementations of frequency dividers utilize digital flip-flops to generate the requisite signals, as shown in Figure 6.42. These circuits typically utilize master/slave flip-flops, where the Q-bar output is fed to the data input to divide the clock frequency by a factor of two. Digital frequency dividers can be classified into static or dynamic types. Static frequency dividers operate over a broad frequency range and use the latching action of the circuit to set the output voltage. Dynamic circuits typically utilize charge storage at critical nodes in the circuit to "hold" the output voltage [36]. These circuits typically have a narrower range of operating frequency (since charge storage at high-impedance nodes has a limited lifetime) although they can often operate at higher frequencies

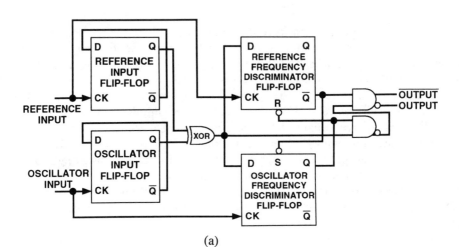

(a)

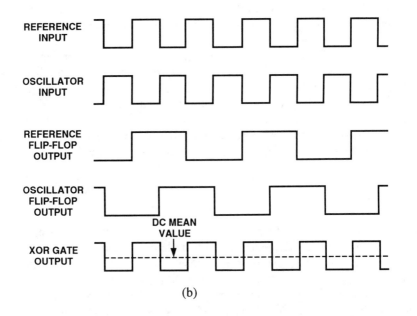

(b)

Figure 6.40 Analog Devices AD9901 edge-triggered phase detector. (a) Circuit diagram, (b) timing waveform at lock, (c) timing waveform (ϕ_{out} leads ϕ_{in}), and (d) timing waveform (ϕ_{out} lags ϕ_{in}).

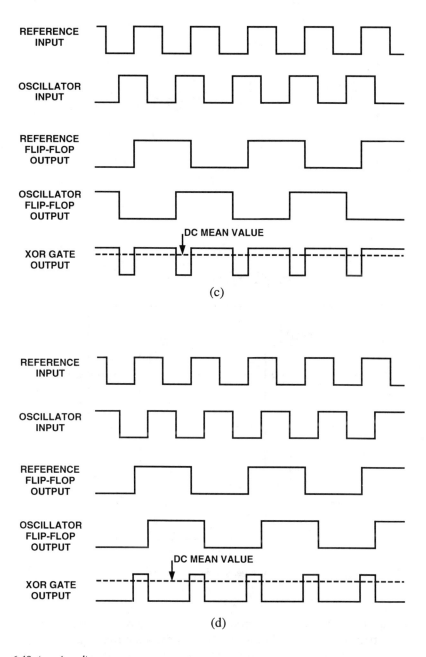

Figure 6.40 (continued)

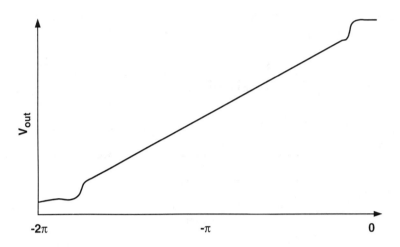

Figure 6.41 Transfer function of edge-triggered phase detector.

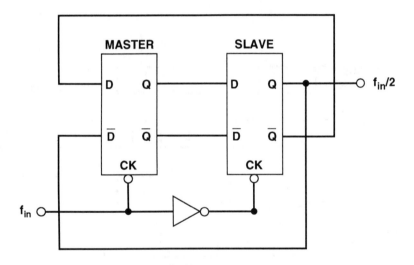

Figure 6.42 Master/slave D-type flip-flop used for frequency division. The circuit is configured as a divide-by-two.

than static dividers. Due to their reputation for improved robustness, static frequency dividers are typically employed more often than dynamic frequency dividers.

An example of a monolithic static D-type flip-flop divide-by-four is shown in Figure 6.43 [37]. This circuit operated at frequencies in excess of 5 GHz, and dissipated less than 30 mW per flip-flop. The maximum frequency of operation of

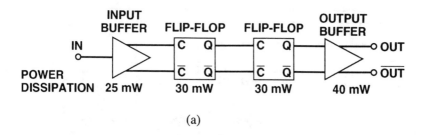

(a)

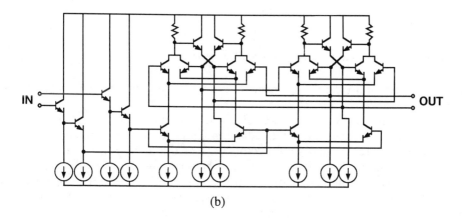

(b)

Figure 6.43 Divide-by-four frequency divider with input and output buffering for communications applications [37]. (a) Block diagram, and (b) flip-flop schematic.

these circuits is typically limited by the f_T and f_{MAX} of the transistors, with an upper limit on the clock frequency lying somewhere between $f_T/2$ and $f_T/3$, depending on the circuit design.

Flip-flops can be embedded in more complicated digital circuits to realize *programmable dividers*, whose output frequency can be altered by changing a digital input word. The division ratios can be integers, as well as fractional integers. A variety of schemes have been proposed to implement these programmable features. An example of a commercially available programmable dividers from Motorola is shown in Figure 6.44 [38], which realizes an 8/9 divider. Other typical circuits include 10/11, as well as 5/6 counters.

A second class of frequency dividers utilizes analog techniques for the realization of frequency division. These so-called *regenerative frequency dividers* use mixing, filtering, and amplification to implement the frequency division function, and a

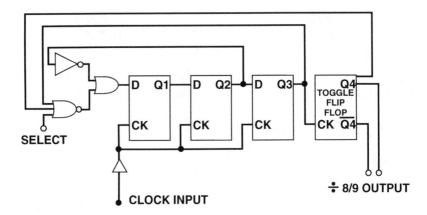

Figure 6.44 Schematic diagram of programmable 8/9 divider [38].

block diagram of a typical implementation appears in Figure 6.45 for the general case of a divide-by-N circuit.

In the case of a divide-by-two circuit, the output can simply be fed back to the mixer input without any intermediate multiplication. The interstage bandpass filter is used to reject the $f_{in} + f_{out}$ component of the mixer output, and pass the $f_{in} - f_{out}$ component. The mixer can often be combined with the amplifier to create a very simple circuit design, as shown in Figure 6.46 [39]. The circuit operated up to 28 GHz in a 40-GHz silicon bipolar process. The filter is accomplished by the addition of some capacitors within the circuit, creating a lowpass response.

Loop Filters

The final component to consider in the implementation of the PLL is the loop filter. Although this is considered the most straightforward, some subtle issues are involved that require consideration.

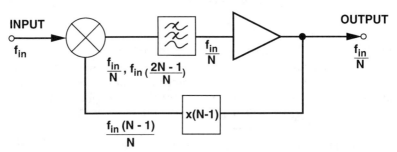

Figure 6.45 Regenerative frequency divider.

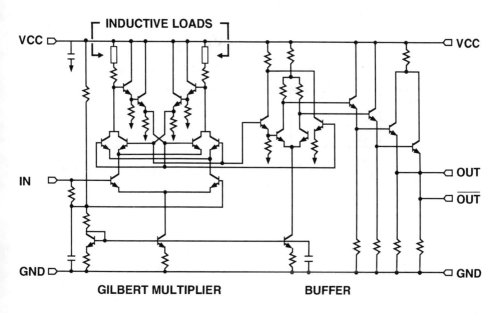

Figure 6.46 Schematic diagram of silicon bipolar regenerative frequency divider [39].

Loop filters are typically implemented using operational amplifiers (op amps), combined with resistors and capacitors to implement the desired transfer function. A typical single-pole, single-zero lowpass filter is shown in Figure 6.47. Referring back to the filter transfer function of (6.62), the necessary element values are defined as follows:

$$K_f = \frac{-R_2}{R_1} \qquad \text{(6.77(a))}$$

$$\tau_Z = C_1 R_1 \qquad \text{(6.77(b))}$$

$$\tau_P = C_2 R_2 \qquad \text{(6.77(c))}$$

One of the major limitations in the design of the loop filter is its bandwidth and dynamic range under high slew-rate conditions; the output of the amplifier must be able to supply the VCO the full voltage required during the clock cycle, without entering slew-rate limited operation.

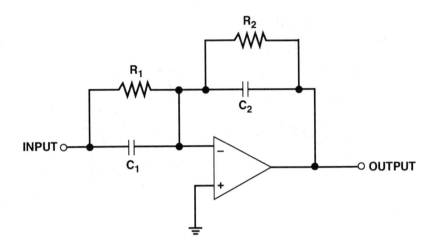

Figure 6.47 Schematic diagram of active op-amp based PLL lowpass filter.

6.5 DIRECT DIGITAL SYNTHESIS (DDS)

An alternative approach for the realization of frequency synthesizers is the use of digital techniques to synthesize the sine wave, and to follow the digital synthesis with a digital-to-analog converter (DAC). The benefits of this approach are due to the well-known advantages of digital signal processing: accuracy, absence of drift, and scalability to increasingly high levels of integration. At the same time, DDS-based systems have historically exhibited higher levels of spurious outputs than their analog counterparts, and these levels must be carefully understood and analyzed before a DDS is used in a system. This section will analyze the architectures, performance tradeoffs, and limitations of the DDS approach to frequency synthesis.[2]

The block diagram of a typical DDS appears in Figure 6.48. It consists of an n-bit accumulator, sine lookup table ROM, DAC, and lowpass filter. The accumulator performs a running calculation of the instantaneous phase angle of the desired sinusoidal wave and delivers its register value R to the sine table. The sine lookup table uses R as an address to find the appropriate sine-wave amplitude, which is stored in the ROM; that is,

$$\sin(\phi) = \sin\left(\frac{2\pi R}{2^n}\right) \tag{6.78}$$

[2]Drs. Ronald Lundgren and Viktor Reinhardt of Hughes Electronics contributed to some of the material in this section.

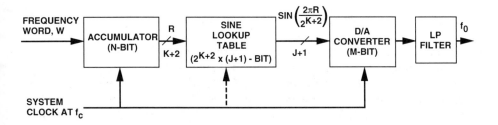

Figure 6.48 Direct digital synthesizer (DDS), utilizing an accumulator, sine lookup table, DAC, and smoothing filter.

The output of the sine table is then sent to an M-bit DAC, which converts the digital waveform into an analog one. The lowpass filter is used to remove the out-of-band harmonics and spurs, and smooth the output waveform. The output of the DAC is given by

$$V(n) = A \; \text{mod}_m[\sin(2\pi nF)] \tag{6.79}$$

where $\text{mod}_m(x)$ truncates x to J bits.

As a result, the ideal output of the DDS is given by

$$V_{\text{out}} = A \; \sin\!\left(\frac{2\pi Wt}{T_c 2^N}\right) \tag{6.80}$$

where W is the N-bit input word, t is time, and T_c is equal to the clock period $(1/f_c)$. As a result, the output frequency can theoretically have 2^N different unique values.

These last two expressions illustrate some of the unique advantages and limitations of the DDS approach to frequency synthesis. The advantage is that the output frequency is discretely variable over a very wide frequency range, from $f_c/2^N$ to $f_c/2$, and this frequency range is set digitally and is solely dependent on the clock frequency, which is intrinsically very accurate.

One disadvantage of the DDS approach is that the finite phase resolution of the input to the sine lookup table creates spurious degradation in the output signal. Generating a sine table to full resolution directly would require a prohibitively large ROM, and spurious errors are created by the periodic excitation of the approximation to an ideal sine wave. In addition, the *output* of the sine ROM is itself a quantized version of an ideal sine wave, further adding to spurious levels. The DAC is not ideal and can exhibit considerable "glitch" energy at major transitions, and this can also degrade the integrity of the output waveform significantly.

6.5.1 DDS Architectures and Design Considerations

The architecture outlined in Figure 6.48 forms the baseline for most basic DDS implementations. It has been noted, however, that there is a theoretical tradeoff between spurious performance and sine ROM and DAC complexity, with higher complexity resulting in better theoretical spurious performance. One major limitation is that the size of the sine lookup ROM must be $2^K \cdot J$ bits, where K is the phase resolution (in bits) and J is the DAC resolution. Various alternative architectures have been proposed that should reduce the complexity of the sine lookup ROM, and these are briefly discussed here.

Hutchinson has proposed a sine-table scheme for reducing the ROM size necessary for a given level of table accuracy [40]. The scheme is based on the approximation

$$\begin{aligned}
\sin(X_K) &= \sin(X_A + X_B) \\
&\approx \sin(X_A) + \sin(X_B)\cos(X_A)
\end{aligned} \tag{6.81}$$

where the input word X_K is separated into the A most significant bits X_A and the least significant bits X_B (i.e., $K = A + B$).

The sine lookup table can now be partitioned into two separate ROM circuits, ROM1 and ROM2, where ROM1 is used to generate $\sin(X_A)$ and ROM2 is used to generate $\sin(X_B)\cos(X_A)$. The output of the two ROMs are then added together to produce a complete output. The size of the ROM table is reduced, because ROM1 must store 2^A words of J-bit length, and ROM2 must store 2^B words of $(J - A + 1)$ bit length. Therefore, the total storage requirements of the ROM are now reduced from $2^K \cdot J$ to $(2^A J + 2^K(J - A + 1))$ bits.

The total phase error that results from this scheme is $(\pi^2/2(2A + 3))$, so for a J-bit accurate phase, $A \geq (J + 0.3)/2$, and

$$A_{\min} = K - \log_2(J ln(2)) \tag{6.82}$$

For example, in the case where $K = 10$ and $J = 10$, $A_{\min} = 7$, and the number of bits required is 5,376, half of the 10,240 bits required for the brute-force approach. A block diagram of an architecture that implements this algorithm is shown in Figure 6.49.

Spurious Responses

Spurious responses arise from a number of factors, and are a significant limitation to the overall performance of DDS systems. They can be further subdivided into phase quantization errors, amplitude quantization errors, and DAC nonlinearities.

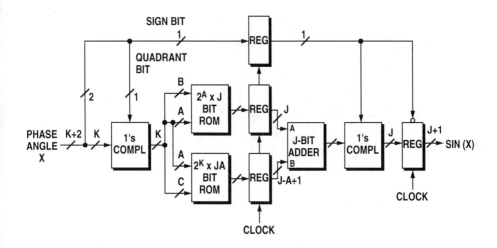

Figure 6.49 Block diagram of memory structure used to implement Hutchinson algorithm.

Phase quantization errors arise from the finite resolution of the input address for the sine lookup ROM. In particular, when the frequency select word is not an integer power of 2, the accumulator does not overflow to zero for each cycle of the output wave. As a result, residual phase increments add up in the accumulator until an overflow occurs after multiple output cycles. This long-term phase error accumulation produces spurious responses at a fraction of the clock frequency.

A variety of techniques have been proposed to minimize this effect by introducing a pseudo-random code into the accumulator with the intent of creating a small amount of phase randomization in the input phase function, with the hope of "spreading out" the resulting spur energy into a wider bandwidth [41, 42].

Amplitude quantization errors result from the finite resolution of the DAC. As a result, it occurs even when the DAC is "perfect," since even an ideal DAC will have finite resolution. This quantization noise is often assumed to be "white" (i.e., uniformly distributed in the frequency domain), but in fact is often harmonically related to the input signal, especially when the input signal is a submultiple of the clock frequency. A rough estimate for the worst case dynamic range with a perfect DAC is 6 dB/bit. So, for example, an 8-bit DAC could have a relatively modest 48-dB dynamic range.

Finally, there are nonlinearities associated with the DAC, which are different from quantization error. These nonlinearities include gain errors, offset errors, non-monotonicity, differential nonlinearity, integral nonlinearity, and major transition glitches. An excellent discussion of the definition of all of these appears in [43], but their effect on DDS performance is not simple to analyze. In effect, the errors can be viewed as generators of harmonics and intermodulation products. Any nonlinearities in the transfer curve will produce these to varying degrees.

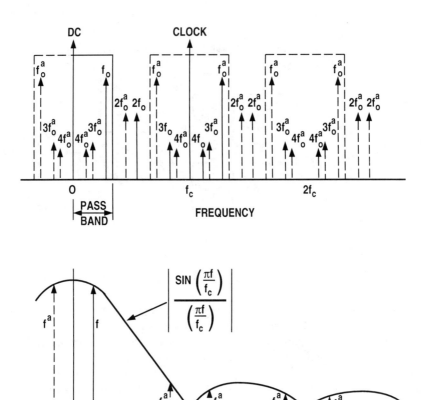

Figure 6.50 (a) Harmonics and aliased harmonics of sampled DDS output waveform. (b) The effect of lowpass filter on the spurious components of DDS output.

Major carry "glitches" create particular problems for DDS systems. These occur when the input word to the DAC changes from, say, 011111 to 100000. In typical implementations of the DAC, this will cause all of the $n - 1$ LSB current sources to turn off and the single MSB current source to turn on. In such a case, there may be a brief period where *all* of the current sources are off, or on, in which case the output swings dramatically before settling to its correct value. One measure of this problem is so-called *glitch energy* which is the integrated area under the voltage curve during the transition period. It is often specified in $pS - V$, or $nS - V$, with a value less than $1pS - V$ being a very good result for a high-speed DAC.

Spurious responses are especially a problem for high-speed DDS systems. For example, if a 500-MHz clock is used to produce a 150-MHz signal, then a second and third harmonic will also be generated at 300 and 450 MHz, which are aliased back to 200 MHz and 50 MHz, creating an in-band spurious response.

Aliasing

A DDS System that incorporates a DAC can be viewed as the convolution of a sampling process and a hold process. The sampling of a signal at frequency f at f_{clock} causes the spectrum to be duplicated at $mf_{clock} \pm f$, where m is an integer from $-\infty$ to $+\infty$. An example of this is shown in Figure 6.50(a). The hold multiplies the amplitude of these frequencies by $(\sin(\pi f/f_{clock})/(\pi f/f_{clock}))$, as shown in Figure 6.50(b). This process would not be a problem if it were not for the harmonics of the fundamental signal, which are aliased back into the band of interest.

References

[1] J. Obregon, and P. S. Khanna, "Exact Derivation of the Non-Linear Negative Resistance Oscillator Pulling Figure," *IEEE Trans. Microwave Theory Tech.*, Vol. MTT-30, July, 1982, pp. 1109–1111.

[2] A. S. Sedra and K. C. Smith, *Microelectronic Circuits*, Philadelphia, PA: Saunders College Publishing, 1991, pp. 841–844.

[3] J. K. Hardy, *High Frequency Circuit Design*, Reston, VA: Reston Publishing, 1979, pp. 259–260.

[4] H. L. Krauss, C. W. Bostian, and F. H. Raab, *Solid State Radio Engineering*, New York, NY: John Wiley and Sons, 1980, pp. 151–152.

[5] J. Smith, *Modern Communication Circuits*, New York, NY: McGraw-Hill, 1986, pp. 252–261.

[6] K. L. Kotzebue and W. J. Parrish, "The use of large signal S-parameters in microwave oscillator design," *Proc. of the Int'l Microwave Symposium on Circuits and Systems*, 1975.

[7] M. Vehovec, L. Houselander, and R. Spence, "On oscillator design for maximum power," *IEEE Trans. on Circuit Theory*, Sept. 1968, pp. 281–283.

[8] B. K. Kormanyos and G. M. Rebeiz, "Oscillator design for maximum added power," *IEEE Microwave and Guided Wave Letters*, Vol. 4, No. 6, June 1994, pp. 205–207.

[9] A.P.S. Khanna, "Oscillators," book chapter in *GaAs MESFET Circuit Design*, ed. R. Soares, Norwood, MA: Artech House, 1988, pp. 361–364.

[10] S. A. Maas, *Nonlinear Microwave Circuits*, Norwood, MA: Artech House, 1988, pp. 463–467.

[11] J. Y. Lee and U. S. Hong, "Voltage controlled dielectric resonator oscillator using three-terminal MESFET varactor," *Electronic Letters*, Vol. 30, No. 16, Aug. 1994, pp. 1320–1321.

[12] E. A. Allamando, "X-band MMIC voltage-controlled oscillator using HEMTs," *Microwave and Optical Technology Letters*, Vol. 5, No. 8, July 1992, pp. 395–398.

[13] A. A. Sweet, *MIC and MMIC Amplifier and Oscillator Circuit Design*, Norwood, MA: Artech House, 1990, pp. 183–189.

[14] A. A. Sweet, "A general analysis of noise in Gunn oscillators," *Proc. of IEEE*, Vol. 60, Aug. 1972.

[15] J. S. Yuan, "Modeling the bipolar oscillator phase noise," *Solid State Electronics*, Vol. 37, No. 10, Oct. 1994, pp. 1765–1767.

[16] U. Rohde, "Oscillator design for lowest phase noise," *Microwave Engineering Europe*, May 1994, pp. 35–40.

[17] A. Howard, "Simulate oscillator phase noise," *Microwaves and RF*, Vol. 32, No. 11, Nov. 1993, pp. 65–70.

[18] L. D. Cohen, "Low phase noise oscillator with flicker $(1/f)$ noise suppression circuit," IEEE MTT-S Int'l Microwave Symposium Digest, June 1992, pp. 1081–1084.

[19] A. Buchwald, and K. Martin, "High-Speed Voltage Controlled Oscillator with Quadrature Outputs," *Electronics Letters*, Vol. 27, No. 4, 1991, pp. 309–310.

[20] S. K. Enam and A. A. Abidi, "NMOS ICs for Clock and Data Regeneration in Gb/s Optical-Fiber Receivers," *IEEE J. of Solid State Circuits*, Vol. 27, No. 12, 1992, pp. 1763–1774.

[21] R. Razavi, and J. J. Sung, "A 60 GHz 60 mW BiCMOS Phase-Locked Loop," *IEEE J. of Solid-State Circuits*, Vol. 29, No. 12, Dec. 1994, pp. 1560–1565.

[22] A. Buchwald, K. Martin, A. Oki, and K. Kobayashi, "A 6-GHz Integrated Phase-Locked Loop Using AlGaAs/GaAs Heterojunction Bipolar Transistors," *IEEE J. of Solid-State Circuits*, Vol. 27, No. 12, Dec. 1992, pp. 1752–1762.

[23] A. Abidi, and R. Meyer, "Noise in Relaxation Oscillators," *IEEE J. of Solid-State Circuits*, Vol. SC-18, No. 6, Dec. 1983, pp. 794–802.

[24] H. Bellescize, "La Reception Synchrone," *Onde Electr.*, Vol. 11, June, 1932, pp. 230–240.

[25] W. Lindsey, and M. Simon, (Eds.), *Phaselocked Loops and Their Application*, New York, NY: IEEE Press, 1978.

[26] F. Gardner, *Phaselock Techniques*, 2nd Edition, Wiley Interscience, 1979.

[27] Motorola MECL Data Book, Chapter 6, "Phase-Locked Loops," 1993.

[28] B. Sklar, *Digital Communications: Fundamentals and Applications*, New York, NY: Prentice-Hall, 1988.

[29] J. G. Proakis, *Digital Communications*, New York, NY: McGraw-Hill, 1983.

[30] A. Hodisan, Z. Hellman, and A. Brillant, "CAE Software Predicts PLL Phase Noise," *Microwaves and RF*, Nov. 1994, pp. 95–100.

[31] V. F. Kroupa, "Noise Properties of PLL Systems," *IEEE Trans. on Communications*, Oct. 1982, pp. 2244–2251.

[32] U. L. Rohde, and T.T.N. Bucher, *Communications Receivers: Principles & Design*. New York, NY: McGraw-Hill, 1988.

[33] U. L. Rohde, *Digital PLL Frequency Synthesizers—Theory and Design*, New York, NY: Prentice-Hall, 1983.

[34] W. F. Egan, *Frequency Synthesis by Phase Lock*, Wiley Interscience, 1981.

[35] P. Gray, and R. Meyer, *Analysis and Design of Analog Integrated Circuits*, New York, NY: John Wiley and Sons, 1989.

[36] H. Ichino et al., "18 GHz 1/8 Dynamic Frequency Divider using Si Bipolar Technologies,"

[37] I. Kipnis et al., "A Wideband Low-Power, High Sensitivity And Small-Size 5.5 GHz Static Frequency Divider," *IEEE 1988 Bipolar Circuits and Technology Mtg.*, pp. 88–91.

[38] Motorola MECL Data Book, 1993.

[39] M. Kurisu et al., "A Si Bipolar 28-GHz Dynamic Frequency Divider," *IEEE J. Solid-State Circuits*, Vol. 27, No. 12, Dec. 1992, pp. 1799–1804.

[40] B. H. Hutchison, "Contemporary Frequency Synthesis Techniques," *Frequency Synthesis and Applications*, Ch. 1, New York, NY: IEEE Press, 1975.

[41] V. S. Reinhardt, "Direct Digital Synthesizers," *Proc. of 17th Annual Precise Time and Time Interval Applications and Planning Mtg.*, Washington, DC, Dec., 1985.

[42] C. E. Wheatley, "Digital Frequency Synthesizer with Random Jittering for Reducing Discrete Spectral Spurs," U.S. Patent 4,410,954 (Oct. 1983).

[43] *Analog Devices Data Converter Reference Manual*, Vol. 1, 1993.

CHAPTER 7
▼▼▼

AMPLIFIERS FOR WIRELESS COMMUNICATIONS

Cynthia Baringer
Hughes Research Labs
Malibu, CA

Christopher Hull
Rockwell International
Newport Beach, CA

7.1 INTRODUCTION

Signal amplification is a fundamental function of the radio portion in all wireless communication systems. The receiver must take the smallest possible signal level at its input and amplify that signal to the minimum level that can be detected by the demodulator. Correspondingly, the transmitted signal level must be amplified sufficiently so that despite loss inherent in wireless transmission, the signal can be received by a nearby base station.

There are several types of amplifiers that serve specific functions in the radio. Amplifiers in the receiving chain that are closest to the antenna contribute directly to the noise performance. These low-noise amplifiers (LNAs) must provide enough gain so that the noise contributions from the rest of the receiving components are minimal. The amplifiers closest to the antenna in the transmit path are power amplifiers (PAs). They must amplify the signal linearly and efficiently. Noise is not a primary consideration.

Along the receiver chain, there are variable-gain amplifiers (VGAs). The function of these amplifiers is to help keep the receiver output constant over the range of expected signal intensity. The noise and linearity requirements of VGAs are not as strict as for LNAs and PAs, respectively. However, VGAs must have reasonable dynamic range, which is a measure of an amplifier's ability to handle both small and large signals.

Higher frequency bands are being sought to avoid congestion. Therefore, wireless systems are requiring higher frequency and wider bandwidth amplification. The bandwidth of an amplifier may ultimately be limited by the technology. Low-cost objectives and higher integration often prohibit the use of higher performing technologies. Techniques used to maximize the bandwidth of an amplifier, for a given technology, are introduced in Section 7.2.

In the next three sections of this chapter, amplifiers widely used in current wireless systems are presented. Section 7.3 discusses the benefits of using a simple cascode for use in low-noise applications. Section 7.4 compares the dynamic range of several VGAs. The design of a PA is covered in Section 7.5. Issues of practical concern, such as package limitations, are emphasized.

7.2 WIDEBAND AMPLIFIER TECHNIQUES

Operating frequencies of many commercial wireless systems are in the low gigahertz range. Amplifiers must provide linear gain at these frequencies.

Feedback is a common technique used in wideband amplification to provide gain stability over process and supply variations, wider bandwidth, and lower distortion. Other techniques for broadbanding, not discussed in this section, include a Gilbert-cell amplifier approach [1] and feedforward techniques such as the Cascomp circuit [2].

The bandwidth of amplifiers is fundamentally limited by the transistor gain-bandwidth product. Applying feedback to an amplifier does not change this product, but rather, allows the designer to trade gain for bandwidth. Compound stages, which replace single transistors with multiple transistors that behave like a single transistor, can exceed the single device gain-bandwidth product. This section will conclude with a discussion of practical design issues in wideband feedback amplifiers, including a discussion of parasitic capacitance and impedance matching.

7.2.1 Gain-Bandwidth Product

A useful measure of the performance of a wideband amplifier is its gain-bandwidth product. It can be used to compare the effective utilization of topologies for a given application. In an N-stage amplifier, the gain-bandwidth is arbitrarily defined as

$$(GBW)_N = (Gain)^{1/N} BW \tag{7.1}$$

where N is the number of stages, Gain is the total midband gain, and BW is the overall bandwidth.

In general, amplifiers with fewer gain stages will have better gain-bandwidth because the bandwidth shrinks as stages are added. Consider an amplifier that is a cascade of N identical stages each with a single-pole role-off response given as

$$\frac{v_o}{v_i} = \frac{G}{1 + j\left(\dfrac{\omega}{B}\right)} \tag{7.2}$$

The gain-bandwidth product of each stage is $G \cdot B$. (G is the gain and B is the bandwidth of each individual stage.) The overall transfer function is

$$H(j\omega) = \frac{G^N}{\left(1 + j\dfrac{\omega}{B}\right)^N} \tag{7.3}$$

The total midband gain is Gain = G^N and the overall bandwidth is when $|H(j\omega)| = \dfrac{G^N}{\sqrt{2}}$. This gives

$$BW = B\sqrt{2^{\frac{1}{N}} - 1} \tag{7.4}$$

and using (7.1), the amplifier gain-bandwidth product is

$$(GBW)^N = GB\sqrt{2^{\frac{1}{N}} - 1} \tag{7.5}$$

So, as the number of stages increases, the amplifier gain-bandwidth is reduced.

Only single-stage and two-stage amplifiers are considered in the following sections. The overall gain-bandwidth product of single-stage and two-stage amplifiers are denoted $(GBW)_1$ or $(GBW)_2$, respectively. In the discussion of two-stage amplifiers, the individual stage gain-bandwidth products are denoted GB_1 and GB_2 for the first and second stages, respectively.

Single-Stage Amplifiers

Amplifier gain-bandwidth products are ultimately limited by the unity-gain frequency f_T of the transistor. The device f_T is the frequency at which the transistor current gain equals one. This can be found from a high-frequency model of a transistor, shown in Figure 7.1. The input impedance is predominantly capacitive and the transconductance g_m is approximately independent of frequency. The unity-gain frequency can be found to be

$$f_T = \frac{g_m}{2\pi C_{in}} \tag{7.6}$$

To first-order, the expression predicts the unity-gain frequency of a FET when $C_{in} = C_{gs}$ and of a bipolar device when $C_{in} = C_\pi$. Analysis with a more accurate transistor model yields a two-pole transfer function. In most cases, there is a dominant pole and while the presence of the second pole does not ideally degrade the gain-bandwidth product, it does contribute excess phase in the circuit. For low source impedance, the gain-bandwidth is degraded by resistive parasitics [3].

The amplifier gain-bandwidth of a simple single-stage amplifier, such as a common-emitter stage, is identical to the transistor gain-bandwidth product. That is, $(GBW)_1 = 2\pi f_T$. By reducing the load resistance, voltage gain can be exchanged for bandwidth, which is a technique referred to as resistive broadbanding. Lower impedances also reduce the interaction between stages by minimizing the Miller effect on device feedback parasitics. These parasitic capacitances can lead to narrowbanding. For these reasons, wideband circuits are always low-impedance circuits.

The gain-bandwidth product of a single-pole feedback amplifier is also identical to the transistor gain-bandwidth product. Though there are only two circuit topologies to consider, series feedback (as in emitter-degeneration) and shunt feedback, a general approach will be taken to find the gain-bandwidth product. In a feedback amplifier, the closed-loop gain is given in terms of the *open-loop gain* and *loop gain.* Open-loop gain is the gain of the basic amplifier when the feedback network is not present and the loop gain is the product of the open-loop gain and the feedback. The closed-loop gain of the single-pole feedback amplifier is [4]

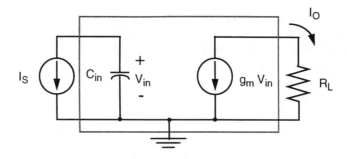

Figure 7.1 Simplified high-frequency small-signal transistor model.

$$A(s) = \frac{a(s)}{1 + T(s)} \tag{7.7}$$

For an amplifier with a single pole and gain-bandwidth of ω_T, the open-loop gain is expressed as $a(s) = \dfrac{a_o}{1 + a_o\dfrac{s}{\omega_T}}$.

If the feedback network is made up of resistors, then the feedback factor f_o is independent of frequency and the loop-gain can be expressed as $T(j\omega) = f_o a(s)$. Substituting the expressions for open-loop gain and loop-gain into (7.7) gives (after some manipulation)

$$A(s) = \frac{\dfrac{a_o}{1 + a_o f_o}}{1 + \left(\dfrac{s}{\omega_T}\right)\left(\dfrac{a_o}{1 + a_o f_o}\right)} \tag{7.8}$$

Referring to (7.8), the low-frequency closed-loop gain is obtained by setting $s = 0$ and the bandwidth is obtained by finding the frequency at which the denominator equals zero. The gain-bandwidth is calculated to be

$$G \cdot B = \left(\frac{a_o}{1 + a_o f_o}\right)\left(\frac{1 + a_o f_o}{a_o}\omega_T\right) \tag{7.9}$$

or

$$G \cdot B = \omega_T \tag{7.10}$$

Thus, ideal resistive feedback does not change the gain-bandwidth product of a single-pole amplifier.

Two-Stage Amplifiers

Improved performance can often be obtained by applying feedback around two stages rather than a single stage. The drawback is that amplifier stability must be addressed in the design. The negative feedback at low frequencies becomes positive feedback at high frequencies, reducing or eliminating any phase margin in the system.

The problem of inadequate phase margin can be overcome by either dominant-pole compensation or by creation of a feedback zero. In the dominant-pole technique, a single pole is introduced at a sufficiently low frequency that the magnitude of the loop gain drops to less than unity before the phase shift reaches 180 deg [4]. The disadvantage of dominant-pole compensation is a significant gain-bandwidth reduction. For wideband amplification, the feedback-zero technique is preferred since little or no gain-bandwidth is lost.

An example of a feedback configuration with a feedback zero is shown in Figure 7.2. The amplifier is shown schematically as an operational amplifier. The capacitor in parallel with R_F increases feedback at high frequencies, giving extra phase lead to compensate for the phase lag in the amplifier. Excessive peaking can be observed in the two-stage amplifier without the capacitor. The feedback zero can eliminate the peaking and eventually cause an over-damped response characteristic.

The closed-loop response of the two-stage feedback amplifier may be written as

$$A(s) = \frac{K}{s^2 + \dfrac{\omega_o}{Q}s + \omega_o^2} \qquad (7.11)$$

where K is the product of the low-frequency open-loop gain $a(0)$ and the two open-loop pole frequencies p_1 and p_2. The closed-loop response is characterized by a resonance frequency, $\omega_o = \sqrt{p_1 p_2(T_o + 1)}$ and a quality factor, $Q = \dfrac{\omega_o}{p_1 + p_2 + \dfrac{p_1 p_2 T_o}{z}}$ where T_o is the low-frequency loop gain and z is the frequency of the feedback zero. The value of z can be chosen to give a desired value of Q. For $Q = \dfrac{1}{\sqrt{2}}$, the bandwidth *is* the resonant frequency and is given as

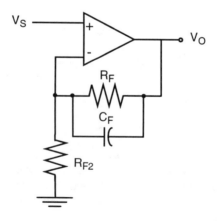

Figure 7.2 Amplifier with feedback zero.

$$\text{BW} = \sqrt{(T_o + 1)p_1 p_2} \tag{7.12}$$

The low-frequency gain is

$$A_v = \frac{a(0)}{1 + T_o} \tag{7.13}$$

Combining (7.12) and (7.13) yields the expression

$$\sqrt{\text{Gain}} \cdot \text{BW} = \sqrt{p_1 p_2 a(0)} \tag{7.14}$$

If there are two gain stages with gains $a_1(j\omega)$ and $a_2(j\omega)$, then (7.14) can be rewritten as

$$\sqrt{\text{Gain}} \cdot \text{BW} = \sqrt{p_1 p_2 a_1(0) a_2(0)} \tag{7.15}$$

$$= \sqrt{\text{GB}_1 \cdot \text{GB}_2} \tag{7.16}$$

where $\text{GB}_1 = p_1 a_1(0)$ and $\text{GB}_2 = p_2 a_2(0)$.

The product $\sqrt{\text{Gain}} \cdot \text{BW}$ is conserved and is independent of the loop gain. Because a two-pole amplifier generally is made up of two gain stages, a reduction in bandwidth by a given factor increases the gain by the square of that factor. The reason for this is that each of the two stages increases in gain when the bandwidth is reduced. Hence $\sqrt{\text{Gain}} \cdot \text{BW}$, as described in the beginning of this section, is the proper performance measure for a two-stage amplifier.

The gain-bandwidth will only be achieved if a sufficiently high Q is obtained. A necessary and sufficient condition for $Q > \frac{1}{\sqrt{2}}$ is

$$T_o > \frac{1}{2}\left(\frac{p_1}{p_2} + \frac{p_2}{p_1}\right) \tag{7.17}$$

Thus, the loop-gain must be at least unity. For increasing loop gains, further initial pole position spreads are permitted. Because of the unwanted feedback zero introduced through capacitive parasitics (such as the collector-base junction capacitance of a bipolar transistor), the loop gain must be larger than the right-hand side of (7.17) for optimal gain-bandwidth.

Compound-Transistor Stages

The use of compound transistors in place of single bipolar transistors is common in wideband amplifier design because they can achieve higher current gain for a given frequency. A common configuration is common-collector/common-emitter, referred to as a Darlington configuration. The first transistor can be biased either resistively as shown in Figure 7.3(a) or with a diode as shown in Figure 7.3(b). The former has seen more implementation in circuit design so far, but the latter configuration exhibits superior phase characteristics [5, 6].

The current gain of the resistive Darlington has a two-pole characteristic, causing the compound transistor to appear more like a two-stage amplifier. There is a zero in the transfer characteristic, which can be used to achieve extremely wideband designs [7]. Substituting the resistive Darlington for a single transistor in a parasitic-sensitive circuit is not advantageous, since the base resistance and output capacitance is double that of the single transistor.

At low frequencies, the diode Darlington (Figure 7.3(b)) appears as a single transistor with twice the current gain and twice the effective f_T. At very high frequencies, the pole-zero pair will add extra phase to the current gain. The input impedance is simply double that of a simple transistor and the transconductance is identical to a simple transistor at low frequencies, though it is slightly lower at high frequencies (near f_T). The impact of degraded transconductance is even more evident when the source impedance is low.

The use of the diode Darlington over the resistive Darlington is preferable. The resistive Darlington has excellent current-gain performance but the diode Darlington maintains double the current gain of a single transistor to very high frequencies. The resistive Darlington exhibits poor phase characteristics; phase shifts of 180 deg at frequencies one decade below f_T may occur. Another key advantage of the diode Darlington is that it can be substituted for a simple transistor with only minor

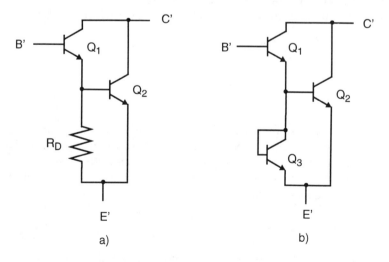

Figure 7.3 Resistive Darlington (a) with resistor feedback, (b) with transistor feedback.

changes in bias and compensation for the slight extra phase shift while the resistive Darlington may require an entire redesign of the circuit.

7.2.2 Gain-Bandwidth of Practical Circuits

Gain-bandwidth concepts are applied to practical circuits in this section. Although the circuit equations given in this section are formulated for bipolar transistors, the results apply equally well to FETs with little modification. Because of the low transconductance of FETs, at lower bias currents some configurations are not suitable to use with FETs.

Transistor Model

To simplify the analysis of the practical amplifier designs presented in the following sections, a simplified equivalent circuit of a bipolar transistor is shown in Figure 7.4. The elements in Figure 7.4 can be related to the elements of the small-signal hybrid-π model [3] shown in Figure 7.5. The small-signal model is equivalent to

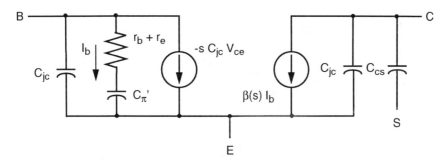

Figure 7.4 Simplified model of bipolar transistor.

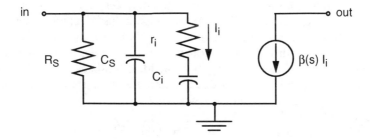

Figure 7.5 Equivalent circuit hybrid-pi model.

the commonly used version, with the exception that feedforward through the collector-base junction capacitance has been neglected. The mapping of the amplifier elements to a bipolar transistor is given as

$$C_s = C_{jc} + C_{source} \tag{7.18}$$

$$R_s = R_{source} \tag{7.19}$$

for input stages and for other stages

$$C_s = C_{jc} + C_{out} \tag{7.20}$$

$$R_s = R_{out}, \tag{7.21}$$

where $C_{out} = C_{cs} + C_{jc}$ is the output capacitance of the previous stage and R_{out} is the output resistance of the previous stage. For all stages,

$$C_i = C'_\pi \tag{7.22}$$

or

$$C_i = \frac{C_\pi}{1 + g_m r_e} \tag{7.23}$$

$$r_i = r_b + r_e \tag{7.24}$$

The emitter resistance r_e is typically small and $C_i \approx C_\pi$ and $r_i \approx r_b$.

The ideal feedback source $(-j\omega V_{ce} C_{jc})$ in Figure 7.5 is dealt with separately in the mapping from the multistage amplifier to the equivalent hybrid-π model. Usually, this ideal feedback creates a feedback zero, which helps to stabilize the amplifier. However, if the magnitude of the feedback zero becomes too low, the Q will be reduced and the bandwidth may drop from its optimum value, as discussed in Section 7.2.1. In the terminal stage of an amplifier, the ideal feedback source has the effect of Miller-multiplication of the feedback capacitance without the compensating benefit of pole-splitting. Hence, a large feedback capacitance in the output stage reduces the gain-bandwidth product.

An FET small-signal model can be manipulated into a form similar to the equivalent model in Figure 7.5. The element and node names change, but the structure is identical. This is quite convenient since circuit equations developed for bipolar transistors generally apply equally well to field-effect transistors.

Current-Feedback Pair

One of the most commonly used wideband amplifiers is the current-feedback pair shown in Figure 7.6. Shunt feedback is used to give a low input impedance (typically 50Ω). The output can be taken from either the emitter or collector of Q_2. Taking the output from the collector gives the feedback loop a better degree of isolation from the loading of subsequent stages. A shunt feedback capacitance stabilizes the amplifier by introducing a feedback zero.

In the first stage, the important pole is contributed by the base capacitance of Q_1. The parasitic capacitance at the base of Q_1 is equal to the sum of the source capacitance, feedback capacitance, and collector-base capacitance. The collector-base capacitance has no effect (ideally) on the gain-bandwidth product but, rather, it adds to the shunt feedback capacitance C_F and lowers the feedback-zero frequency. The gain-bandwidth product of the first stage is

$$GB_1 = \frac{\omega_{T_1}}{1 + \dfrac{r_{b_1}}{R_F} + \dfrac{C_s + C_{jc_1} + C_F}{C_{\pi_1}}} \tag{7.25}$$

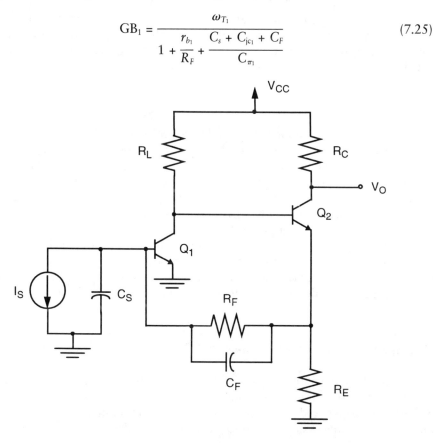

Figure 7.6 Bipolar amplifier current-feedback pair.

where $\omega_{T_1}(= 2\pi f_{T_1})$ is the gain-bandwidth product of transistor Q1. The pin and input bonding pad capacitance can be lumped into the source capacitance C_S.

The form of the second-stage gain-bandwidth product is similar to the first stage. The differences are that there is emitter degeneration by R_E and Miller multiplication of C_{jc2}. Both affect the gain-bandwidth product, which is given as

$$GB_2 = \frac{\omega_{T_2}}{1 + \dfrac{R_E + r_{b_2}}{R_L} + \dfrac{C_x}{C_{\pi_2}}(1 + g_{m_2}R_E)} \qquad (7.26)$$

where $\omega_{T_2} = (2\pi f_{T_2})$ is the gain-bandwidth product of transistor Q2 and $C_x = C_{cs_1} + C_{jc_1} + C_{jc2}\left(1 + \dfrac{R_{cc}}{R_E}\right)$.

Since the current-feedback pair is a two-stage feedback amplifier, the analysis in the previous section can be applied. The overall gain-bandwidth performance is equal to the geometric-mean of the gain-bandwidth product of each of the two individual stages as expressed in (7.16) and is given as

$$(GBW)_2 = \sqrt{GB_1 \cdot GB_2} \qquad (7.27)$$

Parasitic effects can limit the useful bandwidth of the circuit. The existence of base ohmic resistance in series with the input capacitance gives one parasitic pole for each stage. These poles do not appreciably affect the magnitude of the open-loop gain characteristic, but they do add excess phase. To compensate for the excess phase, the feedback-zero location must be lowered by increasing C_F. For extremely wideband amplifiers with bandwidths approaching the device f_T, it may be impossible to compensate for the extra phase shift added by the parasitic poles.

Although bipolar transistors are shown, the current-feedback pair can be designed in an FET technology. In a BiCMOS process, better noise performance can be achieved through replacement of transistor Q_1 with a MOSFET [8]. The equations can be adapted to a different technology by correct substitution of device parameters (e.g., C_{gs} for C_{π}).

Current-Feedback Pair With Darlington

An improved version of the current-feedback pair is shown in Figure 7.7. By replacing transistor Q_2 with a diode Darlington, a significant increase in the gain-bandwidth product can be observed. The diode Darlington configuration can be used with only minor design changes since it adds less than 10 deg of extra phase to the current gain response.

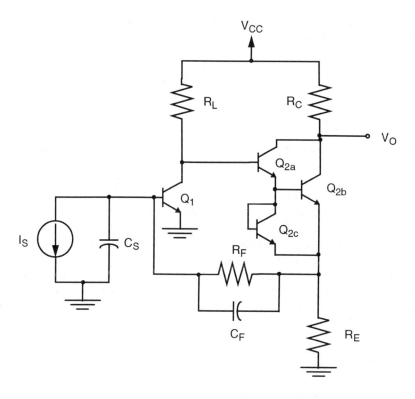

Figure 7.7 Bipolar amplifier current-feedback pair with mirror Darlington.

The diode Darlington has a couple of advantages that help to improve the performance of the current-feedback pair. As shown previously, the compound transistor has a higher current gain than the single transistor. The Darlington configuration has the additional advantage that the extra V_{be} drop in Q_2 provides an extra voltage drop across the collector-emitter of Q_1. The collector-base capacitance of Q_1 is then lowered, which increases the device f_T. The extra voltage drop also allows for a larger output swing.

There is usually little advantage in using a compound in the first stage. The impedance at the input node at high frequencies is fairly low due to the source capacitance and the gain-bandwidth of the stage is more dependent on resistive parasitics. A disadvantage to using a compound stage is the increase in noise. Finally, the advantages of having an extra diode drop in the second stage are lost if the first stage is also a Darlington.

If the excess phase in the current gain of the diode Darlington (due to the pole-zero pair near f_T) can be neglected, then the diode Darlington appears to be identical to a single transistor with twice the current gain (and twice the input impedance). The gain-bandwidth of the improved second stage is

$$GB_2 = \frac{2\omega_{T_2}}{1 + 2\left[\dfrac{R_E + r_{b_2}}{R_L} + \dfrac{C_x}{C_{\pi_2}}(1 + g_{m_2}R_E)\right]}$$

(7.28)

The gain-bandwidth product is doubled but the effects of the parasitics are also doubled. In a situation where parasitics dominate, little advantage is gained by the use of the Darlington compound structure. In practice, however, about 35% improvement in gain-bandwidth can be expected.

The resistive Darlington is also widely used as the second stage in the current-feedback pair [9]. Because of the high collector impedance of Q_1, the Darlington is slightly modified to avoid any Miller multiplication of the junction capacitance of Q_2. (The collector of the input transistor Q_{2a} is connected to V_{CC}.) Unfortunately, current gain near f_T drops off and this, combined with the large phase shift (in the absence of dominant-pole compensation), causes the resistive Darlington/current feedback pair configuration to be less than ideal for applications requiring bandwidths beyond about 10% of f_T.

Voltage-Feedback Pair

While the current-feedback pair provides a low input impedance and a high output impedance, the voltage-feedback pair shown in Figure 7.8 does the reverse. Using series-shunt feedback, the voltage-feedback pair gives a high input impedance and a low output impedance. The circuit approximates an ideal voltage gain stage.

The gain-bandwidth of the stages are

$$GB_1 = \left(\frac{R_{CC}}{R_S}\right)\frac{\omega_{T_1}}{1 + \dfrac{r_{b_1} + R_E}{R_S} + \dfrac{C_{jc_1} + C_F}{C_{\pi_1}}}$$

(7.29)

and

$$GB_2 = \frac{\omega_{T_2}}{1 + \dfrac{r_{b_2}}{R_L} + \dfrac{C_{cs_1} + C_{jc_1} + C_{jc_2}\left(1 + \dfrac{R_{CC}}{R_E}\right)}{C_{\pi_2}}}$$

(7.30)

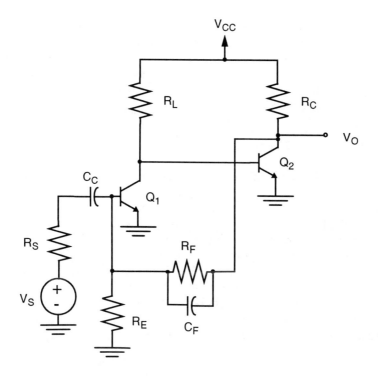

Figure 7.8 Bipolar amplifier voltage-feedback pair.

and, again, (7.16) can be used to find the system gain-bandwidth product.

There are some drawbacks to this design. The frequency response of the voltage-feedback pair is quite sensitive to the source resistance. Unless the source resistance is well-known, and the parasitics at the input are well-controlled, the amount of peaking becomes difficult to control. Additionally, the high input impedance of this configuration makes it unsuitable as an input stage for bandwidths beyond a few hundred megahertz. The high input return ratio can be lowered by applying imped-ance-matching techniques, as discussed in the next section.

The voltage-feedback pair need not employ bipolar transistors. Because the circuit is not suitable for ultrawideband applications, there is little to gain by using GaAs FETs. However, the circuit is suitable for MOSFETs, JFETs, and even vacuum tube triodes.

7.2.3 Wideband Matched-Impedance Amplifiers

There are advantages to matching integrated circuit terminal impedances as well as designing for matched impedances internal to the chip. Terminal impedances matched to a specified system impedance (often 50Ω or 75Ω) will prevent standing waves in transmission lines that connect amplifiers separated by many wavelengths. A major advantage to matching impedances on the integrated circuit is to reduce sensitivity to parasitics. Using impedances that are too high makes a circuit sensitive to capacitive parasitics, and using impedances that are too low makes a circuit sensitive to series resistive and inductive parasitics.

Practical Single-Stage Design

The most straightforward way to achieve matching is by using a shunt resistor at the input and output of a high-impedance amplifier. While sometimes used, this brute-force approach reduces gain and power output capability, and degrades noise performance. A better approach to impedance matching is to use shunt feedback (as was used in the current-feedback pair) since it causes only slight degradation in noise performance and power output capability.

A simple matched-impedance amplifier is shown in Figure 7.9. The design uses shunt feedback around a voltage gain stage to achieve the match. Series feedback in the form of emitter degeneration is used to control the gain. The bipolar transistor may be replaced by an FET, though no source resistance should be included in the FET design because the device transconductance is already low and would be further

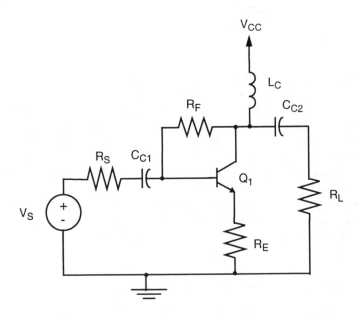

Figure 7.9 Simple matched-impedance bipolar transistor amplifier.

reduced. To achieve reasonable gain, the FET must be run at very high currents (usually tens of milliamps). For this reason, bipolar transistors are usually preferred for amplifiers of this type. GaAs FETs are used only when the widest bandwidth is required.

A small-signal analysis with a simplified model can be applied to the circuit to arrive at the characteristic equations. The effect of finite beta and loading of the feedback network have been ignored. The amplifier is matched if the loop gain of the shunt feedback is unity. The condition for matching is then

$$R_F \approx G_m R_S R_L \qquad (7.31)$$

where $G_m = \dfrac{g_m}{1 + g_m R_E} - \dfrac{1}{R_F}$

The gain and gain-bandwidth product under matching conditions are

$$A_v = -\frac{1}{2} G_m R_L \qquad (7.32)$$

$$\text{GBW} = \frac{\omega_T}{1 + \dfrac{r_b + R_E}{R_S \| R_F} + \dfrac{C_x}{C_{\text{in}}}} \qquad (7.33)$$

where $C_x = C_{\text{jc}}(1 + G_m R_L)$ and $C_{\text{in}} = G_m / \omega_T$.

Two-Stage Matched-Impedance Amplifiers

While one-stage matched-impedance amplifiers are very simple, they often do not provide adequate gain. Two-stage amplifiers offer superior gain for a given bandwidth and have flatter bandpass characteristics. For FET designs, it is even more essential to use two gain stages since the low transconductance of the FET limits the gain that can be achieved in a single stage.

The most straightforward way to create a two-stage matched-impedance amplifier is with a cascade of single stages, as shown in Figure 7.10. The shunt feedback around the first stage is effectively global since it encompasses two fundamental poles (one at the base of Q_1 and the other at the base of Q_2). The global feedback has a loop gain of unity (required for matching) and, as discussed in Section 7.2.1, the optimal gain-bandwidth is difficult to meet for loop gains less than or equal to unity.

Figure 7.11 is a design that is a variation of two cascaded single stages [10]. The design, referred to as the Kukielka configuration, contains a voltage-gain stage (common-emitter/common-collector) with shunt feedback around it. Series-series

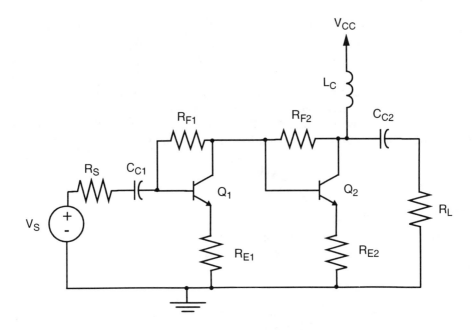

Figure 7.10 Two-stage matched-impedance bipolar transistor amplifier.

feedback, in the form of emitter degeneration in Q_1, gives a high input impedance (in the absence of R_{F1}). Local shunt feedback around Q_2 provides a low input impedance node at its base. Then shunt feedback is applied around this voltage amplifier via R_{F_i} to achieve matching.

The circuit can be designed to achieve optimal gain-bandwidth by appropriate positioning of the poles. As before with the single stage, the loop gain of the shunt feedback must be unity to achieve matching. Therefore, to meet the condition for optimum gain-bandwidth as specified in (7.17), the two open-loop pole positions must be nearly equal for unity loop gain. Under the assumption that $G_{m_2} \gg 1$, $R_{E_1} \gg R_S$ and $R_L C_{jc2} \gg \tau_2$, the condition for equal pole positions is

$$R_{F_2} = G_{m_1} R_S R_L \qquad (7.34)$$

The condition for matching is

$$G_{m_2} R_{F_1} = G_{m_1} R_{F_2} \qquad (7.35)$$

and the gain when matched is

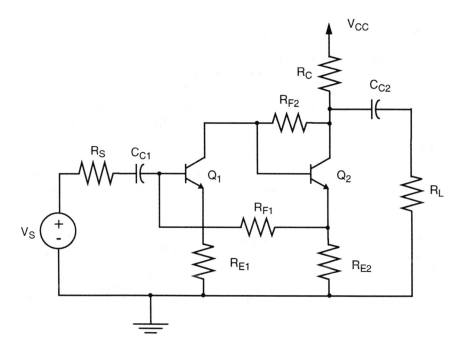

Figure 7.11 Kukielka configuration.

$$A_v = \frac{1}{2} G_{m_1} R_{F_2} \tag{7.36}$$

$$= \frac{1}{2} G_{m_2} R_{F_1} \tag{7.37}$$

Combining the conditions for matching and pole position gives the first-order design equations

$$R_{E_1} \approx \sqrt{R_L R_S} - \frac{1}{g_{m_1}} \tag{7.38}$$

$$R_{F_2} \approx \sqrt{2A_V R_L R_S} \tag{7.39}$$

$$R_{F_1} \approx \frac{2A_V}{G_{m_2}} \tag{7.40}$$

The choice of R_{E_2} is unconstrained by these conditions. A smaller value will increase the bandwidth because it reduces sensitivity to parasitics at the base of Q_2. On the other hand, a large value of R_{E_2} will give better noise performance since R_{F_1} will be forced to be larger by the third design equation.

The gain-bandwidth performance of the Kukielka configuration is essentially identical to that of the current-feedback pair. The optimum gain-bandwidth is not easily met because of parasitics. The circuit tends to give an over-damped response since parasitics decrease Q from its maximum value of $\dfrac{1}{\sqrt{2}}$.

A diode Darlington can replace Q_2. The extra phase shift will help raise the Q of the circuit and increase the bandwidth. Use of the resistive Darlington has also given good results [7].

Another interesting two-stage variant first proposed by Meyer [11] is depicted in Figure 7.12. The circuit can be viewed as a hybrid between the voltage-feedback pair and the current-feedback pair. Like the Kukielka configuration, the Meyer configuration is a voltage-gain stage (a voltage-feedback pair in this case) with shunt feedback around it for matching. The ability to adjust the circuit Q through use of a feedback zero or by design is a significant advantage of this design.

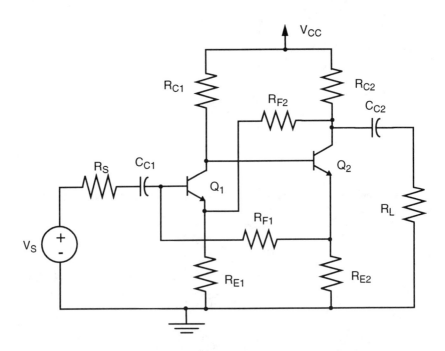

Figure 7.12 Meyer configuration.

7.3 LOW-NOISE AMPLIFIERS

Noisy components in the beginning of the receiver chain have the most significant impact on the overall noise performance. For this reason, a large emphasis is placed on low-noise operation of the first RF amplifier, also referred to as the low-noise amplifier (LNA).

The design of an LNA involves numerous tradeoffs. The amplifier must have sufficient gain to overcome mixer noise, but not so much to cause mixer overload. Good noise characteristics are desired while achieving desired input and output matching. The matching networks at the input and output of the LNA are often a compromise between optimum gain matching and optimum matching for noise figure. Selecting a technology and package involves tradeoffs of cost and level of integration.

The apparent simplicity of an LNA is misleading—the design should be easy because there are relatively few components but the tradeoffs complicate the design. A handful of guidelines are provided in the next section that may facilitate the design. The design of an LNA commonly used in wireless systems is presented.

7.3.1 Guidelines for Low-Noise Amplifier Design

Noise figure (NF) is the most common method used to specify the noise performance of an amplifier. Noise figure is defined as ten times the logarithm (base 10) of the noise factor, F. The noise factor is defined as the ratio of the total output noise power of an amplifier to the output noise power due to the source alone.

A two-port network with noise sources can be represented by the same network with the noise sources removed and with a noise voltage $\overline{v_i^2}$ and current generator $\overline{i_i^2}$ connected at the input. The noise factor can be expressed in terms of these equivalent input noise generators. It is

$$
F = 1 + \frac{\overline{v_i^2}}{4kTR_s\Delta f} + \frac{\overline{i_i^2}}{4kT\dfrac{1}{R_s}\Delta f}
\tag{7.41}
$$

The design of an LNA is strongly influenced by the source resistance, R_S. For high source resistance, $\overline{i_i^2}$ is the dominant equivalent noise source and, conversely, for low source resistance, $\overline{v_i^2}$ is the dominant equivalent noise source. When the source resistance is large, FET devices offer superior performance to bipolar devices at low frequencies because the noise-current generator is determined by gate leakage current, which is small. The noise-voltage generator of a bipolar device is typically smaller than that of an FET, so for low source impedances, a bipolar transistor often has noise performance superior to that of an FET.

From (7.41), a value of R_S giving minimum F can be found. It is

$$R_{Sopt} = \frac{\overline{v_i^2}}{\overline{i_i^2}} \tag{7.42}$$

The optimum source resistance is unlikely to be the actual source resistance (typically, 50Ω or 75Ω), so a matching network is implemented using integrated or off-chip passive components. (The package parasitics can be incorporated in the matching network if the package has been well-characterized.) Often, the optimum source resistance is much larger than the actual source resistance and the matching network cannot easily be realized. Therefore, minimum noise figure cannot be achieved.

For a bipolar transistor, the optimum source impedance and the minimum noise factor are given as

$$R_{Sopt} = \frac{\sqrt{\beta}}{g_m}\sqrt{1 + g_m r_b} \tag{7.43}$$

$$F_{min} \cong 1 + \frac{1}{\sqrt{\beta}}\sqrt{1 + g_m r_b} \tag{7.44}$$

Increasing the current will increase g_m and decrease the optimum source impedance. Optimum matching conditions are more likely achievable, however, the noise figure increases from its minimum value as the current increases. Advantages for both optimum source resistance and minimum noise figure come from using larger input devices. The base resistance r_b decreases, thereby reducing the optimum source impedance and lowering the minimum noise figure.

For a FET transistor, the optimum source impedance and the minimum noise factor are given as

$$R_{Sopt} \cong \frac{1}{\omega C_{gs}} \tag{7.45}$$

$$F_{min} \cong 1 + \frac{4}{3}\frac{\omega C_{gs}}{g_m} \tag{7.46}$$

It has been assumed that the frequency of operation is such that flicker noise and gate leakage current are negligible. Choosing a larger device will increase C_{gs} and reduce R_{Sopt}. (Note that at low frequencies, $R_{Sopt} \rightarrow \infty$.) The effect on minimum noise of a larger device is negligible since both capacitance and transconductance increase. Biasing the input transistor at high currents reduces the minimum noise figure because of an increase in the transconductance.

In general, whether FET or bipolar technology, larger input transistors biased at large drain (or collector) currents are commonly used in low-noise amplifiers. Often, the designs include just the single transistor with input and output matching networks because fewer active components means fewer noise sources. For ultralow noise applications, the gate (or base) is biased through an RF choke.

Feedback is used in many LNAs to set bias and achieve matching. Shunt-resistive feedback is commonly used though the resistor adds noise at the input. If the feedback is minimal, as in [12], the noise performance is not degraded by much.

The use of reactive feedback for matching is preferable. A technique used in some LNA designs [13] is adding inductance (e.g., from the bond wire) at the emitter of a bipolar transistor. The small-signal input impedance becomes small at high frequencies due to the shunting effects of the input capacitance C_π and the Miller effect due to C_μ. Ultimately, the input impedance approaches the base resistance, which is small ($\sim 10\Omega$), because a large input device is used. Emitter inductance can increase the input impedance by the amount $Z_i = \omega_T L_E$. An input match of 50Ω can be achieved without raising the noise figure. The practical design discussed in the next section uses this technique.

7.3.2 Practical Design of a Low-Noise Amplifier

A low-noise amplifier design that uses a cascode arrangement as shown in Figure 7.13 offers performance advantages in wireless applications over other configurations. A single common-emitter amplifier will typically not provide the gain required to overcome the noise figure of successive stages. A cascade of two common-emitter stages may be used, however, the required dc bias current on the second stage will have to be quite high to handle amplified versions of the input blocking signal (unless additional filtering is used between the two stages). A cascode LNA provides high gain and low noise. Also, the cascode is a more power-efficient approach, as the same bias current flows through both Q_1 and Q_2.

The gain of the cascode amplifier is higher than a common-emitter stage because Miller feedback (due to the collector-base junction capacitance of Q_1) is eliminated. The cascode arrangement limits the voltage gain from the base to the collector of Q_1 to unity. The effect is to raise input and output impedances without lowering the current gain. Hence, increased power gain can be obtained.

An additional advantage of the cascode is its ability to handle larger supply voltages. A common-emitter stage can generally be operated only up to the *collector-emitter* breakdown voltage of the process (typically only 4V in a 25-GHz f_T process). However, the cascode arrangement with its common-base output can be operated up to the *collector-base* breakdown of the process (typically greater than 10V for a 25-GHz f_T process).

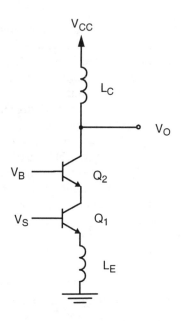

Figure 7.13 Cascode bipolar transistor amplifier.

The cascode arrangement has three main disadvantages. First, the available output-voltage swing is decreased by a diode drop (or at least a collector-emitter saturation voltage). This loss does not limit the cascode performance much; the circuit can still be operated on voltage supplies as low as 1.8V (2 AA cells at end-of-life) for a bias current around 4 mA and a collector impedance of 200Ω. The collector impedance should be no more than $\sim200\Omega$ to easily match to an external 50Ω impedance.

The second disadvantage of the cascode is a somewhat higher noise figure than a cascade of two common-emitter stages. For example, an achievable noise figure at 900 MHz for a cascode amplifier is 1.5 dB compared to the cascaded common-emitter with 1.2-dB noise figure.[1] For systems where noise is more critical than blocking performance, such as GPS, a cascade arrangement is preferable to the cascode. For handheld cellular systems, the cascode offers an excellent tradeoff between gain, noise, and blocking performance.

The third main disadvantage of the cascode is that the input 1-dB compression point and third-order intercept point will be lower than that of a single common-emitter stage because of the increased gain.

An example of a practical circuit that uses a cascode is shown in Figure 7.14. The circuit splits the signal current between Q_3 and Q_5. Transistor Q_3 is much larger

---

[1] The process for comparison has $f_T = 25$ GHz.

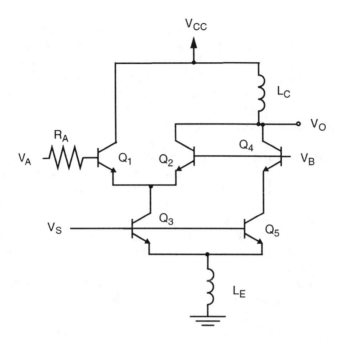

Figure 7.14 Low-noise bipolar transistor amplifier.

than Q_5 and carries most of the signal current. When the voltage V_A is low, the signal current in Q_3 flows through Q_2 and to the output. This is the high-gain state of the amplifier. When V_A is high, all of the signal current in Q_3 is diverted from Q_2 to Q_1 and only the signal current that flows from Q_5 through Q_4 makes it to the output. This is the low gain state [14].

The ability to change gain stages is quite useful for cordless and cellular systems. The feature prevents overload of the mixer and subsequent IF stages in the presence of a large input signal. For example, typical radio systems can have input signals as high as −15 dBm incident at the antenna. If the LNA has 15-dB gain, the mixer would be required to handle 0 dBm of signal (neglecting filter losses). However, with a gain reduction in the LNA from 15 dB to 3 dB (a 12-dB step), the mixer need only handle −12 dBm at the input (a much easier design task). In addition, the voltage swing at the collector will be reduced in the low-gain setting. Furthermore, different gain stages in the LNA reduce the required gain-control range of the variable-gain amplifiers.

7.4 VARIABLE-GAIN AMPLIFIERS

In a variable-gain amplifier (VGA), an applied dc control voltage causes the amplifier gain to change. The gain varies monotonically with increasing (or decreasing) control voltage. A common application of a VGA is within an automatic gain control (AGC) environment such as used in receivers. In many systems, the receiver output signal strength is monitored. A control voltage, which is a function of the output signal strength, is fed back to the VGA to vary the gain for a constant receiver output. Digitally controlled VGAs are becoming increasingly popular. Rather than varying over a continuum of gains, the amplifiers take on either a high-gain or low-gain state. Cascading several VGAs will give finer increments of gain variation for the receiver chain.

The vast number of designs can be classified in a handful of circuit configurations. Most VGAs can be categorized as either emitter-driven, where a current signal is injected at the emitters of diff pairs (differential pairs), or base-driven, where a voltage signal is applied to the bases of the diff pairs. In emitter-driven designs, current is steered between pairs of diff pairs, called a quad pair. Unlike emitter-driven designs, which are inherently linear, local feedback is used in base-driven designs to linearize the circuit. Gain is typically varied in these designs by effectively altering the feedback.

VGAs used in wireless communication systems typically operate at IF signal frequencies and must attain a large gain-control range (30 to 40 dB). Currently, the technology most often used in commercial systems is silicon bipolar and, consequently, is emphasized in this section. Several of the designs can easily be translated to FET technologies. Some attention is given to BiCMOS, which is the technology probably best suited to the design of variable-gain amplifiers; the bipolar devices have good bandwidth and gain characteristics while the FETs can be used as variable resistors.

How well a circuit can handle large and small signals is a measure of its dynamic range. The dynamic range of the emitter- and base-driven circuits are explored first. Drawing from the results of the emitter-driven circuit, three VGA designs are compared. The last part of this section discusses the merits of several gain-control circuits (not restricted to emitter-driven and base-driven configurations).

7.4.1 Dynamic Range

An important performance metric of a variable-gain amplifier is its dynamic range. A definition for dynamic range is the ratio of maximum allowable input signal to minimum detectable signal. The minimum detectable signal is the signal at which the signal-to-noise ratio (SNR) is deemed acceptable (typically $\cong 10$ dB). The maximum allowable input signal is often limited by distortion.

In gain-control amplifiers, the output SNR is a more useful measure of circuit noise than the noise figure because the noise figure depends on the circuit gain, which is variable. At maximum gain, the amplitude of the input signal reaches its lowest value but it does not imply that the SNR is at a minimum. In many circuits this is the case, but often the minimum SNR is found at half-gain when all devices in the amplifier are on and contributing to the noise level. For a specific minimum requirement in SNR over the whole dynamic range, only the amount of noise at the worst case (e.g., either maximum gain or half-gain) has to be examined.

Of particular concern in variable-gain amplifiers used in wireless communication systems is third-order intermodulation distortion. Intermodulation distortion can severely degrade receiver performance because unwanted signals falling within the RF passband encounter nonlinearities in the circuit. A third-order product can be created that for all practical purposes appears to be a legitimate signal because it falls within the desired signal bandwidth.

There is no single satisfactory measure of intermodulation performance. The ratio of the third-order intermodulation product magnitude to the fundamental product magnitude is third-order intermodulation distortion (IM_3). The problem with this is that the value specified for IM_3 is signal-level dependent. Instead, the concept of intercept point is widely used as a measure of amplifier performance. The intercept point is signal independent, but it is a theoretical not actual measure of performance.

The third-order intercept point (IP_3) is the power level (either input or output is specified) at which the extrapolated curves of the linear response and the third-order intermodulation product intersect [15]. Extrapolation is necessary since as the amplifier approaches compression, the actual distortion curves no longer follow the predicted cubic behavior. Another problem with the intercept point is that it only has meaning for amplifiers that can be described as a cubic system. The power in the third harmonic does not increase at three times the rate of that of the fundamental in all systems.

7.4.2 Emitter-Driven and Base-Driven Gain Control

Often the linearity and noise performance of a gain-control amplifier are constrained by the driver design. Two possibilities are the base-driven and emitter-driven designs [16]. In the base-driven design of Figure 7.15(a), the signal is applied to the bases of the diff pair Q_1 and Q_2. A gain-control voltage V_C varies the amount of tail current I_{EE} flowing through the pair. Interchanging the positions of input signal and gain control of the base-driven circuit results in an emitter-driven circuit, shown in Figure 7.16(a). The gain-control voltage steers the current between Q_1 and Q_2.

Intermodulation distortion performance of the emitter-driven circuit is superior to the base-driven circuit. The output current of the base-driven circuit is a nonlinear function (tanh) and thus, the output is distorted with respect to the input. On the other hand, in the emitter-driven circuit the output current is linearly related to the input signal current i. Thus, to first-order, no distortion is produced by the diff pair in the emitter-driven circuit.

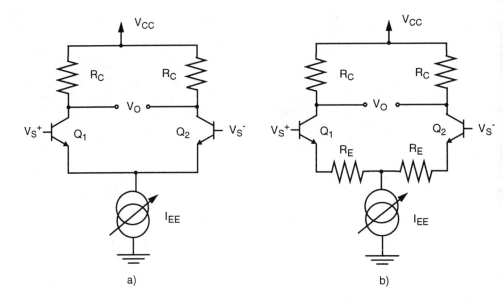

Figure 7.15 Base-driven transistor amplifier: (a) basic circuit approach, (b) practical design.

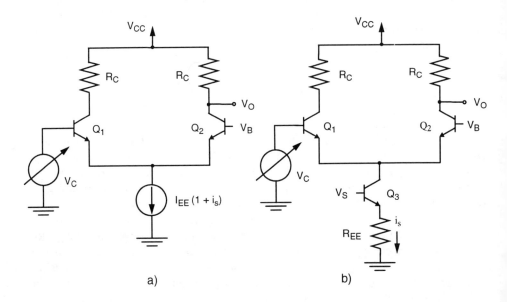

Figure 7.16 Emitter-driven transistor amplifier: (a) basic circuit approach, (b) practical design.

More practical versions of emitter- and base-driven circuits are shown in Figures 7.15(b) and 7.16(b), respectively. When the practical versions are compared, the results are not as dramatic, though the performance of the emitter-driven circuit still exceeds that of the base-driven circuit. Nonlinearities in the emitter-driven pair introduced as the input voltage v_s are converted to current i_s. Since the distortion performance of the base-driven circuit is not satisfactory for most wireless communication systems, the linear range is extended by the presence of emitter and base resistors, either inherent in the transistor or added intentionally. These resistors introduce local negative feedback. The output currents flow in these resistors and are converted by them into voltages that subtract from the input voltage. The smaller effective input signal reduces the distortion.

At low frequencies, a power series can be used to estimate the amount of distortion for a given output signal level [17]. For the base-driven circuit with emitter degeneration,

$$\text{IM}_3 = \left(\frac{v_o}{I_{EE}R_L}\right)^2 \left(\frac{1}{1 + \dfrac{I_{EE}R_E}{2V_T}}\right) \tag{7.47}$$

The effect of emitter degeneration is a reduction in third-order intermodulation of approximately the loop gain, A_L, of the circuit. For a given input, the third-order intermodulation reduces by $(1 + A_L)^3$, which appears as a significant reduction in distortion; however, the gain is reduced by $(1 + A_L)^2$. Distortion in the emitter-driven circuit is

$$\text{IM}_3 = \left(\frac{v_o}{I_{EE}R_L}\right)^2 \left[\frac{\left(1 - \dfrac{2I_{EE}R_E}{V_T}\right)}{8\left(1 + \dfrac{I_{EE}R_E}{V_T}\right)^2}\right] \tag{7.48}$$

It has been assumed that the circuit is operating at maximum gain and Q_2 is acting as a cascode. Equations (7.47) and (7.48) are expressions for third-order intermodulation distortion (the ratio of the third-order intermodulation product and the fundamental component). The output third-order intercept point is found by setting $\text{IM}_3 = 1$, solving for v_o, and converting the resulting signal voltage to an output power level. A numerical example shows that the emitter-driven circuit has better distortion than the base-driven with emitter degeneration. An output signal of 325 mV can be tolerated in the emitter-driven circuit before IM_3 reaches 1%. When the base-driven circuit reaches 1% third-order intermodulation, the output is only 110 mV. These calculations were arrived at using $R_L = 250\Omega$, $I_{EE} = 2$ mA, and $R_E = 100\Omega$.

When noise is compared in these two circuits, it is found that again the emitter-driven circuit has better performance than the base-driven circuit with emitter degeneration. Dominant noise sources for both circuits are the base resistance and collector current noise of transistors Q_1 and Q_2. However, the emitter resistors used for linearization further increase the peak noise in the base-driven circuit. The input noise voltage for the emitter-driven circuit is

$$v_i = \sqrt{4kT\Delta f \left(2r_b + \frac{2V_T}{I_{EE}} \right)} \tag{7.49}$$

and for the base-driven circuit it is

$$v_i = \sqrt{4kT\Delta f \left(2r_b + \frac{2V_T}{I_{EE}} + 2R_E \right)} \tag{7.50}$$

Another numerical example will illustrate the noise performance advantage of the emitter-driven circuit. With the parameter values used in the distortion example and $r_b = 35\Omega$, $\dfrac{v_i}{\sqrt{\Delta f}} = 1.3 nV/\sqrt{Hz}$ and $2.2 nV/\sqrt{Hz}$ for the emitter-driven and base-driven designs, respectively. The output noise of a base-driven circuit with emitter degeneration may actually be lower than the emitter-driven circuit because the gain is less in the former.

7.4.3 Basic Circuit Configurations

Three variable-gain amplifiers commonly used as AGC circuits are referred to as the *AGC amplifier, multiplier,* and as *Gilbert's quad.* Each of these designs offers wideband performance and wide dynamic range operation. Input and output ports are differential and input and output impedances are independent of amplifier gain. A summary of the comparisons are provided in this section. A more detailed analysis can be found in [18].

The AGC amplifier and the multiplier are realized using the emitter-driven pair described in the previous section. They are formed by balancing two emitter-driven circuits. In the AGC amplifier (Figure 7.17), the control voltage V_C causes the currents in Q_4 and Q_5 to rise or fall together. Maximum gain (no attenuation) occurs at a large positive control voltage and causes all the tail current I_{EE} to flow through these transistors. Minimum gain (maximum attenuation) occurs at a large negative control voltage.

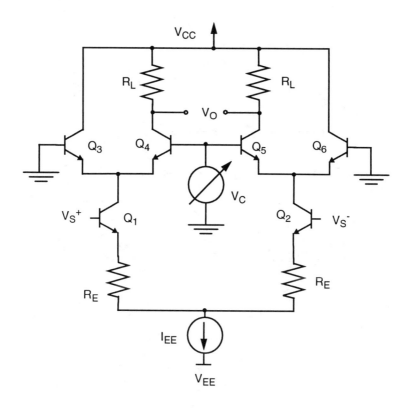

Figure 7.17 AGC transistor amplifier.

There are only slight differences between the AGC amplifier and the multiplier (Figure 7.18). The outputs of the multiplier are cross-coupled and maximum attenuation occurs when $V_C = 0$, at which point the currents through the quad pair (Q_3-Q_6) are equal. Minimum attenuation occurs at both large positive and large negative values of the control voltage.

A different variable-gain quad is based on Gilbert's wideband amplifier technique [1]. The quad is shown in Figure 7.19. Under ideal conditions, there exists a perfectly linear relationship between the input current i_i and the output current i_o. A technique known as *predistortion* is applied to achieve the linear relationship. The nonlinear relationship of current to voltage in transistors Q_1 and Q_2 is complimentary to the nonlinear relationship of voltage to current in transistors Q_3 and Q_4. Thus, ideally, a linear relationship between input and output currents exists. The nonlinearity in the circuit is introduced by converting the input voltage to the input current.

For the AGC amplifier and multiplier circuit configurations, gain is varied by steering the tail current away from some quad transistors and towards others. In

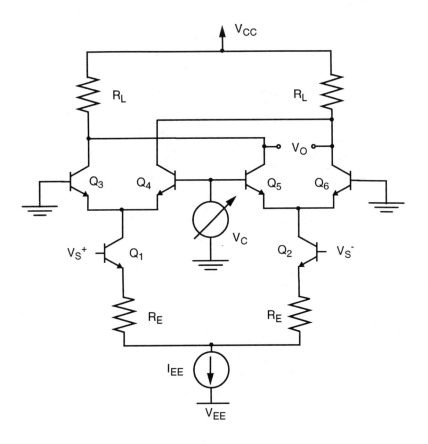

Figure 7.18 Bipolar multiplier.

Gilbert's quad, gain variation is achieved by varying the ratio of the tail currents I_{E1} and I_{E2}. The voltage gain of each configuration is summarized:

$$A_{v_{\text{AGC}}} = \frac{R_L}{R_E} \left(\frac{1}{1 + \exp - \dfrac{V_C}{V_T}} \right) \tag{7.51}$$

$$A_{v_{\text{mult}}} = \frac{R_L}{R_E} \left(\frac{1 - \exp - \dfrac{V_C}{V_T}}{1 + \exp - \dfrac{V_C}{V_T}} \right) \tag{7.52}$$

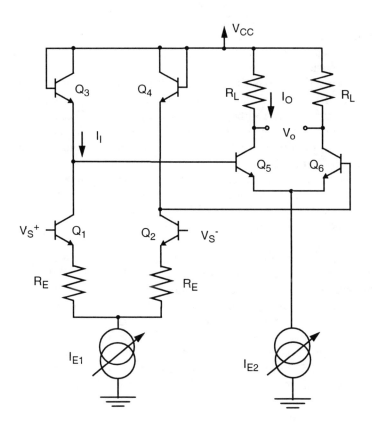

Figure 7.19 Gilbert's quad.

$$A_{v_{\text{quad}}} = \frac{R_L}{R_E} \frac{I_{E2}}{I_{E1}} \tag{7.53}$$

The mechanisms of distortion in the variable-gain amplifiers are the same as in the emitter-driven circuit. Distortion occurs due to base and emitter resistance in the quad devices, causing deviation from exact exponential characteristics. Linearity is preserved if the voltage drops across the base resistors are equal and the voltage drops across the emitter resistors are equal. The effect is observed as a null in the distortion at specific attenuation levels. A null exists at half-gain in the AGC amplifier and multiplier. When the tail currents I_{E1} and I_{E2} of Gilbert's quad are equal, distortion is significantly reduced.

In practice, the AGC amplifier is found to have the lowest distortion. The performance at low attenuation is limited by the nonlinearity introduced by the base resistance. The performance of Gilbert's quad is affected more by the base resistance

[16] than the AGC amplifier. At high attenuation, the multiplier theoretically has the best performance because the currents in all transistors are nearly equal and the effect of base resistance is reduced because of more equal voltage drops. However, the imbalances in a practical multiplier circuit implementation cause the fundamental to cancel at a different point than the distortion. Thus, distortion increases significantly at high attenuation and becomes worse than the AGC circuit.

The noise results of the emitter-driven pair presented in the previous section can be applied to the three variable-gain circuits. The total output noise power in the AGC amplifier and multiplier is twice that for the emitter-driven pair. The noise peaks at half-gain (6-dB attenuation) in the AGC circuit and decreases for both increasing and decreasing attenuation. Though the multiplier has the same peak noise power as does the AGC amplifier, the noise does not decrease for higher attenuation, and so has the disadvantage that the worst SNR is the value obtained over most of the attenuation range. In Gilbert's quad, the maximum output noise voltage due to base resistance equals $2\sqrt{2}$ times the noise peak for the other circuits if base resistance and quad current are assumed to be equal. Peak noise occurs at maximum gain and decreases with increasing attenuation.

The choice of gain-control circuit configuration is dependent on application. For large gain-control variation (~40 dB), the AGC amplifier is found to have the greatest dynamic range performance. The multiplier and quad have good performance for small gain-control ranges (~10 dB) and offer some advantages. The multiplier has a constant dc output voltage and polarity inversion of the control voltage, which is attractive in some applications. Maximum distortion at high frequencies in Gilbert's quad increases only linearly with frequency whereas in the AGC amplifier, the maximum distortion increases twice as fast. For high frequencies, Gilbert's quad has a better distortion performance than the AGC amplifier. However, direct feedthrough in Gilbert's quad is worse than in the AGC amplifier.

7.4.4 Variable-Gain Circuit Configurations

Variable-gain amplifiers used in wireless systems are not restricted to the class of designs based on the emitter-driven circuit. Many designs extend the dynamic range of the base-driven circuit by employing variable resistance. Current trends in VGA design include the use of FETs as variable resistors in the feedback path of the circuit. A sampling of these designs is now discussed. The first designs are improvements to the AGC amplifier and Gilbert's quad.

A proposed improvement [18] to the AGC amplifier increases the current in the lower differential pair and decreases the current in the upper quad transistors. The circuit is shown in Figure 7.20. More current through the degeneration resistors improves distortion. Noise will also improve. The optimum dynamic range and flat transfer characteristics can only be realized at fairly low quad currents. Additional dynamic range at high attenuation is achieved through proper sizing of the quad

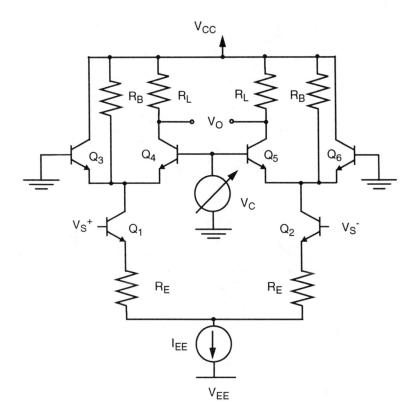

Figure 7.20 Improved AGC circuit.

devices. Making Q_3 and Q_6 a matched pair of large devices minimizes r_b and creates a more ideal exponential junction relation. The other quad devices, Q_4 and Q_5, should be a matched pair of small devices to minimize feedthrough.

Feedback is often used to linearize a circuit at the expense of increased circuit noise. This was observed in the base-driven circuit. Alternative techniques, such as *predistortion* used in Gilbert's quad, achieve linearity by canceling circuit nonlinearities with complimentary nonlinearities. Another example of this is a variation on Gilbert's quad shown in Figure 7.21 [1].

The circuit can achieve a slightly larger gain-bandwidth product. The current gain for this circuit is $A_v = 1 + \dfrac{I_E}{I_B}$.

Further advancing an alternative to emitter degeneration is the *multitanh* doublet shown in Figure 7.22 [19]. It is comprised of two differential pairs with their inputs and outputs in parallel. The junction area of one emitter in each pair is

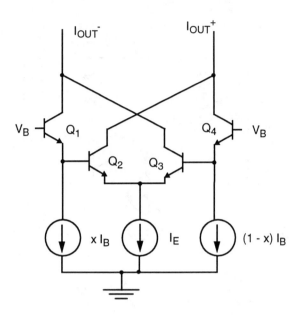

Figure 7.21 Variation on Gilbert's quad.

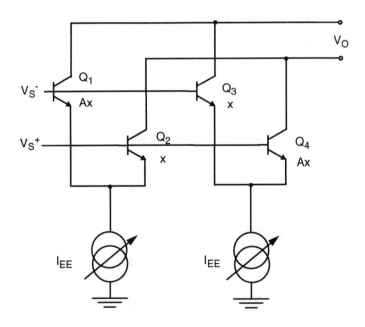

Figure 7.22 Multitanh doublet.

arranged to be A times larger than the opposite emitter. The gain of the circuit is the gain of a basic differential pair reduced by $\dfrac{4A}{(1 + A)^2}$ where A is the ratio of area of Q_1 to Q_2 and Q_4 to Q_3. The gain can be varied by simultaneously increasing or decreasing the equal tail currents of the pairs.

The effect of the emitter area ratio is to shift the peak of each transconductance by an equivalent offset voltage, resulting in an overall g_m that is much flatter than the simple diff-pair. This leads to improved linearity. There is an optimum value of A; if too low, the g_m curve is still a hump and if too high, there is a double hump. For best performance, A is taken to be equal to 4. The gain is thus 0.64 times that of the basic diff pair. Even though the input-referred noise of the multitanh doublet is 25% higher than the basic differential pair, the dynamic range is much better. There is also a significant improvement in dynamic range of the multitanh doublet over the differential pair with emitter degeneration.

The multitanh is a base-driven circuit that uses transconductance offset to linearize the circuit. The next few circuits revisit the idea of base-driven variable-gain circuit configurations. In the dual-gain-state differential amplifier shown in Figure 7.23 [20], current is switched between two diff pairs with connected bases and collectors. The pairs are degenerated differently with emitter resistors R_L and R_H of low and high values, respectively. At maximum attenuation, all current flows through R_H, and conversely for minimum attenuation, all current flows through R_L. Therefore, using this topology, the lowest gain state allows a large undistorted input signal, while the highest gain state has a lower input noise, resulting in a very large dynamic range. This particular design is an example of a digitally controlled VGA as discussed earlier.

A similar approach to achieving wide dynamic range as was used in the previous design is used in the diode-bridge variable-gain amplifier shown in Figure 7.24 [21]. It is another base-driven design that operates with small emitter resistance for high-gain, low-noise operation and large emitter resistance for low-gain, low-distortion operation. Unlike the dual-gain-state amplifier, there is only one diff pair and the emitter resistance is varied by controlling the current, I_C, through the diode bridge. For minimum current, the diode bridge with the additional resistors R_D has a very high impedance, roughly 10 times greater than $2R_E$. Therefore, the amplifier has minimum gain for minimum I_C. With maximum current, the diode bridge has minimum impedance about four times smaller than $2R_E$; therefore, the amplifier has maximum gain.

A BiCMOS technology is desirable when designing a variable-gain amplifier. The bipolar devices can be used to achieve high gain and the FETs can be used as resistors in feedback paths. For example, the diode-bridge network in the circuit just described can be replaced by a single MOSFET. The control voltage can be applied to the gate voltage to vary the transconductance of the transistor. Another

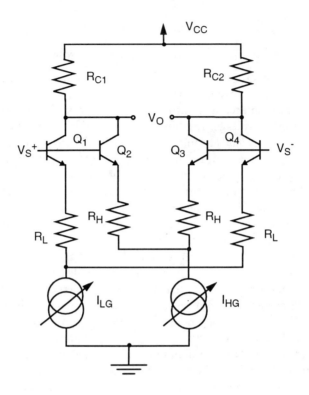

Figure 7.23 Dual-gain-state differential amplifier.

BiCMOS variable-gain amplifier is shown in Figure 7.25 [22]. FETs are used as local shunt feedback in the first and second stage. The advantage of this design above the other designs that vary feedback is that the input impedance remains constant over gain variation. Local shunt feedback with a high impedance load makes the input impedance independent of the feedback. Dynamic range in this circuit is achieved by low base resistance and high collector current in the first stage. Increasing current along the signal path will help to minimize distortion.

7.5 POWER AMPLIFIERS

Power amplifiers (PAs) are in the transmitting chain of a radio in a wireless system. They are the final amplification stage before the signal is transmitted, and therefore must produce sufficient output power to overcome transmission loss between the transmitting and receiving units. The amount of output power is highly dependent on the application cell size (how closely units are arranged), ranging from milliwatts in indoor picocell applications to watts for broader regional coverage.

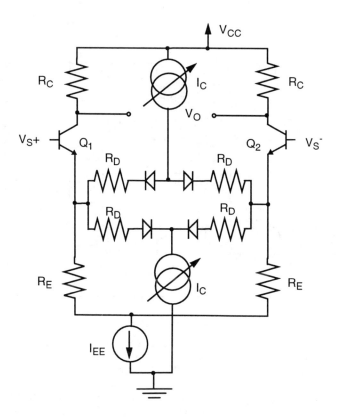

Figure 7.24 Diode-bridge gain control.

The PA is the primary consumer of power in the radio portion of a wireless unit. Thus, a major design issue is how efficiently the PA can convert dc input power to RF output power. Another important characteristic of a PA is linearity; the input-output relation must be linear to preserve signal integrity. A primary consideration in the design of a PA involves tradeoffs of efficiency and linearity.

The choice in system and corresponding modulation method defines the RF waveform and directly affects the linearity requirements of the PA. For highly linear applications, efficiency may be sacrificed. Modulation methods (such as DQPSK in direct sequence systems) that have waveforms with high peak-to-average ratios must have greater linearity. Often, this is at the expense of efficiency. On the other hand, there are modulation techniques (such as GFSK in frequency-hopping systems) that result in more constant envelope signals, allowing the PA to operate in saturation (a nonlinear regime) so higher efficiency can be achieved.

Power amplifiers can generally be classified by mode of operation. A single power device operates continuously in a Class A amplifier. In Class B designs, the

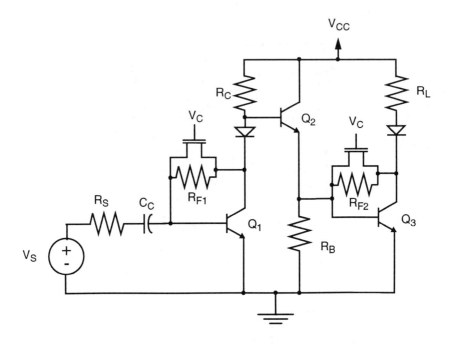

Figure 7.25 BiCMOS RF variable-gain amplifier.

power device is operating for half of the time. A common implementation of a Class B design is in a push-pull fashion, where two power devices operate half of the time in an alternating fashion. Class AB amplifiers fall between Class A and Class B PAs; the efficiency is better than Class A and worse than Class B operation. There are also Class AB push-pull power amplifiers that operate like Class B push-pull PAs, except there is a small overlap of time when both devices are on. Higher class power amplifiers distinguish themselves either by how often the transistors are on, which affects efficiency, or by how hard they are driven (i.e., some act like switches).

This section covers Class A and Class B operation. Design fundamentals are presented as well as practical issues in PA design. In particular, the effect of the package is treated in the design of a Class A power amplifier. Amplifier stability, input-output matching, and package thermal conductivity are also addressed.

7.5.1 Class A Power Amplifier Design

At RF frequencies, Class A amplifiers are usually designed with a common-emitter structure (or common-source for FETs), as shown in Figure 7.26. The collector is biased with a choke to allow the collector to swing from ground to twice the supply rail. The result is that the maximum efficiency doubles from 25% to 50%.

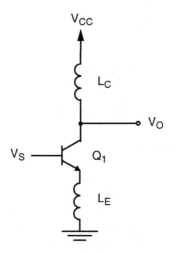

Figure 7.26 Basic Class A amplifier.

A common figure of merit used to compare different PA configurations is efficiency. Efficiency, denoted η, is defined as the ratio of output power to dc input power. Power-added efficiency (PAE) is another measure of efficiency for transistors that discriminates against low-gain devices in which a significant fraction of the output power comes from the RF input power [23]. The latter measure of efficiency is more specific to FET devices, which have lower gain performance than bipolar transistors. Expressions for efficiency and PAE are

$$\eta = \frac{P_{RF}}{P_{DC}} \tag{7.54}$$

$$PAE = \left(1 - \frac{1}{G}\right)\frac{P_{RF}}{P_{DC}} \tag{7.55}$$

where G is the RF gain of the device at the particular value of P_{RF} being achieved.

Maximum efficiency in Class A amplifiers can easily be found by making simplifying assumptions. First of all, the bias current I_C is approximately equal to the average supply current I_{CC}. Next, the maximum current delivered to the load equals the bias current. Finally, the peak voltage delivered to the load is the supply voltage V_{CC}. Consequently, the input power is

$$P_{DC} = I_C V_{CC} \tag{7.56}$$

the optimum load to present to the collector of the power device is

$$R_{Lopt} = \frac{V_{CC}}{I_C} \tag{7.57}$$

the rms power delivered to the load is then

$$P_{RF} = \frac{1}{2} V_{CC} I_{CC} \tag{7.58}$$

and from (7.54), the efficiency is

$$\eta = 50\% \tag{7.59}$$

Maximum efficiency in PAs can rarely be achieved in practice. The output swing is limited by collector-emitter saturation. As the collector falls below ~200 mV, the output signal is no longer linear. Another major limitation is that the emitter is not at an ideal ground; voltage drops develop across emitter resistance and package inductance, which further limit the output swing.

The problem of emitter voltage moving from 0V is commonly known as "ground-bounce" and is a major design issue in PAs. The ground-bounce voltage, or the voltage at the emitter referred to ground, is given by

$$V_{GB}(\omega) = j\omega L_E I_C(\omega) + R_E I_C(\omega) \tag{7.60}$$

The inductance introduced by the package (L_E) can be as little as 100 pH for an RF power package and as much as 4 nH for an inexpensive plastic package such as a thin quad flat pack (TQFP). Emitter resistance (R_E) increases ground bounce, but is useful for temperature compensation (an issue discussed later in this section).

Note that the ground bounce due to the inductor is in quadrature to the signal current, while the ground bounce due to emitter resistance will be in phase with the signal current.

An expression for the efficiency of a Class A amplifier in the presence of ground bounce is readily found. For maximum output power, the collector current will have an amplitude equal to the bias current. If the collector current is approximately sinusoidal, it can be written as

$$I_C(t) = I_C(1 + \sin\omega t) \tag{7.61}$$

The collector-emitter potential of the device can be written as

$$V_{ce}(t) = V_{CC} - R_L I_C \sin\omega t - \omega L_E I_C \cos\omega t - R_E I_C(1 + \sin\omega t) \tag{7.62}$$

The optimum load to present to the collector of the power device is

$$R_L = \frac{V_O}{I_C} \tag{7.63}$$

where V_O is the maximum output voltage amplitude. An expression for the maximum output voltage is found by assuming that the square root of the sum of the squares of inductive ground bounce and collector-emitter potential plus resistive ground bounce must not be larger than the supply voltage. It is given as

$$V_o = \sqrt{(V_{CC} - V_{CE_{sat}} - 2I_C R_E)^2 - (\omega L_E I_C)^2} \tag{7.64}$$

Assuming that all of the signal current is delivered to the load, the rms output power is

$$P_{RF} = \frac{1}{2} V_O I_C \tag{7.65}$$

and the efficiency is given by

$$\eta \cong \left(\frac{V_O}{V_{CC}} \right) 50\% \tag{7.66}$$

The output amplitude is limited by ground bounce and collector-emitter saturation voltage. If both effects are negligible, V_O approaches V_{CC} and a maximum efficiency of 50% is achieved.

The value of the optimum load impedance is invariably different (much lower) from the value required to achieve a good impedance match, which is typically 50Ω or 75Ω. An impedance transformation network can be used to meet both requirements. However, for low supply voltages and high output powers, the required step-up ratio can be quite large.

A numerical example will illustrate how to approach the design of a Class A power amplifier. The goal is to design a 1W power amplifier operating at 1 GHz with a supply voltage of $V_{CC} = 5V$. It is easiest to begin by neglecting ground inductance, thus simplifying (7.64). The bias current can be calculated from

$$I_C = \frac{2P_{out}}{V_O} \tag{7.67}$$

$$\approx \frac{2P_{out}}{V_{CC} - V_{CE_{sat}} - 2I_C R_E} \tag{7.68}$$

When the resistive drop $I_C R_E$ is taken to be 0.1V (for ease of calculation) and the saturation voltage is $V_{CE_{sat}} = 0.2V$, the bias current is approximately 435 mA.

(When ground inductance is considered, the bias current will have to increase to achieve the desired output power. The output voltage decreases and therefore, by (7.67), bias current must increase.)

The presence of ground inductance will reduce the efficiency of the amplifier. A reduction in efficiency as a function of inductance is given as

$$\frac{\eta|_{\text{with L}}}{\eta|_{\text{w/o L}}} = \sqrt{1 - \left(\frac{\omega L_E I_C}{V_{CC} - V_{CEsat} - 2I_C R_E}\right)^2} \tag{7.69}$$

Rewriting (7.69) for L_E will give a maximum inductance that can be tolerated for a given reduction in efficiency. In this example, if the efficiency is to be reduced by no more than 4%, the inductance can be no larger than 146 pH. A low-inductance RF package would be needed to meet this requirement.

Using (7.64), the output voltage is found to be ~4.6V and the efficiency from (7.66) is ~46% (as expected by design). The output power is found from

$$P_{RF} = \frac{1}{2} V_O I_C \tag{7.70}$$

and is calculated to be 978 mW. The output power is slightly lower than the target of 1W because of the inductive ground bounce. The output power can be increased to meet the specification by increasing the bias current by roughly 5%. The optimum load impedance is found (by (7.63)) to be 10.6Ω. This impedance can be matched to 50Ω with a single or two-stage L-section matching network. As the output power increases, the optimum load impedance will drop and an impedance match will become more difficult to achieve. Increasing the supply voltage is a potential solution if the device used is not limited by breakdown or there are no battery voltage constraints as would be the case in a handheld unit.

The design example was optimistic. In practice, a Class A amplifier will typically achieve less than 40% efficiency since the design must tolerate process, temperature, and supply variations.

7.5.2 Higher Efficiency Power Amplifiers

In low power receivers, Class B operation has significant advantages over Class A operation. One major advantage is that there is essentially zero power dissipation with zero input signal in a Class B amplifier. Unlike with Class A operation, the power dissipated in a Class B power amplifier is not wasted when the unit switches to a standby (or idle) mode. Another advantage of Class B amplifiers is that the efficiency is much higher than Class A amplifiers.

The difference between a single-ended Class B amplifier with resistive load and the preceding Class A amplifier is that the quiescent bias point is changed. In the Class B amplifier, the device is biased on the edge of conduction. The resulting output current waveform is a half sine-wave. When Fourier-analyzed, the output current waveform is found to contain a dc component that is a function of the RF power output, a component at the input signal frequency, and only even-order harmonics.

A Class B amplifier with resistive load is rarely used in practice because all the power in the harmonics is also dissipated in the load. An improved design has a tuned load. The load is a shunt parallel-tuned circuit that provides a short-circuit path for all the harmonics so that only the fundamental current component flows into the load. This results in the voltage across the device being a pure sinusoid at the fundamental frequency. The downside of this design is that optimum Class B performance from this single-ended circuit is only achieved over a narrow bandwidth.

For wider bandwidth performance, a commonly used Class B amplifier is in a push-pull configuration, as shown in Figure 7.27. Each device conducts for alternate half-cycles. The two transistors are driven 180-deg out of phase from each other. A Fourier expansion of the current waveforms shows that the even-order harmonics (including dc) are in phase and that the fundamental components are out of phase. The outputs can be combined in such a way that the even-order harmonics cancel and the fundamental components sum.

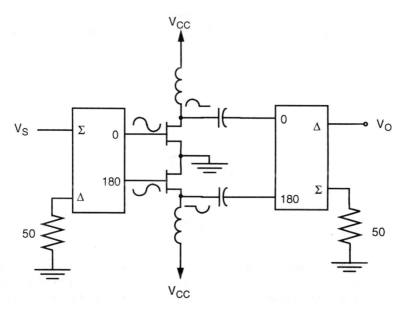

Figure 7.27 Push-pull Class B amplifier.

The power delivered to the load is the same in Class A and Class B operation. However, efficiency improves in Class B amplifiers because the average power drawn from the supplies is smaller than in Class A amplifiers.

The average supply current in each of the transistors is found by integrating the half-sinusoid over one period. It is found to be

$$I_D = \frac{1}{\pi} I_O \tag{7.71}$$

$$= \frac{1}{\pi} \frac{V_O}{R_L} \tag{7.72}$$

where I_O and V_O are the peak output current and voltage amplitudes. Since each supply delivers the same current magnitude, the total average power drawn from the two supplies is

$$P_{DC} = 2I_O V_{CC} \tag{7.73}$$

$$= \frac{2}{\pi} \frac{V_O}{R_L} V_{CC} \tag{7.74}$$

Note that in the case of the Class B amplifier, the average power drawn from the supplies does vary with signal level and is directly proportional to V_O. The average power delivered to the load is

$$P_{RF} = \frac{1}{2} \frac{V_O^2}{R_L} \tag{7.75}$$

and, from the definition of circuit efficiency,

$$\eta = \frac{\pi}{4} \left(\frac{V_O}{V_{CC}} \right) \tag{7.76}$$

$$= \left(\frac{V_O}{V_{CC}} \right) 78.5\% \tag{7.77}$$

Many recently reported designs have been targeted at the commercial wireless market. In [24], a GaAs Class B power amplifier is designed for use in cellular phone systems. The amplifier achieves a PAE of 65% for an input power of 20 dBm from a 3.5V supply. The output power delivered to the load is 1.3W. Another design targeted for the cellular market was reported in [25]. This design is a two-stage

Class AB power amplifier achieving a PAE of 60% from a 6V supply with an input power of 7 dBm and an output of 1.2W. A third design recently reported was developed for portable wireless LAN (local area network) applications [26]. The design is a two-stage Class B amplifier. The output power is 24 dBm for an input power of 7 dBm. The supply voltage is 5V and the operating frequency is at 2.4 GHz.

7.5.3 Maximum Power Gain and Stability

The maximum power gain of a packaged device is typically limited by bond-wire inductance. The bond wire creates series-series feedback in the emitter and the parasitic collector-base capacitance of the device creates shunt-shunt feedback. The two effects combined will generate a narrowband matched-impedance condition that is analogous to the series-series/shunt-shunt resistive broadbanding technique for matched impedance discussed in Section 7.2. With simultaneous input and output match, the optimum power gain is

$$A_p = \frac{1}{\omega^2 L_E C_{jc}} \tag{7.78}$$

where L_E is the bond-wire inductance and C_{jc} is the parasitic collector-base capacitance. The power gain decreases by 20-dB per decade change in frequency.

The collector-base junction capacitance will increase with larger devices that are required for more output power. Therefore, bond-wire inductance L_E must scale inversely with power to keep power gain constant. Recall also that the inductance has to decrease linearly with increasing bias to keep ground bounce constant. A decrease in supply voltage must be accompanied by a decrease in package inductance to maintain power gain and ground bounce.

In choosing an appropriate technology, a compromise must be made between breakdown performance and junction capacitance. As discussed in the next section, low breakdown voltages present difficult problems for the designer. However, it is desirable to have a process with low C_{jc} for higher power gain. For higher frequency applications, a technology with lower parasitic junction capacitance and a low-inductance package should be used. To avoid breakdown problems, the supply voltage should also be fairly low. Applications at lower frequencies can tolerate a higher breakdown technology and a less expensive package.

As output power and frequency of operation increase, distributed effects need to be considered. Higher power levels require multiple devices spread out over large physical areas. The distributed effects of interconnect inductance and device input capacitance cause a phase shift at the input of each device. A similar phase shift can occur at the output of each device. Care must be taken in the device layout so that these phase shifts do not become excessive; otherwise, power output and efficiency will be lost. Specifically, the outputs should be on the opposite side of the inputs so that the phase shifts at the input and output tend to cancel rather than add. This approach is analogous to a traveling wave amplifier, which is widely used in GaAs applications for frequencies in the millimeter wave region [27].

Stability is another crucial issue when designing PAs; the power amplifier must be stable under all possible load impedances. The combination of capacitive shunt feedback and series feedback, which will include both a resistive and inductive part, can lead to instability when driving inductive loads. Instability caused by variations in source impedance is less of an issue in monolithic implementations, because the driver impedance can more easily be controlled.

The source of the stability problem can be examined by calculating the input impedance as a function of load impedance. The input impedance of a device with both series and shunt feedback is given by

$$Z_{in} \cong \frac{\left(Z_e + \dfrac{1}{g_m}\right)Z_F}{Z_L} \tag{7.79}$$

where $Z_E = R_E + j\omega L_E$ and $Z_F = \dfrac{1}{j\omega C_{jc}}$. The load impedance Z_L includes the collector-substrate capacitance and the Early resistance $\left(\dfrac{V_A}{I_C}\right)$, as well as the impedance presented externally. The angle of Z_F equals –90 deg (purely capacitive feedback) and the angle of $Z_E + \dfrac{1}{g_m}$ is less than +90 deg. Therefore, the angle of Z_{in} will be more negative than –90 deg if the load is highly inductive. This implies a negative input resistance and possible instability.

Simple modifications to the PA of Figure 7.26 can be made to prevent instability. The improved design is shown in Figure 7.28. The resistor reduces the angle of the purely capacitive feedback and can be chosen to guarantee stability for any source or load. Resistive shunt feedback has the additional benefit of stabilizing the gain of the device over process, temperature, and supply variations. The shunt feedback will draw power from the output, so R_F should be much larger than R_L.

7.5.4 Practical Design Issues

Device breakdown is a major issue in the design of PAs. As process geometries shrink, the breakdown voltages of the devices decrease (in exchange for higher speed) and this creates a difficult problem for PA design. For example, a bipolar device with an f_T = 25 GHz will have a collector-base breakdown voltage of ~10V. As a

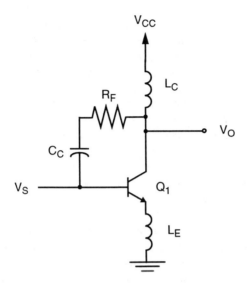

Figure 7.28 Stable Class A power amplifier.

general rule, the breakdown voltage of a power device should be three times the supply voltage. (Although the collector normally doesn't swing more than twice the supply voltage, it may swing higher under transient conditions.) Therefore, the supply voltage is limited to 3.3V. For a lower supply voltage, the bias current must be increased to deliver sufficient power to the output. The impact of increased bias current is more ground bounce and a decrease in the optimum load impedance, making matching more difficult.

The maximum power available from a PA in a bipolar technology is typically less than 1W from a 3V supply. Most cellular applications require PAs to produce 2 to 3W. The amplifiers operate off of four or five batteries, which corresponds to an end-of-life voltage of 4 to 5V. They are constructed in a process with larger geometry. PAs commonly use GaAs MESFETs or silicon DMOS power devices with custom geometries optimized for a high breakdown voltage.

Thermal properties of the package can limit the amount of power dissipated in the amplifier. A critical number for thermal design is the package θ_{J-A}, or thermal conductivity. This figure of merit indicates how poorly the package dissipates heat by specifying the increase in junction temperature from ambient temperature per watt dissipated in the circuit. A good practice is to keep the junction temperature of the power device no more than 50°C above ambient temperature. An inexpensive plastic package may have θ_{J-A}, = 100° C/W, which would limit the amount of power dissipated by the circuit to 500 mW. Assuming maximum efficiency of the Class A power amplifier, the package is required to dissipate at least as much power as delivered to the load. Alternatives are to use a massive heat sink structure to reduce θ_{J-A}, to as low as 2°C/W [4] or use and a more expensive package, which could have a θ_{J-A}, of approximately 40°C/W. For even more output power, arrays of packaged devices may be used and the output power may be combined through a power-combining network. For TDMA cellular communications, the thermal requirements are ameliorated somewhat because the transmit power is only required for a certain fraction of the time.

Another temperature concern in the design of power amplifiers is thermal runaway. When a bipolar device self-heats, its intrinsic base-emitter voltage drops. However, if the bias voltage forces a constant V_{BE}, then the collector current will increase. This increase in current will produce more self-heating and at some point the thermal feedback causes instability or thermal runaway and the device burns up.

Thermal runaway problems can be reduced in two ways. First, the bias voltage should be developed from a reference device that is at thermal equilibrium with the power device. Second, each power device should have some emitter degeneration resistance. The degeneration will act as negative feedback both electrically and thermally. Using separate resistors in each of the multiple power devices ensures that no single device will consume all of the current and create a hot spot. This technique is known as emitter ballasting [28].

References

[1] B. Gilbert, "A New Wide-Band Amplifier Technique," *IEEE Journal of Solid-State Circuits*, Dec. 1968.

[2] P. A. Quinn, Feed-Forward Amplifier, United States Patent #4,146,844, March 1979.

[3] C. Hull, and R. G. Meyer, "Principles of Monolithic Wideband Feedback Amplifier Design," *Int'l Journal of High Speed Electronics*, March 1992.

[4] P. R. Gray, and R. G. Meyer, *Analysis and Design of Analog Integrated Circuits*, Second Edition. New York, NY: John Wiley & Sons, 1984.

[5] C. R. Battjes, Monolithic Wideband Amplifier, United Stages Patent #4,323,119, Nov. 1980.

[6] B. W. Lai, Amplification Beyond f_T Using Non-Inductive Circuit Techniques, Master's Thesis UCB/ERL, University of California, Berkeley, CA, May 1983.

[7] I. Kipnis, J. F. Kukielka, J. Wholey, and C. P. Snapp, "Silicon Bipolar Fixed and Variable Gain Amplifier MMIC's for Microwave and Lightwave Applications up to 6 GHz." *IEEE MTT-S Digest*, 1989, pp. 109–112.

[8] D. Lee, A wideband Low-Noise BiCMOS Transimpedance Amplifier, Master's Thesis UCB/ERL, University of California, Berkeley, CA, May 1990.

[9] R. G. Meyer, and R. A. Blauschild. "A Wide-Band Low-Noise Monolithic Transimpedance Amplifier," IEEE Journal of Solid-State Circuits, Aug. 1986.

[10] J. Kukielka and C. Snapp, *Wideband Monolithic Cascadable Feedback Amplifiers using Silicon Bipolar Technology. Monolithic Microwave Integrated Circuits*, IEEE Press, 1985, pp. 330–331.

[11] R. G. Meyer, and R. A. Blauschild, "A 4-terminal Wide-Band Monolithic Amplifier," *IEEE Journal of Solid-State Circuits*, Aug. 1986.

[12] Avantek Application Note. An-S012: Magic Low Noise Amplifiers.

[13] R. G. Meyer, and W. D. Mack. "A 1-GHz BiCMOS RF Front-End IC," *IEEE Journal of Solid-State Circuits*, March 1994.

[14] I. A. Koullias, et al. "A 900 MHz Transceiver Chip Set for Dual-Mode Cellular Radio Mobile Terminals," *ISSCC 1993 Digest of Technical Papers*, Feb. 1993, pp. 140–141.

[15] H. L. Krauss, C. W. Bostian, and F. H. Raab, *Solid State Radio Engineering*, New York, NY: Wiley & Sons, 1976.

[16] W.M.C. Sansen, and R. G. Meyer, "Distortion in Bipolar Transistor Variable-Gain Amplifiers," *IEEE Journal of Solid-State Circuits*, Aug. 1973, pp. 275–282.

[17] D. O. Pederson, and K. Mayaram, *Analog Integrated Circuits for Communication*, Kluwer, 1991.

[18] W.M.C. Sansen, and R. G. Meyer, "An Integrated Wide-Band Variable-Gain Amplifier with Maximum Dynamic Range," *IEEE Journal of Solid-State Circuits*, Aug. 1974, pp. 159–166.

[19] B. Gilbert, Demystifying the Mixer, Analog Devices Internal Document, 1994.

[20] I. A. Koullias, S. L. Forgues, and P. C. Davis, "A 100 MHz IF Amplifier/Quadrature Demodulator for GSM Cellular Radio Mobile Terminals," *IEEE 1990 Bipolar Circuits and Technology Mtg.*, 1990.

[21] J. Fenk, et al. "An RF Front-End for Digital Mobile Radio." *IEEE 1990 Bipolar Circuits and Technology Mtg.*, 1990.

[22] P. Piriyapoksombut, and R. G. Meyer, "RF Variable Gain Amplifier," *UCB Infopad Retreat*, July 1994.

[23] J.L.B. Walker (editor), *High-power GaAs FET Amplifiers*, Norwood, MA: Artech House, 1993.

[24] M. Maeda, et al., "A 3.5 V, 1.3 W GaAs Power Multi Chip IC for Cellular Phones," *IEEE GaAs IC Symposium*, 1983.

[25] J. Naber, J. Griffiths, and J. Salvey, "A 1.2 W, 60% Efficient Power Amplifier IC for Commercial Applications," *IEEE GaAs IC Symposium*, 1994.

[26] T. Quach, and J. Staudinger, "A High Efficiency Commercial GaAs MESFET Power Amplifier for PCM/CIA Applications at 2.45 GHz," *IEEE GaAs IC symposium*, 1994.

[27] E. Strid, and R. Gleason, "A DC-12 GHz Monolithic GaAs FET Distributed Amplifier," *IEEE Trans. of Microwave Theory and Techniques*, July 1982.

[28] P. R. Gray, "A 15 Watt Monolithic Power Operational Amplifier," *IEEE Journal of Solid-State Circuits*, Dec. 1972, p. 474.

▼▼▼

GLOSSARY

ADPCM	adaptive differential pulse code modulation
AFC	automatic frequency control
AGC	automatic gain control
AM	amplitude modulation
AMPS	advanced mobile phone service
ARDIS	advanced radio data information service
AWGN	additive white Gaussian noise
BER	bit error rate
BPSK	bi-phase shift keying
BT	signal bandwidth × bit time
CDMA	code division multiple access
CDPD	cellular digital packet data
CMOS	complementary metal-oxide semiconductor
CNR	carrier over noise ratio
CIR	carrier over interference ratio
CRC	cyclic redundancy check
CT	cordless telephone (interim ETSI standard)
dBc/Hz	dB with respect to carrier level per hertz

DDS   direct digital synthesis
DECT   digital European cordless telecommunications
DRO   dielectric resonator oscillator
DSMA   digital sensed multiple access
EIA   Electronic Industries Association
EMC   electromagnetic compatibility
ERC   European Radio Commission
ETSI   European Telecommunication Standard Institute
FCC   Federal Communications Commission
FDD   frequency division duplex
FDMA   frequency division multiple access
FET   field effect transistor
FM   frequency modulation
FSK   frequency shift keying
GEO   geosynchronous satellite
GMSK   Gaussian filtered minimum shift keying
GSM   (Groupe Special Mobile) global system for mobile communication
HBT   heterojunction bipolar transistor
HEMT   high-electron mobility transistor
HIPERLAN   high-performance radio LAN
IEEE   Institute of Electrical and Electronics Engineers
IF   intermediate frequency
I and Q   in-phase and quadrature phase
IC   integrated circuits
IP   intermodulation product
ISM   industrial, scientific, and medical
JDC   Japanese digital cellular standard
KTB   Boltzmann's constant × temperature × bandwidth
LAN   local area network
LEO   low-Earth-orbit satellite
LNA   low noise amplifier
LO   local oscillator
MEO   medium-Earth-orbit satellite
MESFET   metal semiconductor field-effect transistor
MOSFET   metal oxide semiconductor field-effect transistor

MSS	mobile satellite service
NF	noise figure
NMT	Nordic Mobile Telephone
NTT	Nippon Telephone and Telegraph
PHS	personal handyphone system (formerly PHP)
PLL	phase lock loop
QCELP	qualcomm coded excited linear predictive coding
QDM	quadrature demodulator
QPSK	quadrature phase shift keying
RES	Radio Expert Systems Group
RF	radio frequency
RMS	root mean square
RPE-LTP	regular pulse excitation-long term predictor
RTMS	radio telephone mobile system
SAW	surface acoustic wave
SFDR	spur free dynamic range
SINAD	signal over noise and distortion
SMR	specialized mobile radio
TACS	total access communication system
TDD	time division duplex
TDMA	time division multiple access
TETRA	trans-European trunked radio system
TIA	Telecommunications Industry Association
UDPC	universal digital personal communications
VCO	voltage controlled oscillator
VSAT	very small aperture (satellite ground) terminal
VSELP	vector sum excited linear predictive coding
WER	word error rate
WLAN	wireless local area network
WLL	wireless local loop

▼▼▼

ABOUT THE EDITOR

Lawrence E. Larson received his B.S. and M.S. degrees in Electrical Engineering from Cornell University, Ithaca, NY, in 1979 and 1980, and his Ph.D. from the University of California - Los Angeles, in 1986, also in Electrical Engineering. He is currently head of the Telecommunications Technology Department at Hughes Research Laboratories, Malibu, CA, where he directs research and development into advanced hardware solutions for emerging communications systems. His past experience includes the development of high-frequency III-V and silicon-based integrated circuits, improved microwave semiconductor devices, and semiconductor device modeling. In the past, Dr. Larson was an Adjunct Associate Professor of Electrical Engineering at UCLA, and has taught numerous short courses in Electrical Engineering at UCLA, UC-Berkeley, Oregon Graduate Research Institute, and Polytechnic University. He is a Senior Member of the IEEE.

▼ ▼ ▼

INDEX

The Artech House Mobile Communications Series

John Walker, Series Editor

For further information on these and other Artech House titles, including previously considered out-of-print books now available through our In-Print-Forever™ (IPF™) program, contact:

Artech House
685 Canton Street
Norwood, MA 02062
781-769-9750
Fax: 781-769-6334
Telex: 951-659
e-mail: artech@artech-house.com

Artech House
Portland House, Stag Place
London SW1E 5XA England
+44 (0) 171-973-8077
Fax: +44 (0) 171-630-0166
Telex: 951-659
e-mail: artech-uk@artech-house.com

Find us on the World Wide Web at:
www.artech-house.com